SPRINGER HANDBOOK OF AUDITORY RESEARCH

Series Editors: Richard R. Fay and Arthur N. Popper

SPRINGER HANDBOOK OF AUDITORY RESEARCH

Volume 1: The Mammalian Auditory Pathway: Neuroanatomy
*Edited by Douglas B. Webster, Arthur N. Popper, and Richard R. Fay*

Volume 2: The Mammalian Auditory Pathway: Neurophysiology
*Edited by Arthur N. Popper and Richard R. Fay*

Volume 3: Human Psychophysics
*Edited by William Yost, Arthur N. Popper, and Richard R. Fay*

Volume 4: Comparative Hearing: Mammals
*Edited by Richard R. Fay and Arthur N. Popper*

Volume 5: Hearing by Bats
*Edited by Arthur N. Popper and Richard R. Fay*

Volume 6: Auditory Computation
*Edited by Harold L. Hawkins, Teresa A. McMullen, Arthur N. Popper, and Richard R. Fay*

Volume 7: Clinical Aspects of Hearing
*Edited by Thomas R. Van De Water, Arthur N. Popper, and Richard R. Fay*

Volume 8: The Cochlea
*Edited by Peter Dallos, Arthur N. Popper, and Richard R. Fay*

Volume 9: Development of the Auditory System
*Edited by Edwin W Rubel, Arthur N. Popper, and Richard R. Fay*

Volume 10: Comparative Hearing: Insects
*Edited by Ronald Hoy, Arthur N. Popper, and Richard R. Fay*

Volume 11: Comparative Hearing: Fish and Amphibians
*Edited by Richard R. Fay and Arthur N. Popper*

Volume 12: Hearing by Whales and Dolphins
*Edited by Whitlow W.L. Au, Arthur N. Popper, and Richard R. Fay*

Volume 13: Comparative Hearing: Birds and Reptiles
*Edited by Robert Dooling, Arthur N. Popper, and Richard R. Fay*

Volume 14: Genetics and Auditory Disorders
*Edited by Bronya J.B. Keats, Arthur N. Popper, and Richard R. Fay*

Volume 15: Integrative Functions in the Mammalian Auditory Pathway
*Edited by Donata Oertel, Richard R. Fay, and Arthur N. Popper*

Volume 16: Acoustic Communication
*Edited by Andrea Simmons, Arthur N. Popper, and Richard R. Fay*

Volume 17: Compression: From Cochlea to Cochlear Implants
*Edited by Sid P. Bacon, Richard R. Fay, and Arthur N. Popper*

Volume 18: Speech Processing in the Auditory System
*Edited by Steven Greenberg, William Ainsworth, Arthur N. Popper, and Richard R. Fay*

Volume 19: The Vestibular System
*Edited by Stephen M. Highstein, Richard R. Fay, and Arthur N. Popper*

Volume 20: Cochlear Implants: Auditory Prostheses and Electric Hearing
*Edited by Fan-Gang Zeng, Arthur N. Popper, and Richard R. Fay*

Volume 21: Electroreception
*Edited by Theodore H. Bullock, Carl D. Hopkins, Arthur N. Popper, and Richard R. Fay*

*Continued after index*

William A. Yost
Arthur N. Popper
Richard R. Fay
Editors

# Auditory Perception
# of Sound Sources

 Springer

William A. Yost
Speech and Hearing Sciences
Arizona State University
Tempe, AZ 85287
USA
William.yost@asu.edu

Arthur N. Popper
Department of Biology
University of Maryland
College Park, MD 20742
USA
apopper@umd.edu

Richard R. Fay
Parmly Hearing Institute and Department
  of Psychology
Loyola University Chicago
Chicago, IL 60626
USA
rfay@luc.edu

*Series Editors:*
Richard R. Fay
Parmly Hearing Institute and Department
  of Psychology
Loyola University Chicago
Chicago, IL 60626
USA

Arthur N. Popper
Department of Biology
University of Maryland
College Park, MD 20742
USA

ISBN-13: 978-0-387-71304-5          e-ISBN-13: 978-0-387-71305-2

Library of Congress Control Number: 2007928313

Printed on acid-free paper.

9 8 7 6 5 4 3 2 1

springer.com

# Contents

# Series Preface

The Springer Handbook of Auditory Research presents a series of comprehensive and synthetic reviews of the fundamental topics in modern auditory research. The volumes are aimed at all individuals with interests in hearing research, including advanced graduate students, postdoctoral researchers, and clinical investigators. The volumes are intended to introduce new investigators to important aspects of hearing science and to help established investigators to better understand the fundamental theories and data in fields of hearing that they may not normally follow closely.

Each volume presents a particular topic comprehensively, and each serves as a synthetic overview and guide to the literature. As such, the chapters present neither exhaustive data reviews nor original research that has not yet appeared in peer-reviewed journals. The volumes focus on topics that have developed solid data and a strong conceptual foundation rather than on those for which a literature is only beginning to develop. New research areas will be covered on a timely basis in the series as they begin to mature.

Each volume in the series consists of a few substantial chapters on a particular topic. In some cases, the topics will be those of traditional interest for which there is a substantial body of data and theory, such as auditory neuroanatomy (Vol. 1) and neurophysiology (Vol. 2). Other volumes in the series deal with topics that have begun to mature more recently, such as development, plasticity, and computational models of neural processing. In many cases, the series editors are joined by a coeditor with special expertise in the topic of the volume.

RICHARD R. FAY, Chicago, IL
ARTHUR N. POPPER, College Park, MD

# Volume Preface

To survive, animals must navigate, find food, avoid predators, and reproduce; and many species survive based on their ability to communicate. All of these crucial behaviors allow animals to function in a crowded world of obstacles, objects, and other animals. Many of these objects vibrate and produce sound, and sound may be used to determine the sources of the sound and to serve as a basis for communication. Sounds produced by different sources are combined in one sound field that must be parsed into information that allows for the determination of the individual sources. This process begins at the level of the auditory receptor organ, but is primarily accomplished by processing of the peripheral code in the brain. Given the variety of sources that produce sound, the complexity of the world in which these sources exist, and the lack of peripheral receptors to analyze sound sources per se, determining the sources of sound presents a significant challenge for the auditory system. At present, not a great deal is known about how the auditory system deals with this challenge. This book reviews several topics that are likely relevant to enhance an understanding of the auditory system's ability to determine sound sources.

Yost, in Chapter 1, provides an overview of the volume and the issues that arise in considering sound source perception. Chapter 2, by Lufti, describes the properties of resonating sources, especially solids, and how the various properties of a resonating sound source (e.g., size, mass, tension) may affect sound source perception. In Chapter 3, Patterson, Smith, van Dinther, and Walters consider the standing-wave properties of sound sources, such as the vocal tract, and how the size of such resonators may determine the perception of the source. Chapter 4, by Demany and Semal, reviews much of the current knowledge about auditory memory, especially as it may relate to sound source perception. In addition to the importance of attending to one source or another to function in our everyday acoustic world, auditory attention may also play a direct role in aiding the auditory system in segregating one sound source from another. Chapter 5, by Hafter, Sarampalis, and Loui, reviews much of the literature related to auditory attention. In Chapter 6, by Kidd, Mason, Richards, Gallun, and Durlach, the topics of masking, especially energetic and informational masking, are reviewed as they relate to sound source perception. This is followed by Chapter 7, by Carylon and Gockel, in which the authors discuss how sources may be perceived and segregated based on a source's fundamental frequency of vibration and its resulting harmonic structure or temporal and spectral regularity.

A great deal is known about how differences in interaural arrival time and interaural level differences are used to locate the position of sound sources. Darwin (Chapter 8) considers the role spatial separation (especially interaural time and level differences) plays in sound source perception and segregation.

In Chapter 9, Sheft discusses temporal patterns of sounds and how these patterns pertain to sound source perception. This is followed by Chapter 10, by Lotto and Sullivan, who consider the speech-perception literature that provides insights into processes that might be considered for a better understanding of sound source perception for any potential sound source. Finally, in Chapter 11, Fay reviews the growing body of literature on how animal subjects other than humans process sound from sources.

Related chapters pertaining to other aspects of sound source perception can be found elsewhere in chapters from the Springer Handbook of Auditory Research series. Several chapters in Volume 15 (*Integrative Functions in the Mammalian Auditory Pathway*) relate to the questions of sound source perception, especially Chapter 9 ("Feature Detection in Auditory Cortex") by Nelken. Lewis and Fay, in Volume 22 (Chapter 2), *Evolution of the Vertebrate Auditory System*, is a treatment of the acoustic variables that could play a role in sound source perception. Volume 24 (*Pitch: Neural Coding and Perception*) contains several chapters that relate to this topic, especially Chapter 8, by Darwin, on "Pitch and Auditory Grouping." Volume 25 (*Sound Source Localization*) is a recent authoritative review of this topic. Volume 28 (*Hearing and Sound Communication in Amphibians*), and especially Chapter 11 on "Sound Processing in Real-World Environments," by Feng and Schul, treats many aspects of sound source perception in amphibians.

WILLIAM A. YOST, Chicago, IL
ARTHUR N. POPPER, College Park, MD
RICHARD R. FAY, Chicago, IL

# Contributors

ROBERT P. CARLYON
MRC Cognition and Brain Sciences Unit, Cambridge CB2 7EF, UK, Email: bob.carlyon@mrc-cbu.cam.ac.uk

CHRISTOPHER J. DARWIN
Department of Psychology, School of Life Sciences, University of Sussex, Brighton BN1 9QG, UK, Email: cjd@sussex.ac.uk

LAURENT DEMANY
Laboratoire Mouvement, Adaptation, Cognition, CNRS and Université Victor Segalen, F-33076 Bordeaux, France, Email: laurent. demany@psyac.u-bordeaux2.fr

NATHANIEL I. DURLACH
Hearing Research Center, Boston University, Boston, MA, 02215, USA

RICHARD R. FAY
Parmly Hearing Institute, Loyola University Chicago, Chicago, IL, 60626, Email: rfay@luc.edu

FREDERICK J. GALLUN
Hearing Research Center, Boston University, Boston, MA, 02215, USA

HEDWIG E. GOCKEL
MRC Cognition and Brain Sciences Unit, Cambridge CB2 7EF, UK, Email: hedwig.gockel@mrc-cbu.cam.ac.uk

ERVIN R. HAFTER
Department of Psychology, University of California at Berkeley, Berkeley, CA, 94720, Email: hafter@berkeley.edu

GERALD KIDD, JR.
Hearing Research Center, Boston University, Boston, MA, 02215, USA, Email: gkidd@bu.edu

ANDREW J. LOTTO
Department of Speech, Language, and Hearing Sciences, University of Arizona, Tucson, AZ, 85721-0071, Email: alotto@email.arizona.edu

PSYCHE LOUI
Department of Psychology, University of California at Berkeley, Berkeley, CA, 94720, Email: psyche@berkeley.edu

ROBERT A. LUTFI
Auditory Behavioral Research Lab, Department of Communicative Disorders, University of Wisconsin – Madison, Madison, WI, 53706, Email: ralutfi@wisc.edu

CHRISTINE R. MASON
Hearing Research Center, Boston University, Boston, MA, 02215, USA

ROY D. PATTERSON
Centre for the Neural Basis of Hearing, Department of Physiology, Development and Neuroscience, University of Cambridge, Cambridge CB2 3EG, UK, Email: rdp1@cam.ac.uk

VIRGINIA M. RICHARDS
Hearing Research Center, Boston University, Boston, MA, 02215, USA

ANASTASIOS SARAMPALIS
Department of Psychology, University of California at Berkeley, Berkeley, CA, 94720, Email: asaram@berkeley.edu

CATHERINE SEMAL
Laboratoire Mouvement, Adaptation, Cognition, CNRS and Université Victor Segalen, F-33076 Bordeaux, France, Email: catherine. semal@psyac.u-bordeaux2.fr

STANLEY SHEFT
Parmly Hearing Institute, Loyola University Chicago, Chicago, IL, 60626, Email: ssheft@luc.edu

DAVID R.R. SMITH
Centre for the Neural Basis of Hearing, Department of Physiology, Development and Neuroscience, University of Cambridge, Cambridge CB2 3EG, UK, Email: drrs2@cam.ac.uk

SARAH C. SULLIVAN
Department of Psychology, University of Texas at Austin, Austin, TX 78712-0187, Email: sullivan@psy.utexas.edu

RALPH VAN DINTHER
Centre for the Neural Basis of Hearing, Department of Physiology, Development

and Neuroscience, University of Cambridge, Cambridge CB2 3EG, UK, Email: rv230@cam.ac.uk

THOMAS C. WALTERS
Centre for the Neural Basis of Hearing, Department of Physiology, Development and Neuroscience, University of Cambridge, Cambridge CB2 3EG, UK, Email: tcw24@cam.ac.uk

WILLIAM A. YOST
Speech and Hearing Sciences, Arizona State University, Tempe, AZ 85287, Email: William.yost@asu.edu

# 1
# Perceiving Sound Sources

WILLIAM A. YOST

## 1. Sound Source Perception

To survive, animals must navigate, find food, avoid predators, and reproduce; and many species survive based on their ability to communicate. All of these crucial behaviors allow animals to function in a crowded world of obstacles, objects, and other animals. Many of these objects vibrate and produce sound, which may be used to determine their sources and as a basis for communication. Thus, evolving an auditory system capable of processing sound provides animals a valuable ability to cope with the world.

Sounds produced by different sources are combined into one sound field that must be parsed into information that allows for the determination of the individual sources. The auditory peripheral receptors did not evolve to process the sources of sound, but instead, they provide the important neural code for the physical attributes of sound (i.e., of the sound field). While the neural code provides crucial information about the sounds from the sources that make up the sound field, the peripheral neural code is not a code for the sources themselves (but see Lewis and Fay 2004). Thus, the neural processes responsible for sound source determination lie above the auditory periphery. Given the variety of sources that produce sound, the complexity of the world in which these sources exist, and the lack of peripheral receptors to analyze sound sources per se, determining the sources of sound presents a significant challenge for the auditory system. At present, not a great deal is known about how the auditory system deals with this challenge. This book reviews several topics that are likely relevant to enhancing our understanding of the auditory system's ability to determine sound sources.

While not a great deal is known about sound source determination, research over the past century has produced a wealth of information about the neural processes involved with the coding and analysis of the physical attributes of sound and the sensations produced by these physical attributes. In the behavioral sciences, it is not uncommon, especially in vision research, to conceptualize sensory processing in terms of sensation and perception. Auditory sensation may be viewed as processing the physical attributes of sound: frequency, level, and time. Auditory perception may be the additional processing of those attributes that allow an organism to deal with the sources that produced the sound. Thus, determining the sources of sound may be viewed as a study of auditory perception.

The perception of "special sources," such as speech and music, may be subtopics of the general topic of the perception of sound sources. Thus, this book is mainly about auditory perception as it relates to sound source determination.

Most of the recent history of the study of hearing has been dominated by the investigation of sensory processing of the physical attributes of sound. This was not the case in the early nineteenth century, when the modern study of biology and behavioral science was emerging (see Boring 1942). The philosopher/scientist of those times viewed the direct investigation of humans' awareness of objects and events in the world as a way to resolve some of the conflicts of mind/body dualism. However, it was recognized that it was the physical attributes of the sound from sources that were being processed. Giants in science, such as von Helmholtz, developed theories for processing acoustic attributes, especially frequency. The investigation of the auditory processing of frequency, level, and time have dominated studies of hearing throughout the past 200 years, as a wealth of information was and is being obtained concerning auditory processing of the physical attributes of sound and its consequences for auditory sensations.

In the late 1980s and early 1990s, several authors wrote about object perception (Moore 1997), perception of auditory entities (Hartmann 1988), auditory images (Yost 1992), and sound source determination (Yost and Sheft 1993). Bregman's book, *Auditory Scene Analysis* (1990), provided a major contribution in restoring a significant interest in sound source perception. These works recognized that others previously had studied issues directly related to sound source perception. For instance, Cherry (1953) coined the term "cocktail party problem" as a way to conceptualize processing sound sources in complex multisource acoustic environments. Cherry proposed that several aspect of sound sources and auditory processing might help solve the "cocktail problem," and he investigated primarily one of these, binaural processing. Bregman's work leading up to *Auditory Scene Analysis* largely involved the study of auditory stream segregation. Thus, the topic of sound source perception has at times been referred to as object perception, image perception, entity perception, identification, the cocktail party problem, streaming, source segregation, sound source determination, and auditory scene analysis.

While at a general level these terms may be synonymous, in many contexts they may refer to different aspects of the study of sound source perception. Images, entities, and objects can refer to either the physical source or the perception of the source, and as such can present an ambiguous description of the problem. Identification implies using a label for the sound from a source. It is not clear that sound source perception requires the use of labels, i.e., that one has to be able to identify a sound in order for it to be processed as a source. Many procedures used to study sound source perception use identification, and we clearly identify (label) the sound from many sources, so that identification plays a role in sound source perception and its study. However, identification is probably not necessary and sufficient for sound source perception.

The cocktail party problem has often been identified with the ability to use binaural processing (or more generally spatial processing) to segregate sound sources. Auditory streaming usually refers to a context of alternating bursts of sound, which does not represent the sound from all sources (e.g., a single breaking twig). And many of these terms emphasize the segregation of sources rather than a more general problem of the perception of a sound source whether it produces sound in isolation or along with the sound from other sources. Thus, this book will attempt to use the term "sound source perception" as the general description of the challenge facing the auditory system.

Figure 1.1 depicts the type of neural information that might be provided to the central nervous system by the auditory periphery for a segment of a musical piece played by a quartet of a piano, bass, drum, and trumpet. In this segment all four instruments are playing together, and each can be identified when one listens to the piece. The figure is based on the computations of the Auditory Image Model (AIM; see Patterson et al. 1995), in which the sound is processed by a middle-ear transfer function, a simulation of basilar membrane analysis (a gammatone filter bank), and a simulation of hair cell and auditory nerve processing (Meddis hair cell). Each horizontal trace reflects a histogram of neural

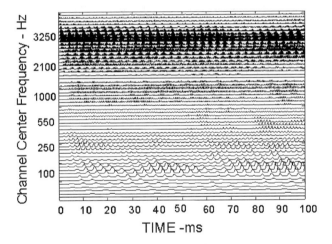

FIGURE 1.1. The neural activation pattern (NAP) for the simultaneous playing of a trumpet, bass, piano, and drum in a 100-ms slice of a piece of jazz music. The NAP represents an estimate of the pattern of neural information flowing from the auditory periphery (see Patterson et al. 1995). Each line in the NAP represents an estimate of the poststimulus time histogram of a small set of fibers tuned to a narrow frequency region indicated on the y-axis. Increased neural activity at different moments in time represents the neural code for temporal modulation, while increased neural activity for one set of tuned fibers over the others represents the neural code for spectral processing. The spectral-temporal pattern of each instrument overlaps that from the other instruments, and as such, the NAP represents the "summed" spectral-temporal pattern of the four instruments. The central nervous system must parse this overall spectral-temporal code into subsets each corresponding to an instrument (sound source) in order for the existence of each instrument in the musical piece to be determined. How this parsing is achieved is the challenge this book addresses.

activity from a small set of tuned fibers with the center frequency of each set of tuned fibers shown on the ordinate and time shown along the abscissa. While this depiction may not represent all details of the analysis performed by the auditory periphery, it does represent the challenge facing the central nervous system. This peripheral code indicates the spectral-temporal neural structure of the complex sound field consisting of the sounds from the four instruments, but it does not provide an obvious depiction of the four sources. That is, how would the central nervous system parse this spectral-temporal neural pattern into four subsets, each representing one of the four instruments?

Two approaches to evaluating the peripheral neural code have been used to account for sound source perception. Many investigators attempt to identify attributes of the sound from sources that are preserved in the neural code that could be used to segregate one source from another. Others have imagined that the problem of sound source perception is aided by the auditory system using information it has gained from experience. The former is a bottom-up approach, and the latter a top-down approach. Bregman (1990) argues that both are necessary to account for sound source perception, and he describes "primitive processes" as the bottom-up mechanisms and "schema-based approaches" as the top-down processes. That is, physical attributes such as temporal onset differences, harmonicity, and common modulation may be used by central nervous system circuits to segregate the neural code to account for sound source perception. At the same time, information gleaned from one's experience in processing sound sources can guide these neural circuits in processing the peripheral neural code. Sheft, in Chapter 9, makes a distinction between segregating neural information to aid sound source perception and extracting information about the segregated neural information as an equally important part of perceiving sound sources.

## 2. Chapters in This Book

Sound sources produce sound based on their vibratory patterns, most often due to the resonance properties of the source. The vibration and resonances produce a sound pressure wave that is propagated from the source to a listener's ears, and this sound wave is characterized by its frequency, level, and timing attributes. A great deal is known about how the auditory system processes these physical attributes. But do the properties of the source itself (e.g., size, mass, tension) have any direct bearing on auditory perception, and, if so, in what way? Chapters 2 (by Lutfi) and 3 (by Patterson, Smith, van Dinther, and Walters) review much of what is known about how the resonant properties of sound sources affect perception. Chapter 2 deals with how the various properties of a resonator, especially solid objects (e.g., a piece of wood that is struck) may affect sound source perception. Lutfi reviews how one characterizes the resonant properties of vibrating sources and several studies concerning the relationship between the physical aspects of resonance and one's auditory perceptions. The chapter introduces several newer psychophysical techniques used to study the perception of complex sounds.

Chapter 3 describes issues of the resonance properties of tubes (e.g., the vocal tract) that resonate due to standing waves. The size of a resonator often has significant biological relevance. In many species (e.g., frog), the female prefers to mate with large males and chooses mates based on the sounds produced by the male. Thus, males need to produce sounds that are correlated with size. In humans, the size of the vocal tract is correlated with the frequency content of speech (e.g., formants). A man and a woman each uttering the same vowel will produce different formant structures due largely to the difference in the size of the vocal tracts. Yet a listener still perceives the same vowel as being uttered by both people. How does the auditory system compensate for the formant differences in order that a constant percept is achieved? These are some of the topics covered in Chapter 3 (and to some extent in Chapter 10).

Sound is a temporal stimulus that is processed over time. The perception of a musical piece, a person speaking, a car driving past requires us to integrate the sound waveform over time. That is, sound that occurred a moment ago has to be retained by the auditory system so that the sound occurring now makes sense. The neural representation of sound must be continually stored and retrieved. How well the nervous system stores and retrieves its neural representations of sound will influence the perceptions of a sound source. The mechanisms of storage and retrieval may directly aid the processes of sound source determination. Researchers must often confront problems of a listener's ability to remember acoustic events when they investigate complex sound processing. Chapter 4, by Demany and Semal, reviews much of the current knowledge about auditory memory, especially as it may relate to sound source perception.

Many sound sources (people talking, music playing, glasses clinking, etc.) occur during a typical cocktail party. Often, the challenge is not just to segregate these sound sources, but to attend to one or more of them (e.g., to follow a friend's conversation). In addition to the importance of attending to one source or another to function in our everyday acoustic world, auditory attention may also play a direct role in aiding the auditory system in segregating one sound source from another. Chapter 5, by Hafter, Sarampalis, and Loui, reviews much of the literature related to auditory attention.

If one sound masks another, then it is unlikely that the masked sound will contribute much to sound source perception. Thus, masking clearly plays an important role in sound source perception, especially source segregation. Masking is typically defined as the elevation in detection threshold (but see Tanner 1958) of one sound (the signal) in the presence of another sound or other sounds (masker or maskers). In the 1980s, Watson and colleagues (see Watson 2005 for a review) showed that when listeners were presented an uncertain stimulus context, they had more difficulty processing sound than if the stimulus conditions were more certain or less variable; the greater the uncertainty, the greater the difficulty in sound processing. Borrowing from earlier work of Pollack (1975), Watson labeled the interference among different sounds that was based only on the sound itself "energetic masking." The additional masking or inter-ference due to making the stimulus context variable and uncertain was labeled "informational masking." Carhart and colleagues (1969) were among the first to show that in studies of speech masking, a masker that was itself speech provided more masking for a speech signal than a nonspeech masker (e.g., noise) when the nonspeech and speech maskers are equated to the extent possible in terms of their acoustic properties. The additional masking that occurs for the speech masker over the nonspeech masker is now also referred to as informational masking, and the masking provided by the nonspeech masker is referred to as energetic masking. In the speech example, it has been suggested that the additional masking that occurs when both the target signal and inferring masker are speech is due to the auditory system's attempt to segregate these two similar information-bearing sound sources. Thus, in the real world of complex sound sources, the challenge is to segregate sound sources that provide information for the listener and occur in uncertain contexts. Chapter 6, by Kidd, Mason, Richards, Gallun, and Durlach, reviews the topics of masking, especially energetic and informational masking, as they relate to sound source perception.

The auditory system has sophisticated biomechanical and neural processes to determine the spectral content of an acoustic event. Thus, it is logical to assume that aspects of a sound's spectrum could be used in sound source perception, especially as an aid in segregating the sound from one source from that from other sources. In simple cases, the spectrum of the sound from one source may not overlap that from another source, and since the auditory system neurally codes for these spectral differences, this neural differentiation could form a basis for sound source segregation. But how does the auditory system segregate sound sources when the spectral content of one source overlaps that of other sources (as indicated in Figure 1.1)?

As Chapters 2 and 3 indicate, resonant sources often have a fundamental frequency with many harmonics. One source is likely to vibrate with a funda-mental frequency that is different from that of another source. The vibratory pattern of a harmonic sound provides a temporal regularity to the waveform as well as spectral regularity. Thus, sources may be perceived and segregated based on the fundamental frequency and its resulting harmonic structure and/or on the resulting temporal and/or spectral regularity. Carylon and Gockel, in Chapter 7, review the current literature on the role harmonicity and regularity play in sound source perception. Often, sounds that are harmonic or contain either spectral or temporal regularity are perceived as having a distinct pitch (complex pitch). Sounds with different harmonic structures (or fundamentals) produce different pitches, so these pitch differences could be a basis for sound source perception. While it is still not certain how the auditory system extracts a pitch from such complex sounds (see Plack et al. 2006), it is clear that such processes involve using information across a broad range of the sound spectrum. Thus, studies of complex pitch also provide useful insights into how the nervous system performs spectral integration.

Cherry (1953) hypothesized that the fact that the source of a sound can be located in space may aid the auditory system in attending to one sound in the presence of sounds from other sources such as occurs at a cocktail party. That is, when different sources are located at different places, this spatial separation may allow the auditory system to segregate the sources. However, spatial separation is not both sufficient and necessary for sound source perception. The many instruments (sources) of an orchestra recorded by a single microphone and played over a single loudspeaker can be recognized; i.e., the sound sources (instruments) can be determined in the complete absence of spatial information.

A great deal is known about how differences in interaural arrival time and interaural level differences are used to locate the azimuth position of sound sources. Thus, it is not surprising that most investigations of spatial separation of sound sources have studied interaural time and level differences. Darwin, in Chapter 8, describes a great deal of the literature on the role of spatial separation (especially interaural time and level differences) for sound source perception and segregation.

Since sound has no spatial dimensions and the auditory periphery does not code for the spatial location of sound sources, the nervous system must "compute" spatial location. It does so based on the interaction of sound with objects in the path of the sound as it travels from its source to the middle ears. For instance, sound from a source off to one side of a listener interacts with the head such that the sound reaches one ear before the other and is less intense (due to the sound shadow caused by the head) at the one ear than at the other. The auditory system computes these interaural time and level differences as the basis for processing the azimuth location of the sound source. However, two sources at different locations that produce exactly the same sound at exactly the same time will not generate two different sets of interaural time and level differences. That is, spatial separation per se is unlikely to be a cue for sound source segregation. The sound from one source must be either spectrally or temporally different from that from another source before the auditory system can compute their spatial locations (e.g., compute different interaural time and level differences). Thus, it is likely that the use of spatial separation of sound sources as a cue for sound source segregation involves an interaction of spectral and/or temporal processing in combination with spatial computations performed by the auditory system.

As already alluded to in this chapter, the temporal pattern of the sound from one source is likely to be different from that of others. These temporal differences provide potential cues for sound source perception and segregation. The sound from one source is likely to start and stop at a different time from that of another source, and the amplitude modulation imparted to one sound by its source's physical properties is likely to differ from the amplitude modulation imparted by a source with different physical properties. These two temporal aspects of sounds—onset characteristics and amplitude modulation—have been extensively studied, and a great deal of this literature as it pertains to sound source perception is presented in Chapter 9, by Sheft.

Speech is clearly the sound of most importance to humans, at least in terms of communication. Speech is generated by the vocal cords and tract, and recognizing one speech sound as different from another requires processing differences in the vocal-tract source. Thus, speech recognition can be viewed as a case of sound source perception, although perhaps a special case. A great deal is known about the relationship between the sound source for speech, the vocal tract, and speech perception. Thus, one might learn about the general problem of sound source perception from studying speech perception. Lotto and Sullivan review in Chapter 10 several topics from the speech perception literature that provide insights into processes that could be considered for a better understanding of sound source perception for any potential sound source. It is clear from speech perception research that one's perception of a sound at one instance in time can be altered by another sound that occurred earlier and in some cases later in time. It is likely that such issues of coarticulation of speech sounds may also play a significant role in the perception of the sounds from other sources.

It is highly unlikely that sound source perception and segregation are unique to a limited number of animal species. It is hard to imagine how any animal with the ability to process sound would not need to deal with processing sound produced by different sources. Thus, sound source perception probably occurs for all animals. Fay, in Chapter 11, reviews some of the growing body of literature on how animal subjects other than humans process sound from sources. This research indicates how animals cope with the challenge of sound source perception, offers a chance to develop an understanding of universal processes that might be used by all animals as opposed to special processes used by only certain animals, and provides a deeper understanding of neural data related to sound source processing that are obtained from different animals. As Fay explains, little research has been devoted to studying sound source perception in animals other than humans. Much of that research uses the streaming paradigm developed by Bregman and colleagues (see Bregman 1990).

The topics covered in this book are likely to play a role in dealing with sound sources both in the laboratory and in the day-to-day world. These topics may play a direct role in how the auditory system processes sound from a source or multiple sources. For instance, everyday experience suggests that auditory attention is required to cope with most complex acoustic scenes. Attention appears to play an important role after the sound sources are processed (segregated), by allowing one to focus on one source as opposed to other sources. It is also possible that attentional mechanisms directly aid the auditory system in segregating sound sources in the first place. That is, attention may play a role in both the segregation process itself and in how the sources are dealt with after they are segregated.

Several topics covered in this book are important for conducting experiments dealing with sound source perception even if they may not always be directly related to general issues of sound source processing. For instance, in many experiments a sound at one instant in time must be compared to a sound occurring at another instant in time. This comparison is based on the earlier sound staying in some form of memory in order for it to be perceptually compared to the later sound. It is possible that the ability to process the sounds in such experiments is related not only to the physical properties of the two sounds and the context in which they exist, but also to memory.

## 3. Other Topics Related to Sound Source Perception

This book and its chapters are somewhat "theoretically neutral" in that the choice of chapters and the presentation in each chapter are not tightly coupled to a particular theory or even a particular theoretical framework. The theories of sound source perception and processing are few. As mentioned previously, Bregman (1990) offers a theoretical framework that helps him organize his views of processing the auditory scene. This framework is based partially on ideas shared with the Gestalt principles of perception. An ecological, or Gibsonian, approach has also been proposed (see Neuhoff 2004) in which sound source perception is viewed as arising from our experience in processing real sounds in the real world. There is also a growing literature on computational models of sound source processing, often referred to as computational auditory scene analysis (CASA); see Wang and Brown (2006).

In addition to not covering a great deal about theories and models of sound source perception, especially the growing use of CASA models, other topics are not covered in great detail in this book. For instance, there is a growing literature of physiological studies of sound source perception. While several chapters in this book review much of this literature, a single chapter was not devoted to physiological processes. The major reason these topics were not covered in detail was the desire to keep the length of the book consistent with the other books in the SHAR series.

But there were other reasons as well. Both the use of CASA models and physiological correlates to studying auditory source processing face a similar challenge. There are very few data on sound source perception per se. That is, there are few data directly related to the actual segregation of sound sources. Thus, it is difficult to provide models or to find physiological measures of actual data indicating how sound sources are segregated. The most common data set that is modeled and studied physiologically is that related to auditory streaming. While the studies of auditory streaming have provided valuable information pertaining to sound source perception, they represent only one class of stimulus conditions. The main streaming paradigm is to alternate two (or more) sounds in time, e.g., two tones of different frequency. Under the proper conditions the sounds are perceived as two pulsating sources (e.g., two pulsating tone bursts) as opposed to one source that alternates in some perceptual attribute (e.g., pitch). Many sound sources are modulated in level over time, and they are still perceived as a single source. Thus, understanding how the pulsating sound from one source provides a perception of a single source rather than different sources is very important to an overall understanding of sound source perception. However, it is probably not the case that the conditions that represent streaming are the same as those for all other conditions that lead to sound source segregation.

Many CASA models are tested by determining whether they produce evidence of segregated sources when the models are provided complex sounds consisting of sounds from two or more sources (e.g., the modeler presents a complex sound mixture to the model and determines whether the model's output produces the perception of different sources). While this can provide valuable insights into how the auditory system might process sounds, without a close tie between the model and actual data on sound source perception, these models are limited in their ability to provide direct evidence for how the actual auditory system may process the sound from sources.

Another topic not covered in great detail is music perception from the perspective of sound source perception. There does not appear to be a large literature on the topic. However, much has been written about the perception of different instruments (e.g., the literature on timbre), and processing timbre is relevant to sound source perception (see Griffith and Todd 1999). However, given the length constraints of the book, a chapter on music perception was not included. While there are not dedicated chapters on CASA models, physiological studies of sound source perception, and music, these topics are covered in many of the book's chapters.

A topic that appears to have great relevance to a better understanding of sound source perception is the development of these processes (e.g., how do young children or infants perceive sound sources?). There is not yet a large literature on the development of sound source perception (but see Winkler et al. 2003). Part of the reason that such studies have not been undertaken might be the complex nature of many of the tasks used to study sound source perception in adults and the difficulty in adapting these paradigms to children or infants. Knudsen's (2004) work on the neural plasticity of binaural processing used by the barn owl offers one example of how neural systems may adapt (especially in the young bird) to environmental factors that influence processing the location of sound sources.

The chapters barely mention how those who are hard of hearing cope with the auditory scene. That is, what changes in sound source perception occur for people with hearing loss and how might various hearing aids and prostheses improve any degraded performance? The reason this topic was not covered is that little research has been conducted that directly investigates sound source processing and segregation in people with hearing loss (but see Grimault et al. 2001). It is well documented that people with hearing loss often have significant difficulty processing sound in noisy or reverberant environments. That is, they cannot process sound well when there are several sound sources presenting sound at about the same time. Such people with hearing loss cannot segregate sound sources as well as those without significant hearing loss. It is still not clear why this is so and how these problems might be ameliorated.

# 4. Summary

Sound source perception plays a crucial role in an animal's ability to cope with the world. This book reviews some of the relevant literature for understanding the processes of sound source perception. The chapter authors and the editors hope that the topics covered in this book will serve two purposes: (1) to provide a review of the literature relevant to understanding sound source perception and (2) to stimulate even more research into what we believe is one of the most important topics in the auditory sciences—sound source perception.

## *References*

Boring EG (1942) Sensation and Perception in the History of Experimental Psychology. New York: Appleton-Century-Crofts.

Bregman AS (1990) Auditory Scene Analysis: The Perceptual Organization of Sound. Cambridge, MA: MIT Press.

Carhart R, Tillman TW, Greetis E (1969) Perceptual masking in multiple sound backgrounds. J Acoust Soc Am 45:694–703.

Cherry C (1953) Some experiments on the recognition of speech with one and with two ears. J Acoust Soc Am 25:975–981.

Griffith N, Todd PM (1999) Musical Networks: Parallel Distributed Perception and Performance. Cambridge, MA: MIT Press.

Grimault N, Micheyl C, Carlyon RP, Arthaud P, Collet L (2001) Perceptual auditory stream segregation of sequences of complex sounds in subjects with normal and impaired hearing. Br J Audiol 35:173–182.

Hartmann WM (1988) Pitch perception and the organization and integration of auditory entities. In: Edelman GW, Gall WE, Cowan WM (eds) Auditory Function: Neurobiological Bases of Hearing. New York: John Wiley & Sons, pp. 623–646.

Knudsen EI (2004) Sensitive periods in the development of the brain and behavior. J Cogn Neurosci 8:1412–1425.

Lewis ER, Fay RR (2004) Environmental variables and the fundamental nature of hearing, In: Manley GA, Popper, AN, Fay RR (eds) Evolution of the Vertebrate Auditory System. New York: Springer, pp. 27–54.

Moore BCJ (1997) An Introduction to the Psychology of Hearing, 3rd ed. London: Academic Press.

Neuoff JG (2004) Ecological Psychoacoustics. San Diego: Elsevier.

Patterson RD, Allerhand M, Giguere C (1995) Time-domain modeling of peripheral auditory processing: A modular architecture and a software platform. J Acoust Soc Am 98:1890–1895.

Plack C, Oxenham A, Fay R, Popper A (2005) Pitch: Neural Coding and Perception. New York: Springer–Verlag.

Pollack I (1975) Auditory informational masking. J Acoust Soc Am 57:S5.

Tanner WP Jr (1958). What is masking? J Acoust Soc Am 30:919–921.

Wang D, Brown G (2006) Computational Auditory Scene Analysis: Principles, Algorithms, and Applications. New York: Wiley-IEEE Press.

Watson CS (2005) Some comments on informational masking. Acta Acust 91:502–512.
Winkler I, Kushnerenko E, Horváth J, Čeponiene R, Fellman V, Huotilainen M, Näätänen R, Sussman E (2003) Newborn infants can organize the auditory world. Proc Natl Acad Sci USA 100:11812–11815.
Yost WA (1992) Auditory image perception and analysis. Hear Res 56:8–19.
Yost WA, Sheft S (1993) Auditory processing. In Yost WA, Popper AN, Fay RR (eds) Human Psychoacoustics. New York: Springer-Verlag, pp. 193–236.

# 2
# Human Sound Source Identification

Robert A. Lutfi

## 1. Introduction

Understanding how we make sense of our world through sound is arguably the most significant challenge for contemporary research on hearing. We rely critically on our ability to identify everyday objects and events from sound to function normally in a world that at any given moment is largely out of view. When this ability is compromised through disordered hearing, the impact on daily function can be profound. Past work has dealt broadly with the problem under different headings: auditory object perception (Hartmann 1988; Moore 1989; Handel 1995), auditory image perception (Yost 1992), auditory scene analysis (Bregman 1990; Ellis 1996), sound source recognition (McAdams 1993), and sound source determination (Yost and Sheft 1993). This chapter avoids much redundancy with these works by focusing on a simple expression of the problem—the identification of a single rudimentary source from the sound of impact. Simplicity is a key reason for this focus, but there are others, as will later become clear.

The chapter is organized into five major sections. The first describes the fundamental nature of the problem, laying the foundation for what follows. The second summarizes the major theoretical approaches, focusing on how these approaches differ in terms of the relative importance attached to different variables. The third reviews some physical acoustics of elementary sound sources undergoing impact and identifies the acoustic information intrinsic to basic physical attributes of these sources. The fourth considers what is known regarding the limits of human sensitivity to this information, while the fifth evaluates its role in identification. Certain parallels will be noted in these sections with those of Chapter 3 of this volume, by Dr. Roy Patterson. There, the focus is on the identification of attributes of the vocal tract from sound, an identical problem in principle but a somewhat different problem acoustically.

## 2. The Problem of Sound Source Identification

Take a moment to listen to the sounds around you, and you are likely to notice a few things. First, you might notice that much of what you hear is hidden from view; the person coughing in the room next door, the fan of the air-conditioning unit, the birds chirping outside. Such examples make clear how much of our perception of the world depends on sound. We tend not to be aware of this simple fact, but it is easily appreciated when what goes undetected is the smoke alarm, the child's call for help, or the sound of unseen oncoming traffic. Continue listening, and you might notice that your perception of the unseen things you hear is also remarkably accurate. A plate drops in an adjacent room, and from its sound you can tell that it was a small ceramic plate; that it bounced once, rolled a short distance, and then came to rest after colliding with the wall. This ability, too, we tend to take for granted, but the best man-made machine-recognition systems developed to date have yet to approach the level of accuracy on the scale that we achieve in everyday listening (cf. Ellis 1996; Martin 1999).

To understand how in everyday listening we might solve the problem of sound source identification and why it is such a challenging problem, it is helpful to begin with a basic understanding of the problem itself. This section describes the fundamental challenges a listener must confront in attempting to identify an arbitrary source based only on the sound it produces. The problem is analyzed on three levels that serve to tie together the different sections of this chapter. The first deals with the acoustic wave front reaching the listener's ears and its ambiguous relation to the physical properties of the sound-producing source, the second with auditory transduction and the associated loss of information at the auditory periphery, and the third with the role of prior knowledge and experience.

## 2.1 The Inverse Problem

The most pervasive challenge of sound source identification is commonly referred to as the *inverse problem*. The term is borrowed from the field of mathematics, where it is used to describe not a single problem but a class of problems that have no unique solution. Consider the algebraic expression $A = x^2y$. If this were to appear in an elementary algebra text, you might be asked to solve for $A$ given certain values of $x$ and $y$. This is a *direct* problem; it has a unique solution obtained by performing the specified arithmetic operations after substitution. The inverse to this problem turns the question entirely around. You are first given a value for $A$ and then asked to solve for $x$ and $y$. In this case, there is no unique solution, since there is an indeterminate number of values of $x$ and $y$ that can yield a given value of $A$. Now, as trivial as this example might seem, it is precisely the kind of problem we confront in everyday listening when attempting to identify an arbitrary source from sound alone. The identifying features of the source—its size, shape and material—and the manner in which it is driven to vibrate dictate the particular sound emitted (direct problem); but the information we might use to separately recover these features from sound (inverse problem) is confounded in the single pressure wave front arriving at the ear. Similar sources can thus produce different sounds, and different sources can produce similar or even identical sounds.

The problem is exacerbated in everyday listening, where there are usually several sources sounding simultaneously. Here, the sound waves emitted by each source sum before reaching the ears; hence, to recover information about any one source, the listener must extract from the sum (in effect, solve the inverse of the problem $A = x + y$). Helmholtz (1954, p. 29) likened the problem to one of viewing the up-and-down motion of waves on the surface of water through a long narrow tube and inferring from these oscillations the various surface disturbances that gave rise to them. Once again, there is no unique solution unless one has, or can in the process acquire, additional information regarding the circumstances surrounding the observation. What that information might be and how the listener might use that information largely distinguishes the major theoretical approaches to sound source identification, as reviewed in Section 3.

## 2.2 Stimulus Variability

Sound source identification would be a simpler task if each time an object was driven to vibrate it produced exactly the same sound. In fact, however, the sound produced by any source varies widely from one instance to the next and from one place to the next for reasons that have little or nothing to do with the properties of the source. Sound is by its very nature stochastic. Fluctuations in the ambient atmospheric pressure resulting from the random motion of air molecules (Brownian motion) ensure that no source will ever produce exactly the same sound twice. Other transient conditions of the atmosphere (wind direction, wind shear, temperature, and humidity) have more profound effects as the sound travels at longer distances from the source. For example, as a sound wave encounters pockets of more or less dense air, it will bend or refract. This can cause a sound source to appear closer at greater distances from the source, or at different times it can cause a sound to appear as if coming from different places.

Other significant sources of acoustic variation result from small unavoidable differences in attempts to replicate the physical action giving rise to a particular sound. This type of variation reflects a general property in which a source's natural modes of vibration are excited to different degrees depending precisely on how energy is injected into the source (cf. Section 4.3). The effects are well known to musicians. The familiar note of a kettledrum, for example, is created by striking the drumhead about one-fourth of the distance from the edge to the center. Strike the drumhead too close to the center, and the result is a dull "thud"; strike too close to the edge, and the result is a hollow "ping." The sound of the note also depends critically on the hardness of the mallet and the precise area and angle at which it comes into contact with the drumhead.

Finally, there is variation in sound transmission from source to receiver due to environmental context. No listening environment is completely transparent. The single note of the trumpet heard in the concert hall is different acoustically from the same note heard on the marching field. This is because the sound reaching the ears depends not only on the emitting source, but also on the obstacles it encounters along the way. Walls, furniture, curtains, and carpets have surface properties that color sound through absorption. Hard walls color sound through reflections that create standing waves and complex patterns of interference in a room. Smaller objects have a similar effect through diffraction. Ambient noise also varies with context, so that one cannot always anticipate what acoustic features of a signal might be masked by the noise in any given situation (see Chapter 6 in this volume, on informational masking, by Kidd et al.). Each listening environment thus imprints its own acoustic "signature" on sound, so that the task of identifying a sound event depends not only on that event but critically on the context in which it occurs.

## 2.3 Information Loss at the Auditory Periphery

Even under ideal circumstances, imperfections of auditory transduction impose certain limits on the amount of information that can be recovered about a source. The transduction process begins when sound energy, transmitted through the ossicular chain, creates a compression wave in the fluids of the cochlea, causing the basilar membrane to vibrate. Along its length, the basilar membrane varies continuously in stiffness and mass, so that the positions of vibration maxima vary systematically with frequency. An active process of the outer hair cells amplifies and sharpens the vibration maxima locally, thus helping to provide a usable *place code* for sound frequency. The place code is then preserved in the average discharge rate of individual nerve fibers that respond selectively to different characteristic frequencies. *Timing* information is also preserved in the group response of fibers synchronized to membrane displacement at each point along the length of the membrane. The result is an effective neural activation pattern (NAP) in frequency and time that preserves many but not all of the details of the sound waveform. Figure 2.1 shows an example NAP generated in response to a "tin can" sound. It was produced using the auditory-image model (AIM) described by Patterson et al. (1995) (Also see Patterson, Smith, van Dinther, and Walters, Chapter 3). Smearing in the NAP reflects limits in the ear's ability to resolve individual partials and to detect changes occurring across frequency and time. Such representations are thus useful for evaluating the information available to a listener when judging source attributes from sound (cf. Ellis 1996; Martin 1999; Lutfi 2001; Tucker and Brown 2003). Later, we shall describe studies suggesting real limits to what a listener might determine about a source from sound because of information loss at the cochlea.

## 2.4 Source Knowledge

Helmholtz considered information loss at the sensory receptors to argue that perception must involve some form of inference—an educated guess, as it were, based on the combination of indefinite information provided by sound and prior knowledge of potential sound sources. Much research has been undertaken to specify the extent and nature of this prior knowledge, but few answers have so far emerged beyond the peculiarities of the individual studies at hand. The difficulty

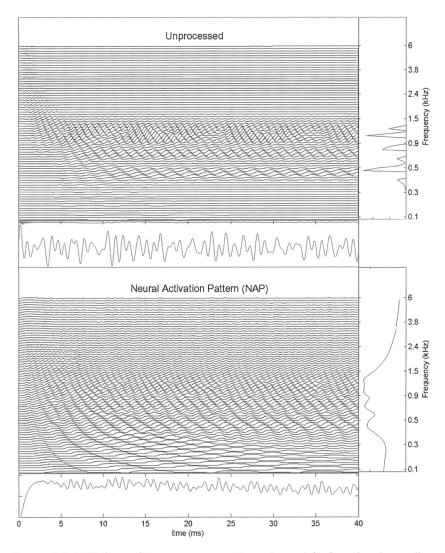

FIGURE 2.1. NAP (bottom) in response to a "tin can" sound (top) produced according to a model of the mechanical response of the basilar membrane and transduction of the inner hair cell (cf. Patterson et al. 1995). The curves to the right of the NAP give the sum of the response of each neural channel over time; those at the bottom give the sum of the response over channels.

of the problem is easily appreciated when one considers the various levels at which recognition can occur and the differences that exist among listeners in terms of their familiarity or experience with different sounds. The rattle of the car engine, for example, is recognized by the car mechanic as a malfunctioning alternator, but to the car's owner it is just a noise indicating needed repair.

Even for a given level of recognition, the redundant ways in which information is represented in sound can allow for individual differences in the manner of recognition based on prior knowledge. The gender of a walker might be as easily recognized from the loudness of the footsteps as from the pace of the gait or the bright click of high heels. Because of the importance of prior knowledge for sound source identification, our review of studies pays close attention to the information that is given beforehand to listeners regarding the properties of the objects and events that produced the test sounds.

## 3. Approaches to Sound Source Identification

The discussion thus far should make clear that the problem of sound source identification is exceptionally complex. Potential ambiguities exist regarding the properties of a sound source even before the emitted sound reaches the listener's ears. Information loss at the auditory periphery imposes constraints on the ability to recover exact information about a source, as potentially does a listener's unfamiliarity or uncertainty regarding the source. How then do we negotiate these challenges, as we evidently do so well in everyday listening? The question has fueled much debate, but that debate has for the most part concerned just three viable theoretical approaches to the problem.

The earliest meaningful perceptual theory, referred to as the *inferential approach*, is attributed to Helmholtz (1954). Fundamentally, Helmholtz argued that the perceptual problem is not well constrained, that the information arriving at the sensory receptors does not generally permit an unambiguous interpretation of the eliciting stimulus event (for all the reasons discussed above). Working from the premise of an *impoverished stimulus*, Helmholtz proposed that perception must involve a process of inferring (unconsciously) the event from knowledge of its likelihood given the incomplete sensory information at hand. He thus viewed the observer as an active participant in perception, using prior knowledge determined innately or from experience to, in a sense, "add" information to the stimulus. The inferential approach has had a profound impact on the perceptual sciences that continues to the present day. Its contemporary influence is perhaps best represented in computational models of vision and hearing in which the perceptual problem is recast as a form of Bayesian inference involving the speci-fication of prior knowledge and stimulus likelihoods in terms of a priori and conditional probabilities (Ellis 1996; Martin 1999; Kersten et al. 2004).

The second major theoretical approach, in chronological order, is the Gestalt psychologists' *organizational approach*. Like Helmholtz, the Gestalt psychol-ogists believed the stimulus alone to be inadequate for perception; however, whereas Helmholtz supposed that perception was determined by the most *likely* alternative given the stimulus, the Gestaltists believed it to be determined by the most *regular* or *simplest* alternative. It has proven difficult to distinguish the predictions of the organizational approach from those of the inferential approach, inasmuch as the simplest interpretation is also very often the most likely. Sounds having common onsets, for example, are simply interpreted as belonging to a single source; however, by virtue of their common onsets they are also most likely to belong to a single source. Still, the surviving popularity of the organizational approach is in large part due to such credible examples, which serve as heuristics to stimulate and guide research. A modern-day expression of the organizational approach to auditory perception can be found in the highly readable text by Bregman (1990).

The third influential theory of perception would appear more than a century later in the form of Gibson's (1950) *ecological approach*. Gibson advocated a radically different view of perception that fundamentally challenged the premise of the impoverished stimulus. The view would evolve from Gibson's insights into how the physical world through its orderly structure constrains the mapping of environment to stimulus. He would provide numerous examples, mostly in optics, of ways in which information about the layout of the environment and the events occurring therein is conveyed by patterned stimulation called *invariants*. Such significance did Gibson attach to these invariant aspects of the stimulus that he proposed that no additional information was required for veridical perception. He thus rejected the need for any inference based on prior knowledge, maintaining instead that perception is *direct*. Gibson's ideas would bring about a shift in theoretical emphasis from the ambiguous to the ordered nature of the stimulus, and this is generally acknowledged to be his greatest contribution. However, his related claims that the stimulus contains sufficient information for perception and that perception is direct have met with skepticism. For a general discussion of these issues the reader is referred to the works of Ullman (1980), Micheals and Carello (1981), and Neuhoff (2004).

Lately, there has emerged what could qualify as a fourth major approach to sound source identification: an *eclectic approach* that borrows freely from each of the three major perceptual theories described above. The approach has been used to great advantage in the area of computational modeling, most notably in the work of Ellis (1996) on computational auditory scene analysis and Martin (1999) on machine recognition of musical instruments. These authors begin with a realistic auditory front end and then incorporate all the central elements of the three major theories to improve the performance of their models. An appeal of these models is that they are scalable to new stimulus inputs (i.e., they can learn) and they can be easily tested. This makes them a potentially invaluable tool for study when their behavior is compared to that of human listeners for the same acoustic inputs and the same identification task. The eclectic approach may hold great promise for advancing our understanding of human sound source identification in the foreseeable future.

## 4. Some Elementary Physical Acoustics

We next review some fundamentals of physical acoustics that might be bought to bear on the problem of identifying a single source from the sound of impact. This section will provide the background necessary for evaluating the types of acoustic information available to the listener and the best identification performance that might be achieved in the studies to be reviewed in Section 6. The focus is on two simple resonant sources—the struck bar and plate. These two sources have received a good deal of attention in the literature for several reasons. First, while their simplicity makes them an attractive choice for study, they represent at the same time a large class of familiar musical instruments and other sound-producing objects commonly encountered in everyday listening (e.g., tuning forks, gongs, woodblocks, dinner plates). Second, as freely vibrating sources they convey less ambiguous information about their physical properties through sound than they would if acted upon continuously by an outside driving force. Third, they are a logical starting point for a foray into the problem, since the relation between their physical and acoustic properties (prior to sound radiation) is grossly captured by a few, relatively simple, equations of motion.

## 4.1 Equations of Motion for the Ideal Homogeneous Bar and Plate

Bars and plates are capable of complex patterns of motion involving at the same time longitudinal (lengthwise), torsional (twisting), and transverse (bending) modes of vibration. A strike of a hammer can hardly excite one of these modes without also exciting the others. The focus here is on the transverse vibrations, since these vibrations typically produce the largest acoustical displacements of air and so largely determine the sound radiated from the bar and plate.[1] The derivation of the equation for transverse motion of the ideal homogeneous bar is given by Morse and Ingard (1968, pp. 175–178). The result is a fourth-order differential equation giving the displacement $y$ over time at each point $x$ along the length of the bar,

$$\frac{\partial^2 y}{\partial t^2} = -\frac{Q\kappa^2}{\rho}\frac{\partial^4 y}{\partial x^4} \tag{2.1}$$

where $Q$ is Young's modulus, a measure of the elastic properties of the material making up the bar, $\rho$ is mass density, also a material property, and $\kappa$ is the radius of gyration, a measure of how the cross-sectional area of the bar is distributed about the center axis. Nominal values of $Q$ and $\rho$ for different materials can be found in Kinsler and Frey 1962, pp. 502–503. Some values of $\rho$ for different cross-sectional shapes are given in Figure 2.2.

Equation (2.1) is the first step in understanding the precise relation between the geometric/material properties of the bar and the emitted sound. The next step is to specify a driving force and boundary conditions related to how the bar is to be suspended in air. The driving force, as has been decided, will be an impact (to be represented as an impulse) applied at some point $x_c$ along the

---

[1] A notable exception is the glockenspiel, which also presents prominent torsional and shearing modes of vibration (Fletcher and Rossing, 1991, pp. 534–535).

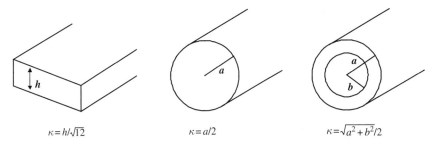

$\kappa = h/\sqrt{12}$          $\kappa = a/2$          $\kappa = \sqrt{a^2 + b^2}/2$

FIGURE 2.2. Radii of gyration $\kappa$ for some different bar shapes.

length of the bar. For boundary conditions we will consider the case in which the bar is allowed to vibrate freely at both ends (e.g., xylophone), is rigidly clamped at one end (e.g., tuning fork), or is simply supported (hinged) at one end (e.g., glockenspiel). These cases are selected because they are useful, although idealized, approximations to models of real resonant objects and because their mathematical properties are well understood. A general solution to Eq. (2.1) is a sum of $N$ exponentially damped sinusoids,

$$y_x(t) = \sum_{n=1}^{N} A_{n,x,x_c} e^{-t/\tau_n} \sin(2\pi f_n t), \qquad (2.2)$$

where $y_x(t)$ represents the wave motion at a point $x$ along the bar, and $A_{n,x,x_c}$, $\tau_n$, and $f_n$ are respectively the amplitude, decay constant, and frequency of the $n$th natural mode of vibration (Morse and Ingard 1968). Note that unlike the modal frequencies and decay constants, the modal amplitudes vary with point of measurement and strike locations. For the different boundary conditions, Fletcher and Rossing (1991, pp. 57–60) give the frequency of the $n$th natural mode of an ideal bar of length $l$ as

$$f_n = \frac{\pi \kappa}{2l^2} \sqrt{\frac{Q}{\rho}} \beta_n^2 \qquad (2.3a)$$

where

$$\begin{cases} \beta_1 = 0.597, \beta_2 = 1.494, \beta_{n \geq 3} \approx n - \frac{1}{2} & \text{Clamped} - \text{Free} \\ \beta_n = n\sqrt{2} & \text{Hinged} - \text{Free} \\ \beta_1 = 1.506, \beta_2 = 2.500, \beta_{n \geq 3} \approx n + \frac{1}{2} & \text{Free} - \text{Free} \end{cases} \qquad (2.3b)$$

Except in the case of the hinged-free bar, the natural frequencies are not harmonic. The bar's stiffness causes the wave velocity to be greater at higher frequencies, where the degree of bending is greater. Higher-frequency partials thus become progressively higher in frequency compared to a simple harmonic progression. The functional relation between the frequency of the lowest mode

$(f_1)$ and the physical parameters of the bar as given by Eq. (2.3) is demonstrated by way of example in Figure 2.3. The curves give constant-$f_1$ contours (higher partials scaled accordingly) for a circular, clamped-free bar 1 cm in diameter and 10 or 11 cm in length (continuous and dashed curves respectively).

Equation (2.3) and Figure 2.3 show that the natural frequencies alone convey a good deal of information, though not completely unambiguous information, about the physical properties of the bar and its manner of support. Note, for example, from Eq. (2.3) that the ratios among the natural frequencies $f_n/f_1$ are unaffected by the material and geometric properties of the bar, but do vary with how the bar is suspended in air. This means that the frequency ratios could possibly be used to identify the manner of support independent of the physical properties of the bar. Note also from Figure 2.3 that a low $f_1$ tends to be associated with bars made up of denser (larger $\rho$), more compliant (smaller $Q$) materials, a possible cue for material identification. The cue is not entirely reliable, however, since $f_1$ additionally varies with the bar's size and width, a low $f_1$ being associated with long and narrow bars (large $l$, small $\kappa$). Indeed, $f_1$ has more often in the literature been implicated as a primary cue for object size rather than material (Gaver 1988, 1993a,b; Rocchesso and Ottaviani 2001). In Section 4.3 we consider how the ambiguity regarding material and size might

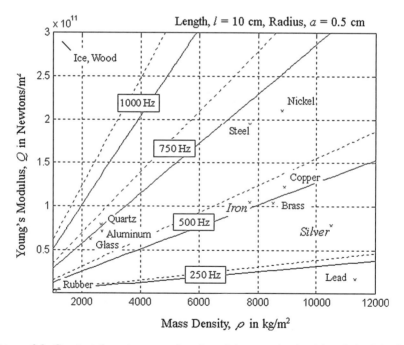

FIGURE 2.3. Constant $f_1$ contours as a function of the mass density ($\rho$) and elasticity ($Q$) of a circular, clamped-free bar 1 cm in diameter and 10 or 11 cm in length (continuous and dashed curves respectively). Values of $\rho$ and $Q$ corresponding to different bar materials are given by small crosses with labels.

be resolved by using the combined information in the frequency $f_1$ and decay constants $\tau_1$ of partials; related perceptual studies are then reviewed in Section 6.

The equation of motion for a plate is rather more complicated than that for the bar. Unlike the bar, a plate tends to curl up sideways when it is bent along its length. This produces an additional strain, which has the same effect as would an increase in the stiffness $Q$ of the plate. The effect enters into the equation of motion as an additional factor known as Poisson's constant, here denoted by $s$ (approximately 0.3 for most materials). Otherwise, the general solution to the equation of motion for the plate is the same as that for the bar, Eq. (2.2), but with the wave motion specified in two dimensions $(x, z)$ rather than one. The natural frequencies depend on the boundary conditions as before, but now they also depend on the shape of the plate. For the simply supported rectangular plate of width $a$ and height $b$, the natural frequencies are

$$f_{mn} = 0.453h \sqrt{\frac{Q}{\rho(1-s^2)} \left[ \left( \frac{m+1}{a} \right)^2 + \left( \frac{n+1}{b} \right) \right]}, \qquad (2.4)$$

where the subscript $mn$ designates the two dimensions of the modes.[2] For the clamped and free rectangular plate, the natural frequencies entail rather lengthy computational formulas and so are not presented here. They can be found, however, along with those of the circular plate, in Fletcher and Rossing (1991, pp. 71–85).

As was the case for the bar, the lowest natural frequency provides information regarding the material and size of the plate, while the ratio of the natural frequencies provides information regarding the manner of support. In contrast to the case of the bar, however, here the ratio of the natural frequencies provides additional information regarding the shape of the plate independent of its material or size. This can be seen, for example, in Eq. (2.4), where for a given shape (given $a/b$), $f_{m0}/f_{10}$ and $f_{0n}/f_{01}$ remain constant despite changes in material properties ($\rho$, $Q$, and $s$) and size (equal scaling of $a$ and $b$). The role of the natural frequencies in the perception of plate shape, material, and size is discussed further in Section 6. Finally, comparing Eqs. (2.3) and (2.4), it is possible to see how the natural frequencies might also mediate the identification of source type: bar or plate. For the plate, the natural frequencies tend to be greater in number and more closely spaced than they are for the bar, this by virtue of the fact that the plate vibrates transversely in two dimensions, the bar in one. Indeed, if one excludes the degenerate case $a = b$, then over the bandwidth $f_{max} - f_{min}$ occupied by the partials, the number of modes $N$ will generally be greater for the plate. The quantity $N/(f_{max} - f_{min})$ is identified here as "modal density." It is a property that is independent of other properties of the bar or plate and so provides a viable cue for the distinction between these two general classes of sound sources.

---

[2]A slight, but perceptually important, modification to Eq. (2.4) is required for anisotropic materials such as wood, which owing to the grain, have different mechanical properties in different directions (cf. Fletcher and Rossing 1991, p. 84).

## 4.2 Damping

After impact, the energy that was initially injected into the object is dissipated over time until the object eventually comes to rest. The damping of motion can generally be ascribed to three causes: (1) viscous drag of the surrounding medium, (2) energy transfer through the supports, and (3) friction created by shearing forces within the object (internal friction). In the idealized representation given by (1), the damping is exponential and given by a single decay constant $\tau$. The combined effect of the three sources of damping is therefore $1/\tau = 1/\tau_a + 1/\tau_b + 1/\tau_c$, where $\tau_a$, $\tau_b$, and $\tau_c$ are the decay constants associated with each factor. Note that $\tau$ in (2) appears with the subscript $n$, which denotes the dependence on frequency associated with each factor. Figure 2.4 shows hypothetical constant decay contours comparable to the constant-frequency contours of Figure 2.3. In this case, the decay is taken to be inversely proportional to the cube root of the frequency (cf. Morse and Ingard 1968, p. 222).

One or more of the three general sources of damping largely determine the value of $\tau$ depending on the type of object, the surrounding medium, and the manner of support. A thin plate, for example, displaces a greater volume of air than a bar of similar length and so is subject to greater damping due to viscous drag by the air. The bar dissipates much of its energy through internal friction, but a greater proportion of that energy may be transferred through the supports if, as in the case of a tuning fork, the bar is coupled to a resonator box.

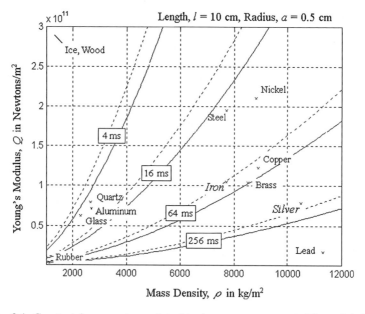

FIGURE 2.4. Constant decay contours plotted in the same manner as in Figure 2.4, for the same bar.

The varied circumstances that give rise to a particular value of $\tau$ can complicate the analysis of how it might ultimately inform a listener of the source. Indeed, of the three general causes of damping, only one, internal friction, is exclusively a property of the source. Notwithstanding these complications, the damping due to internal friction is given special consideration here, since it has been identified as a potential acoustic cue in the identification of materials.

The significance attached to internal friction derives from the fact that it is an intrinsic property of materials; it does not depend on the geometry of the source, the manner of support, or how the source is driven to vibrate. This makes the decay parameter $\tau_c$ associated with internal friction a likely suspect among the potential acoustic cues for material identification. Wildes and Richards (1988) have shown specifically how $\tau_c$ might be used to this end. Their analysis hinges on the fact that for real materials, there is a time dependency between the applied force (stress) and the resulting deformation of the material (strain). To adequately describe this behavior, two elastic moduli are required, one associated with each deformation (note that only one was used in Eq. [2.1]). With the two moduli denoted by $Q_1$ and $Q_2$, the decay constant due to internal friction is

$$\tau_c = \frac{1}{\pi f} \frac{Q_1}{Q_2}, \tag{2.5}$$

where $Q_2/Q_1$ is the measure of internal friction uniquely associated with the material (Fletcher and Rossing 1991, p. 51). Wildes and Richards suggest, based on Eq. (2.5), that a listener who has access to both the frequency $f$ and the decay constant $\tau_c$ of partials could, in theory, recover a measure of internal friction that would allow for a material classification. The measure is derived from a frequency-independent decay time $\tau_c f$ expressed as the number of cycles required for the sound to decay to $1/e$ of its original value. Wildes and Richards are careful to note that the classification of material based on $\tau_c f$ would require a prevailing condition in which internal friction is the predominant source of damping; however, they do not discuss at great length the circumstances under which this is generally expected to be true. Indeed, for many musical instruments, internal friction is not a predominant source of damping. Also, empirical measurements of internal friction have suggested that the functional relation between $\tau_c f$ and $Q_2/Q_1$ is more nearly quadratic than linear, as indicated by (5) (Krotkov et al. 1996). Notwithstanding these qualifications, the role of frequency-independent decay in the human identification of material from sound is the topic of several studies reviewed in Section 6.

## 4.3 Manner of Impact

As noted previously, the physical properties of the striking implement and how it comes in contact with the source can have a profound effect on the resultant sound. Importantly, however, the effect is not to change the natural frequencies of vibration, but rather to change their relative amplitudes. The general rule for point of contact is that each mode of vibration is excited in proportion to how much it is involved in the free vibration at that point. Figure 2.5 demonstrates the principle. The leftmost panels show by vertical displacement the degree to which two different circular modes are involved in the motion of the simply supported circular plate. The two modes are excited maximally when the point of contact is squarely in the middle of the plate, since at this point both modes participate maximally in the motion. In the middle panels, the lower-frequency mode (top) is attenuated relative to the higher-frequency mode (bottom), because the point of contact is near one of the nodes of the higher-frequency mode. In the rightmost panels, the reverse is true, since the point of contact is near a node for the lower-frequency mode. The area of contact works in a similar way, except that the vibration of a given mode is reduced when the area is so large as to cover regions of that mode moving in opposite directions.

Whereas the area and point of contact relate to the spatial dimension of the impact, the hardness of the mallet relates to its temporal dimension. Generally, for a soft mallet the force of impact increases more gradually over time than it does for a hard mallet. There is, as a result, less high-frequency energy in the impact produced by a soft mallet. Less high-frequency energy in the impact simply means that there will be less high-frequency energy showing up in the response of the bar or plate (Fletcher and Rossing 1991, pp. 547–548). For a soft mallet, the resulting "tilt" in the spectrum toward lower frequencies causes the sound to appear muted and less bright than the sound produced by a hard mallet. Freed (1990) identifies an acoustic parameter closely related to spectral tilt as the primary cue underlying auditory perception of mallet hardness. That study is another to be discussed in Section 6.

Finally, the force of impact affects the amplitude of sound in a complex way that depends both on the physical properties of the source and the manner in which it is driven to vibrate. Equation (2.6) shows by way of example the relevant relations that exist between the amplitude of motion, force of impact,

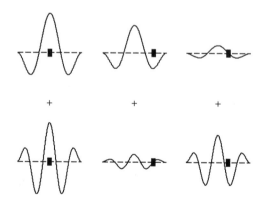

FIGURE 2.5. Top row: Variation in the amplitude of a single mode of vibration of a circular plate for different points of hammer contact indicated by the square symbol. Bottom row: Variation in amplitude for a second, higher-frequency, mode.

and the material and geometric properties of the clamped-free bar struck at the free end:

$$A_n \propto (-1)^{n-1} \frac{Pl}{S\kappa} \sqrt{\frac{1}{\rho Q}} \beta_n^{-2}, \qquad (2.6)$$

where $A_n$ is the amplitude of the $n$th partial as in Eq. (2.2), $P$ is the time integral of the impact force, and the $\beta_n$ are as given in Eq. (2.3b) (Morse and Ingard 1968, p. 185). Note first the relation of amplitudes to the quantities $\rho$ and $Q$ associated with the material of the bar. We know that the natural frequencies grow with the ratio of these quantities (Eq. [2.3a]), but the amplitudes according to Eq. (2.6) decrease with their product; that is, the amplitude of motion (and so the amplitude of sound) is generally less for stiff, dense materials (e.g., iron vs. wood). For a constant applied force, the amplitudes also depend on the geometry of the bar, increasing with length $l$, but decreasing with cross-sectional area $S$ and the modulus of gyration $\kappa$. The latter is expected, since the greatest resistance to bending will be afforded by the bar with the largest cross-sectional area distributed farthest from the center axis of the bar. Figure 2.6 gives constant-amplitude contours generated according to Eq. (2.6), which permit comparison to the constant-frequency and decay contours of Figures 2.3 and 2.4. The complex

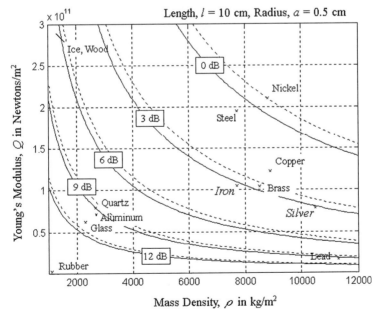

FIGURE 2.6. Constant-amplitude contours plotted in the same manner as in Figure 2.4, for the same bar.

interaction of force of impact with the physical properties of the object and how it is struck makes the determination of force of impact from sound a nontrivial problem and one meriting study (cf. Lakatos 2001).

## 4.4 Sound Radiation

It is common in the literature on sound source identification to treat the sound-producing object as though it occupied a single isolated point in space, that except for intensity, the sound radiating outward is everywhere the same and varies only as a function of time. A point-source approximation to real vibrating objects is acceptable in theory when the object's size is small with respect to the wavelength of the sound (Fletcher and Rossing 1991, pp. 162–163). However, the wavelengths of audible sounds in air are often only a few centimeters (about 17 cm at 2 kHz under normal atmospheric conditions). There are many cases, then, in our common experience for which the point-source representation is not met. These cases are of special interest, since the emitted sound can potentially provide cues about the object's size and shape. The cues may be evident both in terms of the spatial extent of the sound (often associated with the perceptual attribute of volume) and in terms of the efficiency with which sound power is radiated at different frequencies depending on size and shape (Fletcher and Rossing 1991, pp. 156–170). The analysis of these cues is complex and well beyond the scope of this chapter. They are mentioned here only because they will prove pertinent to the discussion of auditory perception of size and shape later, in Section 6. For an in-depth treatment of the topic of sound radiation as it is affected by object size and shape, the reader is referred to the works of Fletcher and Rossing (1991, Chapter 7) and Skudrzyk (1968).

## 5. Some Relevant Auditory Psychophysics

We have so far reviewed some basic physical principles governing the motion and radiated sound of simple resonant objects after impact. In the process, specific acoustic cues have been implicated as those that listeners could potentially use to recover gross physical attributes of these sources and the manner in which they are struck (summarized in Table 2.1). We next evaluate the viability of these cues for source identification. A natural first question is whether these cues vary

TABLE 2.1. Some acoustic features associated with basic source attributes of ideal homogeneous struck bars and plates from Section 4.

| Physical attribute | | Acoustic cue(s) | |
|---|---|---|---|
| Type | bar vs. plate | Modal density | $N/(f_{max} - f_{min})$ |
| Size | $l$ (bar), $a \times b$ (plate) | Frequency, radiation pattern | $f_1, A(x, y, z)$ |
| Shape | $\kappa$ (bar), $a, b$ (plate) | Frequency ratios, radiation pattern | $f_n/f_1, A(x, y, z)$ |
| Material | $\rho, Q, s$ | Frequency-independent decay | $\tau_c f$ |
| Manner struck | $x_c, P(t)$ | Spectral profile | $A_n/A_1$ |

over a sufficient discriminable range to allow meaningful distinctions among sources. The question has not been widely addressed in the literature, but it is not to be ignored. There is evidence at least (reviewed in Section 6) to suggest that in many instances, the cues that would serve to disambiguate sources can be quite subtle, and so they may go undetected even when the sounds themselves are clearly perceived as different. This section then reviews studies whose goal it was to measure the *limits* of a listener's ability to discriminate changes in the cues listed in Table 2.1. Unlike many studies to be discussed in Section 6, these studies make no attempt to simulate "real-world" listening conditions. Rather, to ensure best performance they provide listeners with extensive training in the task, they use forced-choice procedures with trial-by-trial feedback, and they make every attempt to keep stimulus uncertainty to a minimum.

## 5.1 Frequency and Frequency Ratios

The frequencies of partials, or more specifically their scaling by $f_1$, have been implicated here and elsewhere as a possible cue mediating the perception of object size (Gaver 1988, 1993a,b). Clearly, other attributes of objects equally affect frequency; however, to the extent that other attributes are unchanging in a given situation, one should expect listeners to be capable of discriminating quite small differences in the size of objects based on the frequency cue alone. Human sensitivity to changes in frequency is extraordinarily good. At moderate-to-high sound levels (20–80 dB SL) and over most of the dynamic range of hearing (from 125 to 8000 Hz), the just-noticeable difference (JND) in the frequency of a pure tone (expressed as $\Delta f/f$) ranges from 0.2% to 0.9% (Wier et al. 1977). This would translate to a mere 0.4%–1.8% change the length of an ideal bar; cf. Eq. (2.3). The situation is not much different at lower sound levels, since the JND for frequency increases only slightly, to about 1.5%, near the threshold for detection. Similarly small JNDs are obtained for frequency shifts in harmonic complexes (Moore et al. 1984) and synthetic vowels (Flanagan and Saslow 1958).

The frequency ratios among partials have been identified as a cue to object shape and manner of support; however, only one study, to the author's knowledge, has measured differential sensitivity for the frequency ratio between simultaneously presented tones. Fantini and Viemeister (1987) used a two-interval, forced-choice procedure in which the base frequency of simultaneous tone pairs was randomized to encourage listeners to make within-pair comparisons of tone frequencies. Though data from only two listeners were reported, the results indicate that for small frequency ratios, less than 1.4, the discrimination of frequency ratios is quite good. At a base frequency ratio of 1.25, the average JND was just 1.2%, not much greater than that for pure-tone frequency discrimination. In a related pair of studies, good sensitivity was also demonstrated for the detection of inharmonicity of multitone complexes (Moore et al. 1984, 1985). These studies suggest that hearing sensitivity should impose little limit on perceived differences in size, shape, or manner of support wherever frequency differences are a viable cue.

## 5.2 Decay Time

Decay time, or more precisely, frequency-independent decay time, $\tau f$, has been implicated as a cue for material classification. Unfortunately, there are presently few data on the discrimination of decay time. Schlauch et al. (2001) present the most extensive set of data for which an objective forced-choice procedure was used. Their stimuli were exponentially damped bursts of Gaussian noise with base decay times ranging from $\tau = 2$ to 20 ms. Results from four trained listeners were quite similar, with JNDs ($\Delta\tau/\tau$) gradually decreasing from near 35% at a base value of $\tau = 2$ ms to about 12% at a base value of $\tau = 20$ ms. These values, it should be noted, are much larger than the JNDs for changes in frequency or frequency ratio described above. Van Heuven and Den Broecke (1979) obtained similar results for linearly damped sinusoids using a method of adjustment. Only one study, to the author's knowledge, has measured JNDs for frequency-independent decay time (i.e., $\Delta\tau f/\tau f$). Using a same–different forced-choice procedure and synthetically generated sounds of plucked strings, Jarvelainen and Tolonen (2001) report JNDs in the range of 25% to 40% for five listeners whom they described as musically experienced.

## 5.3 Spectral Shape

The relative amplitude of partials (spectral shape) has so far been identified as a source of information regarding manner of impact (point of contact, mallet hardness, and so forth). The discrimination of relative amplitude has undergone extensive study over the last two decades, but the changes to be discriminated have typically been quite rudimentary. In most cases, the listener is simply asked to detect a single bump in an otherwise flat spectral profile created by incrementing the level of one tone of a multitone complex. Overall level is roved to discourage listeners from basing decisions on simple differences in overall level. The results of the different studies are in good agreement, showing best sensitivity (JND of about 13%–15%) for center frequencies of the complex, but progressively poorer sensitivity as the frequency of the incremented component approaches either extreme within the complex (JND from 35% to 50%) (Green and Mason 1985; Green et al. 1987). An important question for the purposes of this chapter is whether such data can be used to predict detection of more complex changes in spectral shape, as might, for example, be associated with differences in manner of impact. Unfortunately, such attempts at prediction have met with only limited success (Bernstein and Green 1987; Green et al. 1987).

## 5.4 Spatial Extent

The spatial pattern of sound power radiating from a source, as has been noted, contains potential information regarding the source's size and shape. This information may be conveyed to a listener by the perceptual impression of the sound's spatial distribution or spatial extent. Unfortunately, like much of the research on auditory perception of size and shape, psychoacoustic research on the spatial attributes of sound has typically treated the sound as originating from a point in space. As a result, most of the psychoacoustic research has been concerned with the perceived location or position of the sound and comparatively little with its perceived spatial extent. The perceived location of sound sources in three dimensions is known to be quite accurate, if not somewhat compressed in the plane of elevation (Wightman and Kistler 1989). This would seem to imply a comparable measure of accuracy in judgments regarding the spatial extent of sound. However, what little research exists on the discrimination of spatial extent suggests quite the opposite. Perrott (1984) found the discrimination of horizontal extent (18.75 ± 12.5°) by trained listeners to be near chance for tones and correlated noise samples played over a stereophonic speaker array. Significantly better than chance performance was obtained only when the signals played over the different speakers were uncorrelated noise samples or tones differing in frequency, and then only when these signals contained energy at frequencies below 1500 Hz. These data do little to suggest that spatial extent is often a viable cue to source size or shape, but in any case, the possibility should not be ruled out without further research.

# 6. Identification of Simple Source Attributes

In the previous two sections we analyzed potential acoustic cues for auditory identification of some simple source attributes and reviewed what is known regarding human sensitivity to these cues. In this section we now review the research investigating the ability of listeners to identify such attributes from sound. The literature is not extensive, and so the criterion for inclusion of studies was intentionally lax. Liberal reference is made to self-described preliminary data presented in meeting abstracts and/or technical reports. Scaling studies, which do not involve identification per se, are included wherever they offer potential insights into the problem. Also, so as not to be overly restrictive, a rather broad definition is adopted of what constitutes a sound source identification experiment. This is judged to be any study in which listeners classify (label) sounds into two or more discrete categories based on some physical attribute or attributes of the sound-producing source. The definition appeals to the broad notion of an identification experiment as described by Luce (1993 p. 112).

## 6.1 Material

Of the various basic attributes of a source one could possibly judge from sound, material composition has received the most attention. Gaver (1988) conducted the earliest experiments. He used sound recordings of wood and metal bars of variable size. The bars were struck with a soft mallet as they rested on a carpeted surface. Listeners were informed beforehand of the two categories of material and were presented by way of example the sound of the smallest and longest bar in each category; no further information was given to aid performance.

In subsequent experimental trials, all 19 participants were able to identify the material of the bar from the recording with near perfect accuracy. Similar results were obtained using synthetic bar sounds. Gaver concluded based on his results that listeners are adept at material classifications from sound. This was also the conclusion later reached by Kunkler-Peck and Turvey (2000), who studied material identification from the airborne impact sounds of large suspended plates. The plates were of three materials (wood, steel, and Plexiglas) and three shapes (square, triangular, and rectangular). The dimensions were chosen so that shapes within each material group had the same mass and across material groups had the same surface area. Listeners saw tokens of each plate type, but did not hear the sounds of the plates prior to experimental trials and were not given feedback during experimental trials. Material identification was near perfect for the seven untrained listeners participating in this study.

Not all studies have reported such high levels of performance for material classification. Tucker and Brown (2003), like Kunkler-Peck and Turvey (2000), had listeners judge material from impact sounds of large suspended plates. The plates were made of wood, aluminum, and plastic and differed is size and shape (square, triangular, and circular). Twelve untrained listeners identified the plate from a sound recording on each trial by selecting from a list of the possible material/shape combinations; no other feedback was given. Plastic and wooden plates were frequently confused but easily distinguished from the aluminum plates. Notably, the confusions between plastic and wood increased when the plates were damped by submerging them in water. Tucker and Brown suggest that the discrepancy of their results with those of Kunker-Peck and Turvey might have been due to the fact that their sounds were recorded. However, similar results have since been obtained by Giordano (2003) and Giordano and McAdams (2006) using "live" sounds. In these studies, rectangular plates of plastic (Plexiglas), wood, glass, and steel were varied in height, width, and area, and were struck with mallets of different hardnesses. Significant confusions were observed for plastic with wood, and for glass with steel, though either member of the one pair was easily distinguished from either member of the other. Damping increased the confusions between plastic and wood, and also caused glass to be confused with these materials.

Lutfi and Oh (1997) conducted an experiment similar to that of Gaver (1988), but with the specific goal of measuring the limits of listeners' ability to classify bar material. The stimuli were synthetic sounds of freely vibrating bars perturbed at random from one trial to the next in mass density and elasticity. To ensure best performance, listeners were given extensive prior training in the task and received trial-by-trial feedback during the course of experimental trials. The physical parameters of the synthesis were then chosen to yield performance levels in the range 70%–90%. The material pairs yielding the targeted performance levels were nominally iron versus silver, steel, or copper; and glass versus crystal, quartz, or aluminum. Though these pairs shared similar physical and acoustical properties, the sounds for each pair were clearly discriminable from one another.

Moreover, a detection-theoretic analysis indicated that performance could have been much better had the information in the sounds been used optimally.

Two interpretations have been given for the less-than-optimal performance observed in the classification of material. The first implicates a lack of sufficient information in the stimulus. Carello et al. (2003), for example, attribute the results of Lutfi and Oh to a loss of information entailed in the use of synthetic stimuli. They do not indicate what this information might be or how similar results could have been obtained by Giordano (2003) using live sounds. The other interpretation emphasizes the listener's failure to use existing information in the stimulus. Lutfi and Oh, for example, found the trial-by-trial judgments of individual listeners to be much more strongly correlated with frequency $f_1$ than with frequency-independent decay $\tau_c f$, even though the latter would theoretically have allowed for more accurate performance (cf. Wildes and Richards 1988).

More recently, studies have placed less emphasis on overall performance and greater emphasis on evaluating the acoustic cues underlying material classi-fication in different conditions. These studies typically undertake finer-grain *molecular* analyses that evaluate the relation between individual stimuli on each trial and the judgment of listeners. Here again the results have been mixed. Results similar to those of Lutfi and Oh were found by Hermes (1998) for both free-response and fixed-category classification of synthesized impact sounds. Giordano (2003) and Giordano and McAdams (2006) also report a stronger corre-lation of judgments with $f_1$ than $\tau_c f$; but only for the confused materials (wood and plastic, glass and steel). Frequency-independent decay was a better predictor of the gross discrimination across these pairs, though this outcome was equally well predicted by other acoustic cues, including most notably overall sound duration and peak loudness. Klatzky et al. (2000), using synthesized sounds of struck, clamped bars, report greater reliance on $\tau_c f$ than $f_1$ for both judgments of material similarity and material classification (rubber, steel, glass, or wood). The correlations, moreover, were in the direction predicted by Wildes and Richards (1988). Similar results were obtained in an informal study by Avanzini and Rocchesso (2001) using synthetic impact sounds and in the study by Tucker and Brown (2003), although Tucker and Brown analyzed only for $\tau_c f$.

In an attempt to reconcile these results, Giordano and McAdams (2006) suggest that $f_1$ may mediate judgments for materials having similar physical properties, whereas $\tau_c f$ may mediate judgments for materials having vastly different physical properties. The idea receives some support in these studies, except that the discrimination between glass and steel, for which $f_1$ was indicated to be the cue, involve two materials that differ greatly in mass density and elasticity (see Figure 2.3). Lutfi (2000, 2001) alternatively suggests that the reliance on $f_1$ may reflect the robust nature of this cue with respect to stimulus ambiguity and sensory noise. The idea is explained further in Section 6.3, on the detection of hollowness.

Taken together, the studies on material identification from sound show a wide range of performance levels in different conditions that do not appear to be simply related to any one acoustic cue. Given the near singular attention to $f_1$ and

$\tau_c f$, correlations of judgments with these cues should probably be interpreted with caution. Their meaning is certainly less clear in cases in which there is little variance in listener judgments that is not already predicted by the correct material categories (i.e., when performance is near perfect). It may be that the apparent discrepancies among these studies will ultimately be understood in terms of the significant stimulus and procedural differences that exist. This, however, will likely require further research.

## 6.2 Size and Shape

Next to material, the physical attributes of size and shape have received the most attention in the literature on sound source identification for inanimate objects (cf. Patterson, Smith, van Dinther, and Walters, Chapter 3). Gaver (1988) and Tucker and Brown (2003) obtained listener estimates of size using the same stimuli and procedures as in their experiments on material classification. Gaver found ratings of bar length to be much compressed and strongly biased by the material of the bar; this in contrast to near perfect material classification by his listeners. The 11 listeners were roughly split in terms of whether they perceived wood or metal bars to be longer, making it appear as though size judgments were independent of material when ratings were averaged across listeners. This is a noteworthy result, given that few of the studies reviewed in this chapter report individual listener data (cf. Lutfi and Liu, 2007). The tendency of listeners to underestimate size increased with the use of synthetic stimuli, though the pattern of results was otherwise similar. Tucker and Brown (2003) largely replicated these results using their recordings of freely vibrating plates of different materials and shapes. Twelve naive listeners judged relative size from sound pairs produced by plates of the same material. Judgments again were less accurate than those for material; listeners grossly underestimated differences in size and often confused the larger of the two plates with the smaller.

The acoustic information that might have mediated judgments of size in these studies is unclear. Gaver (1988, 1993a,b) implicates pitch based on the inverse relation between bar length and frequency (cf. Section 4.1). His listeners could not, however, have judged size exclusively from pitch, since metal bars, which were judged higher in pitch than wood bars, were consistently judged longer in length (Gaver 1988, p. 127). Relevant to this point, Ottaviani and Rocchesso (2004) found pitch judgments for struck cubes and spheres to be highly variable and ordered with size only in constrained conditions. Tucker and Brown (2003) suggest that the combination of frequency and decay may have served as a cue to size in their study. They base their conjecture on informal interviews of listeners and the observation that plates were generally perceived to be smaller when damped. Carello et al. (1998) obtained judgments of the length of rods dropped onto a linoleum floor. They found actual length to be a better predictor of judged length than sound frequency, duration, or amplitude. However, because the rods were allowed to bounce, additional spatial cues and/or cues in the timing between bounces may have been available.

2. Human Sound Source Identification 35

The results from studies on the perception of shape are about as varied as they are for the perception of size. Lakatos et al. (1997) examined the perception of the cross-sectional height/width ratios of metal and wooden bars from sound recordings. On each trial, a pair of sounds accompanied a visual depiction of the corresponding bars in the two possible presentation orders. Listeners selected the visual pair they believed to be in the correct order for each trial. The data were analyzed in terms of a perceived dissimilarity metric, so it is difficult to know just how accurately listeners performed the task; at least some performed significantly below chance. Nonetheless, a multidimensional scaling analysis of the dissimilarity scores showed listeners to be sensitive to differences among sounds associated with differences in the height/width ratios. The two acoustic features found to correlate strongly with both the bar's height/width ratios and the bar's coordinates in the multidimensional solution were the frequencies of the torsional modes of vibration and the ratio of frequencies of transverse modes corresponding to height and width.

Kunkler-Peck and Turvey (2000) had listeners provide analogue estimates of the height and width of large, freely vibrating rectangular plates made of steel, wood, and Plexiglas. Listeners underestimated height and width, consistent with the previously described results for size; yet they were able to judge the proportional relation of height to width with reasonably good accuracy. In a second experiment, listeners were asked to identify the shapes of plates from sound. The listeners were constrained to the labels circular, triangular, and rectangular, but otherwise received no prior information regarding the plates and no feedback during experimental trials. Identification performance was found to be statistically above chance, though not nearly as good as it was for material (cf. Section 6.1). The authors take their results to reflect a dependency of perceived dimensions on the ratios of modal frequencies uniquely associated with plate shape; cf. (4).

Tucker and Brown (2003) conducted a shape classification study very similar to that of Kunkler-Peck and Turvey (2000) involving circular, triangular, and square plates of aluminum, wood, and plastic. In contrast to the results of Kunkler-Peck and Turvey, these authors found shape identification performance to be no better than chance. As a possible cause of the discrepancy the authors point to the fact that Kunkler-Peck and Turvey's live sounds could have contained spatial cues not available in their sound recordings. Only one other study, to the author's knowledge, has reported clearly better than chance shape identification performance for sounds devoid of spatial cues. Rocchesso (2001) reported statistically better than chance identification of shape from the synthesized sounds of resonating spheres and cubes.

## 6.3 Hollowness

The problem of determining whether an object is hollow or solid from sound is of special interest because it is not easily preempted by sight; one often cannot tell simply by viewing an object whether it is hollow or solid. Lutfi (2001) studied the detection of hollowness using sounds synthesized according to the simple equation for transverse motion of the struck, clamped bar (Eq. [2.1]). Listeners were to select from a pair of sounds on each trial the sound corresponding to the hollow bar. The bars were perturbed in length on each presentation. The goal was to measure the limits of the listener's ability to detect hollowness; hence, bar parameters were selected to yield performance in the range of 70%–90% correct after listeners received extensive prior training in the task and feedback throughout all experimental trials. Despite similar performance levels, listeners were clearly split in their approach to the task. A regression analysis of each listener's trial-by-trial responses showed that half of the listeners relied on a specific relation between frequency and decay, a unique solution to the task according to the equations of motion. The other half of the listeners, however, were found to rely on simple statistical differences in frequency. The pattern of results was similar for wood, iron, and aluminum bars.

Lutfi (2001) suggests that the results can be understood in terms of the vulnerability of the relevant acoustic relations to sensory noise. Figure 2.7, which is from that paper, demonstrates the idea. The left panel gives, for different trials, the values of frequency and decay for a single partial of the hollow (o) and solid (x) bar sounds. Here, decisions based on the correct acoustic relation (decision border given by the continuous curve) yields perfect performance, while decisions based on frequency alone (decision border given by the vertical dashed line) produce only somewhat better than chance performance. The right panel shows the effect of sensory noise simulated by adding a small amount of jitter to each acoustic parameter estimated to be in agreement with the jnds for these parameters (as reported in Sections 5.1 and 5.2). Here, the two decision strategies yield nearly equal identification performance. The results of these simulations are noteworthy because they suggest that small variations in the

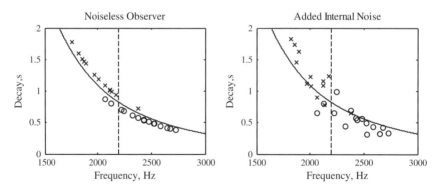

FIGURE 2.7. Simulated effect of internal noise on listener judgments of hollowness (after Lutfi, 2001). Trial-by-trial values of stimulus parameters are given for hollow (o) and solid (x) bars with (right panel) and without (left panel) internal noise jitter. Continuous and dashed curves give, respectively, decision borders for judgments based on intrinsic acoustic information and on frequency alone. See text for further details.

acoustic signal, as would be associated with sensory noise, can have a profound effect on listener judgments even when variation in the acoustic signal is largely dominated by physical variation in the source.

## 6.4 Manner of Impact

As emphasized at the beginning of this chapter, the sound emitted by a resonant object depends as much on the manner in which the object is driven to vibrate as it does on the object itself. This means that to correctly identify any property of the object from sound, the listener must at the same time accurately judge the action giving rise to the sound. Few studies have addressed this latter aspect of the problem, though one particularly relevant study was conducted early on by Freed (1990). Listeners in this study judged the hardness of mallets from recordings of the mallets striking cooking pans of different sizes. Hardness was rated on a scale of 0–9 after listeners were informed about the nature of the sound events and heard examples of the hardest and softest mallet sounds. Listener ratings were well ordered with mallet hardness (as reported, since physical hardness was not measured) and were largely unaffected by changes in frequency $f_1$ related to pan size. A measure of spectral centroid, closely related to spectral tilt, as discussed in Section 4.3, was found to be most predictive of listener ratings. The results were taken as evidence that listeners can perceive properties of the driving force independently of the properties of the resonating source.

More recently, studies have raised questions regarding the generality of this conclusion. Giordano and Petrini (2003) conducted a study involving judgments of mallet hardness similar to that of Freed. Impact sounds were synthesized according to a model having three free parameters: $f_1$ and $\tau_c f$, related to the geometric and material properties of the resonating source, and $k$, a stiffness coefficient, related to mallet hardness. The values of $k$ and $\tau_c f$ had similar effects on listener ratings, suggesting confusion between mallet hardness and material properties of the source. Also, in contrast to the results of Freed, there were large individual differences in listeners' judgments related to the effects of frequency, $f_1$. The scaling results were reinforced by free and forced-choice identification of hardness, which indicated a bias of listeners to perceive variation among stimuli in terms of variation in the source parameters ($f_1$ and $\tau_c f$) rather than variation in impact ($k$).

Similar results have since been obtained using real sounds as stimuli. Grassi (2005) had listeners estimate the size of wooden balls from the sounds they made when dropped from a constant height onto a clay plate. Listeners had no prior knowledge of the ball, the plate, or the height from which the ball was dropped. Though listeners were able to judge ball size with reasonable accuracy, their estimates were influenced by the size of the plate, larger plates yielding larger estimates of ball size. When questioned about what they heard, listeners were able to identify the event as a ball dropping on a plate; however, they more often identified the ball to be made of clay (the material of the plate) than wood, and more often reported the ball to be of different materials than the plate to be of different sizes. Grassi attributes the apparent discrepancy of his results with those of Freed (1990) to what listeners knew beforehand regarding the experiment, in particular, the fact that his listeners were given very little information about the sounds, whereas Freed's listeners were told that the sounds were of mallets striking pans of different sizes and were allowed to hear sound examples of the hardest and softest mallets prior to the experiment.

## 6.5 Synthetic versus Real

In recent years, studies of sound source identification have more frequently employed physically informed models for sound synthesis and analysis. The goal is to exploit what is known regarding the physics of vibrating bodies to achieve a precise analytic representation of the various sources of acoustic information for identification and to exert greater and more reliable control over relevant acoustic variables and their relations. The approach seems likely to gain in popularity as synthesis techniques improve to the point where they can be shown to satisfy basic criteria for psychophysical validity.

Lutfi et al. (2004) considered three such criteria in the evaluation of impact sounds synthesized according to the textbook equations for the transverse motion of struck bars and plates (cf. Section 4). First, in forced-choice comparisons between real-recorded and synthesized sounds, highly trained listeners (professional musicians) were required to identify the real sound with no greater than chance accuracy. Second, in a source identification task, real-recorded and synthesized sounds were required to produce no significant differences in performance or pattern of errors. Third, for a given number of free parameters satisfying the first two criteria, highly practiced listeners were required to be largely insensitive to the changes resulting from further increases in the number of parameters used to synthesize the sounds. All three criteria were satisfied with as few as four physically constrained parameters chosen to capture the gross acoustic effect of object geometry and material, how the object was held, and how it was struck.

Multidimensional scaling techniques provide yet another means of evaluating sound synthesis models. McAdams et al. (2004) took this approach to evaluate a model, similar to that used by Lutfi et al. (2004) for synthesizing the sounds of struck bars. Listeners were instructed to rate the perceived dissimilarity between pairs of the synthesized sounds according to any criteria that seemed to them perceptually salient. The sound pairs were synthesized under different conditions from bars varying in cross-sectional area, length, mass density, and viscoelastic damping coefficient. The perceptual space resulting from the scaling solution had the same dimensionality as that of the original physical parameter space. Moreover, the dimensions of the perceptual space were shown to correlate with the physical parameters after a power-law transformation. These results taken together with those of Lutfi et al. (2004) would seem to go some way in validating the use of physically informed synthesis techniques in studies of sound source identification, at least in regard to these rudimentary resonant sources.

# 7. Summary and Conclusions

Much progress has yet to be made in understanding how we identify the most rudimentary attributes of objects and events from sound. The problem is elusive. The pressure wave front arriving at the ear is a confluence of information regarding the material, geometric, and driven properties of any source, so determining precisely how a listener recovers these attributes from sound requires creative approaches. Measures of performance accuracy have so far demonstrated our capacity for identification, but they have not permitted strong conclusions regarding the basis for identification. New approaches are required that focus on the relation of the listener's response to trial-by-trial variation in stimuli. More work is also needed if we are to understand the factors that limit identification. We are only now beginning to learn of the potentially significant role of limited auditory sensitivity and prior knowledge on performance in simple source identification tasks. Finally, some benefit could be gained by devoting greater effort to understanding individual differences in performance, since these bear fundamentally on questions regarding the role of perceptual inference, prior knowledge, and experience. Though great leaps in our understanding of human sound source identification seem unlikely in the short term, recent advances in the development and validation of analytic sound synthesis techniques, in the formulation of computational recognition models, and in the application of new molecular psychophysical methodologies appear to hold real promise for progress in the foreseeable future.

*Acknowledgments.* The author thanks Dr. Douglas Keefe for many helpful comments on an earlier version of the manuscript. This work was supported by NIDCD grant R01 DC006875-01.

## *References*

Avanzini F, Rocchesso D (2001) Controlling material properties in physical models of sounding objects. In: Proceedings of the International Computer Music Conference, La Habana, Cuba, pp. 17–22.

Bernstein LB, Green DM (1987) Detection of simple and complex changes of spectral shape. J Acoust Soc Am 82:1587–1592.

Bregman AS (1990) Auditory Scene Analysis: The Perceptual Organization of Sound. Cambridge, MA: MIT Press.

Carello C, Anderson KL, Kunkler-Peck AJ (1998) Perception of object length by sound. Psychol Sci 9:211–214.

Carello C, Wagman JB, Turvey MT (2003) Acoustical specification of object properties. In: Anderson J, Anderson B (eds) Moving Image Theory: Ecological Considerations. Carbondale, IL: Southern Illinois University Press, pp. 79–104.

Ellis DPW (1996) Prediction-driven computational auditory scene analysis. PhD thesis, Department of Electrical Engineering and Computer Science, Massachusetts Institute of Technology.

Fantini DA, Viemeister NF (1987) Discrimination of frequency ratios. In: Yost WA, Watson CS (eds) Auditory Processing of Complex Sounds. Hillsdale, NJ: Lawrence Erlbaum.

Flanagan JL, Saslow MG (1958) Pitch discrimination for synthetic vowels. J Acoust Soc Am 30:435–442.

Fletcher NH, Rossing TD (1991) The Physics of Musical Instruments. New York: Springer-Verlag.

Freed DJ (1990) Auditory correlates of perceived mallet hardness for a set of recorded percussive sound events. J Acoust Soc Am 1:311–322.

Gaver WW (1988) Everyday listening and auditory icons. PhD thesis, University of California, San Diego.

Gaver WW (1993a) What in the world do we hear? An ecological approach to auditory event perception. Ecol Psychol 5:1–29.

Gaver WW (1993b) How do we hear in the world? Explorations in ecological acoustics. Ecol Psychol 5:285–313.

Gibson JJ (1950) The Perception of the Visual World. Boston: Houghton Mifflin.

Giordano BL (2003) Material categorization and hardness scaling in real and synthetic impact sounds. In: Rocchesso D, Fontana F (eds) The Sounding Object. Firenze: Mondo Estremo, pp. 73–93.

Giordano BL, McAdams S (2006) Material identification of real impact sounds: Effects of size variation in steel, glass, wood and Plexiglas plates. J Acoust Soc Am 119: 1171–1181.

Giordano BL, Petrini K (2003) Hardness recognition in synthetic sounds. In: Proceedings of the Stockholm Music Acoustics Conference, Stockholm, Sweden.

Grassi M (2005) Do we hear size or sound? Balls dropped on plates. Percept Psychophys, August 6–9, web publication. 67:274–284.

Green DM, Mason CR (1985) Auditory profile analysis: Frequency, phase, and Weber's Law. J Acoust Soc Am 77:1155–1161.

Green DM, Onsan ZA, Forrest TG (1987) Frequency effects in profile analysis and detecting complex spectra changes. J Acoust Soc Am 81:692–699.

Handel S (1995) Timbre perception and auditory object identification. In: Moore BCJ (ed) Hearing. New York: Academic Press.

Hartmann WM (1988) Pitch perception and the organization and integration of auditory entities. In: Edelman GW, Gall WE, Cowan WM (eds) Auditory Function: Neurobiological Bases of Hearing. New York: John Wiley & Sons, pp. 425–459.

Helmholtz H (1954) On the Sensations of Tone as a Physiological Basis for the Theory of Music. New York: Dover, pp. 623–645.

Hermes DJ (1998) Auditory material perception. IPO Annual Progress Report. 33:95–102.

Järveläinen H, Tolonen T (2001) Perceptual tolerances for the decaying parameters in string instrument synthesis. J Audio Eng Soc 49(11).

Kersten D, Mamassian P, Yuille A (2004) Object perception as Bayesian inference. Annu Rev Psychol 55:271–301.

Kinsler LE, Frey AR (1962) Fundamentals of Acoustics. New York: John Wiley & Sons, pp. 55–78.

Klatzky RL, Pai DK, Krotkov EP (2000) Perception of material from contact sounds. Presence Teleoperat Virt Environ 9:399–410.

Krotkov E, Klatzky R, Zumel N (1996) Robotic perception of material: Experiments with shape-invariant acoustic measures of material type. In: Khatib O, Salisbury K (eds) Experimental Robotics IV. New York: Springer-Verlag.

Kunkler-Peck AJ, Turvey MT (2000) Hearing shape. J Exp Psychol [Hum Percept Perform] 26:279–294.

Lakatos S (2001) Loudness-independent cues to object striking force. J Acoust Soc Am 109:2289.

Lakatos S, McAdams S, Causse R (1997) The representation of auditory source characteristics: simple geometric form. Percept Psychophys 59:1180–1190.

Luce D (1993) Sound and Hearing: A Conceptual Introduction. Lawrence Erlbaum.

Lutfi RA (2000) Source uncertainty, decision weights, and internal noise as factors in auditory identification of a simple resonant source. Abstr Assoc Res Otolaryngol 23:171.

Lutfi RA (2001) Auditory detection of hollowness. J Acoust Soc Am 110:1010–1019.

Lutfi RA Liu CJ (2007) Individual differences in source identification from impact sounds. J Acoust Soc Am, in press 122:1017–1028.

Lutfi RA, Oh EL (1997) Auditory discrimination of material changes in a struck-clamped bar. J Acoust Soc Am 102:3647–3656.

Lutfi RA, Oh E, Storm E, Alexander JM (2004) Classification and identification of recorded and synthesized impact sounds by practiced listeners, musicians and nonmusicians. J Acoust Soc Am 118:393–404.

Martin KD (1999) Sound-source recognition: A theory and computational model. PhD thesis, Department of Electrical Engineering and Computer Science, Massachusetts Institute of Technology.

McAdams S (1993) Recognition of sound sources and events. In: McAdams S, Bigand E (eds) Thinking in Sound: The Cognitive Psychology of Human Audition. Oxford: Clarendon Press.

McAdams S, Chaigne A, Roussarie V (2004) The psychomechanics of simulated sound sources: Material properties of impacted bars. J Acoust Soc Am 115:1306–1320.

Micheals CF, Carello C (1981) Direct Perception. Englewood Cliffs, NJ: Prentice-Hall.

Moore BCJ (1989) An Introduction to the Psychology of Hearing. New York: Academic Press.

Moore BCJ, Glasberg BR, Shailer MJ (1984) Frequency and intensity differences limens for harmonics within complex tones. J Acoust Soc Am 75:1861–1867.

Moore BCJ, Peters RW, Glasberg BR (1985) Thresholds for the detection of inharmonicity in complex tones. J Acoust Soc Am 77:1861–1867.

Morse PM, Ingard KU (1968) Theoretical Acoustics. Princeton, NJ: Princeton University Press, pp. 175–191.

Neuhoff JG (ed) (2004) Ecological Psychoacoustics. New York: Academic Press.

Ottaviani L, Rocchesso D (2004) Auditory perception of 3D size: Experiments with synthetic resonators. ACM Transact Appl Percept 1:118–129.

Patterson RD, Allerhand M, Giguère C (1995) Time-domain modelling of peripheral auditory processing: A modular architecture and a software platform. J Acoust Soc Am 98:1890–1894.

Perrott D (1984) Discrimination of the spatial distribution of concurrently active sound sources: Some experiments with stereophonic arrays. J Acoust Soc Am 76:1704–1712.

Rocchesso D (2001) Acoustic cues for 3-D shape information. In: Proceedings of the 2001 International Conference on Auditory Display, Finland, July 29, 2001.

Rocchesso D, Ottaviani L (2001) Can one hear the volume of a shape? In: IEEE Workshop Applications of Signal Processing to Audio and Acoustics. New Paltz, NY, pp. 21–24.

Schlauch RS, Ries DT, DiGiovanni JJ (2001) Duration discrimination and subjective duration for ramped and damped sounds. J Acoust Soc Am. 109:2880–2887.

Scudrzyk E (1968) Simple and Complex Vibratory Systems. University Park, PA: Pennsylvania State University Press.

Tucker S, Brown GJ (2003) Modelling the auditory perception of size, shape and material: Applications to the classification of transient sonar sounds. Presented at the 114th Audio Engineering Society Convention, Amsterdam, the Netherlands, March 22–25.

Ullman S (1980) Against direct perception. Behav Brain Sci 3:151–213.

Van Heuvan VJ, Van Den Broecke MPR. (1979) Auditory discrimination of rise and decay times in tone and noise bursts. J Acoust Soc Am. 66:1308–1315.

Wier CC, Jesteadt W, Green DM (1977) Frequency discrimination as a function of frequency and sensation level. J Acoust Soc Am 61:178–184.

Wightman FL, Kistler DJ (1989) Headphone simulation of free-field listening. II: Psychophysical validation. J Acoust Soc Am 85:868–878.

Wildes R, Richards W (1988) Recovering material properties from sound. In: Richards W (ed) Natural Computation. Cambridge, MA: MIT Press, pp. 356–363.

Yost WA (1992) Auditory image perception and analysis. Hear Res 56:8–19.

Yost WA, Sheft S (1993) Auditory perception. In: Yost WA, Popper AN, Fay RR (eds) Human Psychophysics. New York: Springer-Verlag, pp. 209–237.

# 3
# Size Information in the Production and Perception of Communication Sounds

Roy D. Patterson, David R.R. Smith, Ralph van Dinther, and Thomas C. Walters

## 1. Introduction

This chapter is about the perception of the sounds that animals use to communicate at a distance and the information that these sounds convey about the animal as a source. Broadly speaking, these are the sounds that animals use to declare their territories and attract mates, and the focus of the chapter is the size information in these sounds and how it is perceived. The sounds produced by the sustained-tone instruments of the orchestra (brass, strings, and woodwinds) have a similar form to that of the communication sounds of animals, and they also contain information about the size of the source, that is, the specific instrument type (e.g., violin or cello) within an instrument family (e.g., strings). Animals and instruments produce their sounds in very different ways, and the comparison of these two major classes of communication sounds reveals the general principles underlying the perception of source size in communication sounds.

For humans, the most familiar communication sound is speech, and it illustrates the fact that communications sounds contain information about the size of the source. When a child and an adult say the "same" word, it is only the linguistic message that is the same. The child has a shorter vocal tract and lighter vocal cords, and as a result, the waveform carrying the message is quite different for the child. The situation is illustrated in Figure 3.1, which shows short segments of four versions of the vowel in the word "mama." From the auditory perspective, a vowel is a "pulse-resonance" sound, that is, a stream of glottal pulses each with a resonance showing how the vocal tract responded to that pulse. From the perspective of communication, the vowel contains three important components of the information in the sound. The first component is the "*message*," which is that the vocal tract is currently in the shape that the brain associates with the phoneme /a/. This message is contained in the shape of the resonance, which is the same in every cycle of all four waves. The second component of the information is the glottal pulse rate. In the left column of the

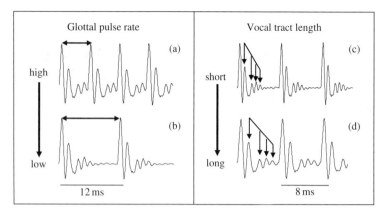

FIGURE 3.1. The internal structure of pulse-resonance sounds illustrating changes in pulse rate (a or b) and resonance rate (c or d).

figure, an adult has spoken the /a/ with a fast glottal pulse rate (a) and then a slow glottal pulse rate (b). The glottal pulse rate determines the pitch of the voice. The resonances are *identical*, since it is the same person speaking the same vowel. The third form of information is the resonance rate. In the right column, the same vowel is spoken by a child with a short vocal tract (c) and an adult with a long vocal tract (d) using the same glottal pulse rate. The glottal pulse rate and the *shape* of the resonance (the message) are the same, but the *rate* at which the resonance proceeds within the glottal cycle is faster in the upper panel. That is, the resonances of the child ring faster, in terms of both the resonance frequency and the decay rate. In summary, the stationary segments of the voiced parts of speech carry three forms of information about the sender: information about the shape of the vocal tract, its length, and the rate at which it is being excited by glottal pulses.

The components of the vocal tract (the nasal, oral, and pharyngeal passages) are tubes that connect the openings of the nose and mouth to the trachea and the esophagus. They are an integral part of the body, and they increase in length as the body grows. The decrease in pulse rate and resonance rate that occurs as humans grow up is a general property of mammalian communication sounds. Section 2 of this chapter describes the form of pulse resonance sounds, and Section 3 describes how information about source size is encoded in these sounds.

The fact that we hear the same message when children and adults say the same word suggests that the auditory system has mechanisms to adapt the analysis of speech sounds to the pulse rate and resonance rate, as part of the process that produces the size-invariant representation of the message. This suggests that there is an initial set of auditory processes that operate like a preprocessor to stabilize repeating neural patterns and segregate the pulse-rate and resonance-rate information from the information about the message. Irino and Patterson (2002) have demonstrated how these processes might work. First, the auditory system adapts the analysis to the pulse rate using "strobed temporal integration."

Then the resulting "auditory image" is converted into a largely scale-invariant Mellin image with the aid of resonance-rate normalization. As a byproduct, the two processes produce a contour of pulse-rate information and a contour of resonance-rate information that the listener can use to estimate speaker size, and to track individual speakers in a multisource environment. Section 4 illustrates the representation of size information in the auditory system.

Section 5 describes recent psychophysical experiments that indicate that the vocal-tract-length information provided by resonance-rate normalization is perceived in terms of source size, and that it functions like a dimension of auditory perception much like pitch. Section 6 describes recent experiments designed to reveal the interaction of pulse rate and resonance rate in the perception of source size.

With regard to the topic of this book, the *auditory perception of sound sources*, the current chapter is restricted to one aspect of the perception of one class of sources, namely, the perceived size of pulse-resonance sources. We focus on this specific problem because we believe that it holds the clue to speaker normalization, that is, the ability of human listeners to recognize the message of speech *independent* of the size of the speaker. Machine recognizers are still severely handicapped in this regard. If we can characterize the transforms that the auditory system uses to perform pulse-rate and resonance-rate normalization, and integrate them with source-laterality processing (e.g., Patterson et al. 2006) and grouping by common onset (e.g., Cooke 2006), the resultant auditory preprocessor might be expected to enhance the performance of automatic speech recognition significantly. We return to the topic of speaker normalization in Section 7, where we compare our perceptual approach to speaker normalization with the more linguistic approach described by Lotto and Sullivan in Chapter 10 of this volume.

## 2. Communication Sounds

Pulse-resonance sounds are ubiquitous in the natural world and in the human environment. They are the basis of the calls produced by most birds, frogs, fish, and insects, as well as mammals, for messages that have to be conveyed over a distance, such as those involved in mate attraction and territorial defense (e.g., Fitch and Reby 2001). They are also conceptually very simple. The animal develops some means of producing a pulse of mechanical energy that causes structures in the body to resonate. From the signal-processing perspective, the pulse marks the start of the communication, and the resonances provide distinctive information about the shape and structure of parts of the sender's body, and thus the species producing the sound. The pulse does not contain much information other than the fact that the communication has begun. Its purpose is to excite structures in the body of the animal that then resonate in a unique way. The resonance has less energy than the pulse but more information; it follows directly after the pulse and acts as though it were attached to it. So the location of the species-specific information is very predictable; it is tucked in behind each pulse.

In human speech, the vocal cords in the larynx at the base of the throat produce a pulse by momentarily impeding the flow of air from the lungs; this pulse of air then excites complex resonances in the vocal tract above the larynx. The mechanism is described in the next section. The mechanism is essentially the same in all mammals, and there is a similar mechanism in many birds and frogs; they both excite their air passages by momentarily interrupting the flow of air from the lungs. Fish with swim bladders often have muscles in the wall of the swim bladder (e.g., the weakfish, *Cynoscion regalis*) that produce brief mechanical pulses, referred to as "sonic twitches" (Sprague 2000), and these twitches resonate in the walls of the swim bladder in a way that makes the combination distinctive. Note that the sound-producing mechanisms in these four groups of vertebrates (fish, frogs, birds, and mammals) probably all evolved separately; the swim-bladder mechanism in the fish did not evolve into the vocal tract mechanism of the land animals, and the vocal tract mechanisms do not appear to have developed one from another. The implication is that this is convergent evolution, with nature repeatedly developing variations of the same basic solution to acoustic communication—the combination of a sharp pulse and a body resonance.

The sustained-tone instruments of the orchestra (brass, strings, and woodwinds) are also excited by nonlinear processes that produce sharp pulses that resonate in the air columns, or air cavities, of the instruments (Fletcher and Rossing 1998); so they also produce pulse resonance sounds (van Dinther and Patterson 2006). Combustion engines produce mini-explosions that resonate in the engine block; so they are also pulse-resonance sounds. They are not communication sounds in the normal sense, but they show that the world around us is full of pulse-resonance sounds, which the auditory system analyzes automatically and effortlessly.

There are also many examples of communication sounds that consist of a single pulse with a single resonance: Gorillas beat their chests with cupped hands, elephants stomp on the ground, and blue whales boom. Chickens and lemurs cluck every few seconds as they search for food in leaf litter. Humans clap their hands to attract attention. The percussive instruments such as xylophones, woodblocks, and drums also produce single-cycle pulse-resonance sounds. One important class of these percussive sounds is the "struck bars and plates," which are described in Chapter 2 of this volume, by Bob Lutfi. These percussive sources produce very different sounds from those of animals and sustained-tone instruments because the resonance occurs within the material of the bar, or plate, rather than in an air column, or air cavity, in an animal or instrument. The materials of the bars and plates (typically metal or wood) are dense and stiff, and so the resonances ring much longer in these sources. Nevertheless, they are pulse-resonance sources, and the principles of sound production and perceptual normalization are similar to those for the sustained tones produced by speech and musical sources.

The variety of these pulse-resonance sounds, and the fact that humans distinguish them, is illustrated by the many words in our language that specify transient sounds; words such as click, crack, bang, thump, and word pairs such as ding/dong, clip/clop, tick/tock. In many cases, a plosive consonant and a vowel are used to imitate some property of the pulse-resonance sound.

Finally, it should be noted that in the world today, most animals produce their communication sounds in pulse-resonance "syllables," that is, streams of regularly timed pulses, each of which carries a copy of the resonance to the listener. The syllables are on the order of 200–800 ms in duration, with a pulse rate in the region 10–500 Hz. The pulse rate rises a little at the onset of the sound, remains fairly steady during the central portion of the sound, and drops off with amplitude during the offset of the sound, which is typically longer and more gradual than the onset. A selection of four of these animal syllables is presented in Figure 3.2; they are the calls of (1) a Mongolar drummer, or Jamaica weakfish (*Cynoscion jamaicensis*), (2) a North American bullfrog (*Lithobates catesbeiana*), (3) a macaque (*Macaca mulatta*), and (4) a human adult saying /ma/.[1] The notes of sustained-tone instruments are like animal syllables with fixed pulse rates and comparatively flat temporal envelopes. Both of these classes of communication sound are completely different from the sounds of inanimate sources such as wind and rain, which are forms of noise. In the natural world,

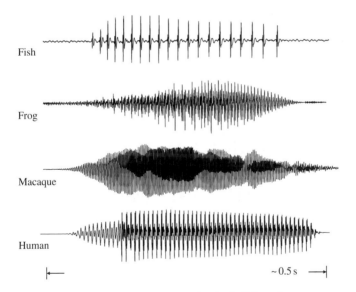

FIGURE 3.2. Communication calls from four animals (a fish, a frog, a macaque, and a human) illustrating that they all use "pulse-resonance" sounds for communication, and the duration of these animal syllables is on the order of half a second.

---

[1]The Mongolar drummer call is available at http://www.fishecology.org/soniferous. The bullfrog and macaque calls were kindly provided by Mark Bee and Asif Ghazanfar, respectively. Many of the sounds presented in this chapter can be downloaded from the CNBH website: http://www.pdn.cam.ac.uk/cnbh/.

the detection of a pulse-resonance sound in syllable form immediately signals the presence of an animate source in the local environment.

## 3. Size Information in Communication Sounds

### 3.1 The Effect of Source Size in Vocal Sounds

In general, as a mammal matures and becomes larger, there is a consistent and predictable decrease in both the resonance rate and the pulse rate of its communication sounds, primarily because they are produced by structures that increase in size as the animal grows. The vibration of the vocal tract of a mammal is often modeled in terms of the standing waves that arise in a tube closed at one end (Chiba and Kajiyama 1942; Fant 1970). The resonances of the vocal tract are referred to as formants, and for present purposes, the relationship between resonance rate and vocal tract length can be taken to be

$$\text{F1}_{\text{voice}} = \frac{c}{4L_{\text{tract}}}, \tag{3.1}$$

where $c$ is the speed of sound in air (340 m/s) and $L_{\text{tract}}$ is the length of the vocal tract, which can be as long as 17 cm in tall men. So the frequency of the first formant for men is on the order of 500 Hz $= \frac{340}{4 \times 0.17}$. The point to note is that the size variable, $L_{\text{tract}}$, is in the denominator on the right-hand side of the equation, which means that as a child grows up into an adult and the length of the vocal tract increases, the resonance rate of the first formant *decreases*. This is a general principle of formants and of mammalian communication sounds.

The vocal cords produce glottal pulses in bursts, and the vibration of the vocal cords can be modeled by the equation for the vibration of a tense string, although the vocal cords are actually rather complicated structures. The glottal pulse rate, $\text{F0}_{\text{voice}}$, is the fundamental mode of vibration of the vocal cords, and the relationship between glottal pulse rate and the properties of a tense string is

$$\text{F0}_{\text{voice}} = \sqrt{\frac{T_{\text{cords}}}{M_{\text{cords}}(4L_{\text{cords}})}} \tag{3.2}$$

where $T_{\text{cords}}$, $M_{\text{cords}}$, and $L_{\text{cords}}$ are the tension, mass, and length of the vocal cords. In this case, there are two physical variables associated with size: They are the length of the vocal cords and their mass; both increase as a child grows up. The point to note is that both the mass and length terms are in the denominator on the right-hand side of the equation, and they combine multiplicatively, so an increase in size, be it length or mass, leads to a decrease in glottal pulse rate in either case. The average $\text{F0}_{\text{voice}}$ for children is about 260 Hz, and it decreases progressively to about 120 Hz in adult men. The reduction in pulse rate with increasing size is also a general principle of mammalian communication sounds.

Thus, when we encounter a new species of mammal, we do not need to learn about the relationship between call and size. If the syllables of one individual have a consistently lower pulse rate and a consistently lower resonance rate than the syllables of a second individual, then we can predict with reasonable confidence that the first individual is larger without ever having seen a member of the species.

Speakers can also vary the tension of the vocal cords and change the pitch of the voice voluntarily. They do this to make prosodic distinctions in speech; for example, in many European languages, speakers raise the pitch of the voice at the end of an utterance to indicate that it is a question. This is also how singers change their pitch to produce a melody. The voluntary variation of tension makes the use of pulse rate as a size cue somewhat complicated. But basically, for a given speaker, the long-term average value of the voice pitch over a sequence of utterances is size information rather than speech information, whereas the short-term changes in pitch over the course of an utterance are speech information (prosody) rather than size information.

Finally, note that in pulse-resonance sounds, the frequency of the resonance is always greater that the pulse rate; this is one of the defining characteristics of the sounds used by mammals for communication.

## 3.2 The Effect of Source Size in Musical Instrument Sounds

The instruments of the orchestra are grouped into "families" (brass, strings, woodwinds, percussion). The members of a family (e.g., trumpet, French horn, and tuba) have similar construction, and they produce similar sounds; they differ primarily in their size. The mechanisms whereby sustained-tone instruments (the brass, string, and woodwind families) produce their notes are quite different from one another, and quite different from the way mammals produce syllables. Nevertheless, the excitation in sustained-tone instruments is a regular stream of pulses (Fletcher and Rossing 1998), each of which excites the body resonances of the instrument. As a result, sustained-tone instruments produce pulse-resonance sounds (van Dinther and Patterson 2006), and the sounds reflect the size of the source both in their pulse rate and their resonance rate, albeit in rather different ways than for the voice.

The French horn illustrates the form of the size information. It is a tube closed at one end like the vocal tract, and so the equation that relates fundamental frequency to tube length is the same as the one used to specify the frequency of the first formant of the voice,

$$\text{F0}_{\text{hom}} = \frac{c}{4L_{\text{hom}}} \qquad (3.3)$$

where $L_{\text{hom}}$ is the length of the brass tube when it is unrolled. However, in brass instruments, the length of the tube is associated with the *pulse rate* of the note rather than the frequency of the lowest body resonance. So the F0 is associated with the pitch of the note that the instrument is playing rather than its brassy timbre. The relationship between the F0 of the instrument and the pulse rate at any particular moment is complicated by the fact that the pulse rate is also affected by the tension of the lips, and the fact that it is not actually possible to excite the instrument with a pulse rate as low as its F0. The length of the French horn is about 3.65 m, so its F0 is about 23.3 Hz. This is actually below the lower limit of melodic pitch (Krumbholz et al. 2000; Pressnitzer et al. 2001). If for the sake of this illustration, however, we take this F0 to be c1, then the instrument can be made to produce pulse rates that are harmonics of C1, beginning with C2, that is, C2, G2, C3, E3, G3, etc., by increasing the tension of the lips. The point of the example, however, is that the equation for pulse rate in brass instruments contains a size variable, e.g., $L_{\text{horn}}$, and as the size of the instrument increases, the pulse rate decreases because the length of the tube is in the denominator on the right-hand side of the equation.

The broad mid-frequency resonance that defines the timbre of all brass instruments is strongly affected by the form of the mouthpiece. The mouthpiece can be modeled as an internally excited Helmholtz resonator (Fletcher and Rossing 1998). The vibration of a Helmholtz resonator is much more complicated than that of a tube, but it is nevertheless instructive with respect to the effects of source size on the acoustic variables of the French horn sound. If we designate the resonance frequency $F1_{\text{horn}}$ by analogy with $F1_{\text{voice}}$, then the resonance rate of the formant is

$$F1_{\text{hom}} = \frac{c}{2\pi} \sqrt{\frac{A_{\text{stem}}}{L_{\text{stem}} V_{\text{bowl}}}} \qquad (3.4)$$

Here $A_{\text{stem}}$ and $L_{\text{stem}}$ are the area and length of the stem that connects the bowl of the mouthpiece to the tube, and $V_{\text{bowl}}$ is the volume of the bowl of the mouthpiece. This is a much more complex equation involving three size variables, and the balance of these variables is crucial to the sound of a brass instrument. For present purposes, however, it is sufficient to note that the most important size variable is the volume of the bowl, and it is in the denominator; so once again, the rate of the body resonance decreases as the size of the bowl increases.

Similar relationships are observed in the other families of sustained-tone orchestral instruments, such as the woodwinds and strings; as the size of the components in the vibrating source and the resonant parts of the body increases, the pulse rate and the resonance rate decrease. This is the form of the size information in the sounds that animals use to communicate at a distance, and it is the form of the size information in the notes of sustained-tone instruments. Musical notes have a more uniform amplitude envelope and a more uniform pulse rate than animal syllables. Nevertheless, the size information has a similar form because of the basic properties of vibrating sources; as the components get larger in terms of mass, length, or volume, they oscillate more slowly.

The same physical principles also apply to the percussive sources described in Chapter 2, which produce single-cycle pulse-resonance sounds. For example, in the equation that specifies the natural frequencies of a struck bar (Chapter 2, Eq. [2.3a]), the length term is in the denominator, so the natural frequencies decrease as bar length increases. Similarly, in the equation for the natural frequencies of a struck plate (Lutfi, Chapter 2, Eq. [2.5]), the length and width terms are both in the denominator. Thus, size information is ubiquitous in mechanical sound sources. We turn now to the perception of source size in speech sounds and musical sounds.

## 4. The Form of Size Information in the Human Auditory System

The representation of size information in the auditory system has been illustrated by Irino and Patterson (2002) using a pair of /a/ vowels like those in the right-hand column of Figure 3.1. The two vowels were simulated using the cross-area function of a Japanese male saying the vowel /a/ (Yang and Kasuya 1995). In one case, the vocal tract length was that appropriate for an average male (15 cm); in the other, the length was reduced by one-third (10 cm), which would be appropriate for a small woman. The glottal pulse rate (GPR) was the same in the two vowels as it is in Figures 3.1c and 3.1d. The auditory image model (AIM, Patterson et al. 1992, 1995) was used to simulate the internal representation of the two vowels; the resulting "stabilized auditory images" are shown in Figure 3.3 which is a modified version of Figure 3 in Irino and Patterson (2002). Briefly, a

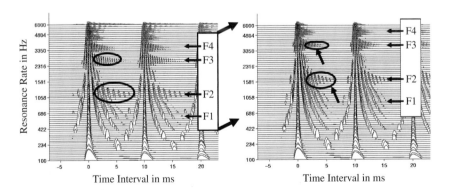

FIGURE 3.3. Auditory images of the vowel /a/ produced by a person with a long vocal tract (left column) and a short vocal tract (right column), showing the form that a change in source size takes in the internal auditory representation of these pulse-resonance sounds. The formants (F1–F4) move up as a unit on the tonotopic axis (the ordinate); that is, the resonance rates of the formants increase proportionately, and they have proportionately shorter duration.

gammatone auditory filterbank is used to simulate the basilar membrane motion produced by the vowel, and the resulting neural activity is simulated by applying half-wave rectification and adaptive compression separately to each channel of the filterbank output (Patterson and Holdsworth 1996). The repeating waveform of the vowel sound produces a repeating pattern of neural activity in the auditory nerve. In AIM, the pattern is stabilized by (1) calculating time intervals from the neural pulses produced by glottal pulses to the neural pulses produced by the remaining amplitude peaks within the glottal cycle, and (2) cumulating the time intervals in a dynamic, interval histogram (one histogram for each channel of the filterbank). The result of this "strobed temporal integration" (Patterson et al. 1992; Patterson 1994) is an array of dynamic, interval histograms that represents the auditory image; it is intended to simulate the first internal representation of the sound of which you are aware. The stabilization mechanism is assumed to be in the brainstem or thalamus.

The GPR of the synthetic /a/ vowels was 100 Hz, so the time between glottal pulses was 10 ms in both cases. The glottal pulses excite most of the channels in the filterbank, and so there are peaks at 0 ms and multiples of 10 ms in each channel, and these peaks form vertical ridges in the auditory image (Figure 3.3). This is the form of voice pitch in the auditory image: a vertical ridge that moves left as pitch increases and right as pitch decreases.

The rightward-pointing triangles on the vertical ridges are the formants of the vowels in this representation (marked by F1–F4 and arrows). They show that the vocal tract resonates longer at these frequencies. The overall shape of the patterns is quite similar, since it is the same vowel, /a/. The formants in the right-hand auditory image are shifted up, as a unit, along the quasi-log-frequency dimension, and a comparison of the fine structure of the formants in the corresponding ellipses shows that the formants ring faster in the auditory image of the vowel from the shorter vocal tract. This is the form of a change in vocal tract length in the auditory image: the resonances move up as a group (that is, the resonance rates increase), and the resonances decay away faster, so that the pattern shrinks in width. The same form of change occurs when the body resonators of musical instruments are reduced in size and when the struck bars and plates described in Chapter 2 are reduced in size. In the latter case, the resonance structure is attached to the 0-ms vertical, and the resonance structure extends across the full width of the image, because the density and stiffness of bars and plates means that the resonances ring much longer than those of the vocal tract or sustained-tone instruments.

The dimensions of the auditory image are both forms of frequency; the ordinate is acoustic frequency, which for narrow resonators is the resonance rate; the abscissa is the reciprocal of pulse rate. So the auditory image segregates the two components of size information and presents them, as frequencies, in a simple orthogonal form. The time interval between the vertical ridges in the auditory image is directly related to the size of the vocal cords, and the period of the individual resonances is directly related to the size of the resonators in the body of the source. Thus, in this image, changes in the size of the excitation source are reflected in proportional changes in the time intervals between the vertical ridges, and changes in the size of the vocal resonators are reflected in proportional changes in the time intervals in the triangular structures that represent the formants. Irino and Patterson (2002) and Turner et al. (2006) have demonstrated how the auditory image can be converted into a Mellin image in which the pattern of the "message," /a/, is truly scale-invariant. This aspect of size processing is beyond the scope of the current chapter.

Finally, note that whereas strobed temporal integration preserves the details of the resonances as they arise in basilar membrane motion, pitch mechanisms based on autocorrelation and the autocorrelogram do not (Licklider 1951; Slaney and Lyon 1990; Meddis and Hewitt 1991; Yost et al. 1996). Autocorrelation averages periodicity information over the glottal cycle. Whenever the resonance period is not an integer divisor of the glottal period, the periodicity information provided by the autocorrelation differs from the resonance rate of the formant. Thus, although autocorrelation can be used to predict the pitch and pitch strength of a vowel with great accuracy, the calculation smears the fine structure of the formant information (cf., for example, Figures 3.2c and 3.3c of Patterson et al. 1995), and consequently, it reduces the fidelity of any subsequent size-invariant representation of the message.

The pulse rate and resonance rate of a sound do not describe the size of a source in absolute terms. They are acoustic variables that describe properties of the sound wave as it travels from the sender to the listener. The acoustic variables change in a predictable way as the resonators in the sender's body grow. However, the brain does not have the equations required to convert a pulse rate into a mass or a length, and even if it had the equations, there would still be difficulties. The information about all of the physical variables involved in the production of the sound has to be transmitted to the listener via only two acoustic variables: pulse rate and resonance rate. These acoustic variables often vary with the product of several physical variables like mass and length, so a given pulse rate could be produced by many different combinations of mass and length. So what the listener receives is one pulse-rate value that summarizes the aggregate effect of all of the physical variables on the vibration source, and one resonance-rate value that summarizes the aggregate effect of another set of physical variables on resonance rate.

Moreover, the brain is not actually interested in the mass, length, or volume of the physical components of the sounder, such as the size of the vocal cords or the length of the vocal tract. What matters to the listener is the size of the sender's body: some perceptual and/or cognitive combination of the sender's height, mass, and volume, and within a species, whether one sender is much bigger or smaller than another. In order to estimate the sender's body size, a more central mechanism must combine the pulse-rate and resonance-rate information with some form of stored knowledge about the structure of the sender and/or a body of experience with a range of individuals from the specific population.

This is a complex problem, to which we return in Section 6. The next section is concerned with the much simpler problem of comparing the relative size of two individuals from the same species, or two musical instruments from the same family.

# 5. The Perception of Relative Size in Communication Sounds

Broadly speaking, the resonators in animals maintain their shape and composition as the animal grows, because the resonators are part of the sender's body. So within a population of senders, the function that relates the physical variables describing resonator components to the acoustic variables remains the same, and the constants maintain their fixed values. Thus, the changes are typically limited to the specific values of a small number of size-related variables, whose growth patterns are correlated and whose effects all go in the same, predictable, direction. As a result, differences in pulse rate and resonance rate provide useful information about the relative size of individuals within a population of senders. In this section, we describe perceptual experiments designed to demonstrate that listeners perceive the size information provided by the resonance rate and the pulse rate, and that they can discriminate relatively small changes in resonance rate as well as pulse rate. The results support the hypothesis that resonance rate is a dimension of auditory perception like pitch, and that together, resonance rate and pulse rate largely determine our perception of the relative size of animals and musical instruments.

## 5.1 Discriminating Speaker Size from Changes in Vocal Tract Length

Recently, two high-quality voice processing systems have been developed that make it possible to dissect segments of natural speech and manipulate the vocal tract length (VTL) and glottal pulse rate (GPR) information without changing the other qualities that specify the message and the speaker's identity. These voice coders, or vocoders, are referred to by the acronyms STRAIGHT (Kawahara et al. 1999; Kawahara and Irino 2004) and PRAAT (Boersma 2001), and they have made it possible to perform experiments on the perception of size information in natural speech with precise stimulus control. PRAAT has the advantage that it can extract formant contours as well as the voice pitch from utterances. The advantage of STRAIGHT is that the spectral envelope of the speech that carries the vocal-tract information is smoothed, as it is extracted, to remove the interference that occurs between the harmonic structure associated with the glottal-pulse rate of the stimulus, and the transfer function of the analysis window in the short-term Fourier transform. This helps to avoid the problem that LPC analysis has with the first formant when the GPR is relatively high and there are only one or two harmonics of the voice pitch to define the first formant. When operating on speech, both PRAAT and STRAIGHT can produce resynthesized utterances of extremely high quality, even when the speech is resynthesized with pulse rates and vocal tract lengths beyond the normal range of human speech.

Smith et al. (2005) used STRAIGHT to control VTL in an experiment designed to measure a listener's ability to discriminate speaker size from differences in resonance rate. If acoustic scale functions as a dimension of sound as suggested by Cohen (1993), then we might expect to find that listeners can readily make fine discriminations in VTL, and thus speaker size, just as they can for the intensity of sound (loudness) or light (brightness). Moreover, if this is a general mechanism of auditory perception, we should expect to find that listeners can make size judgments even when the speech sounds are scaled to simulate humans much larger and smaller than those that the listeners have ever encountered.

Smith et al. (2005) prepared a set of "canonical" vowels /a/, /e/, /i/, /o/, /u/ from recordings made of author RP saying the vowels in natural /hVd/ sequences, i.e., "haad, hayed, heed, hoed, who'd." The vowels were edited to a common length of 600 ms by extracting the central sustained portion of the vowel and gating them on and off with a smooth cosine-squared envelope. The vowels were normalized to the same intensity level and the GPR was scaled to 113 Hz, which is near to the average for men. The VTL of these vowels was then scaled using STRAIGHT, which is actually a sophisticated speech-processing package that dissects and analyzes an utterance at the level of individual glottal cycles. It performs a "pitch synchronous" spectral analysis with a high-resolution fast Fourier transform (FFT), and then the envelope is smoothed to remove the zeros introduced by the position of the Fourier analysis window relative to the time of the glottal pulse. The resultant sequence of spectral envelopes describes the resonance behavior of the vocal tract in a form that is largely independent of pitch.

Once STRAIGHT has segregated a voiced sound into a GPR contour and a sequence of spectral envelope frames, the frequency dimension of the spectral envelope can be expanded or contracted independently of the GPR, and vice versa. Then the vowel can be resynthesized with its new GPR and VTL. The operations are largely independent, with the restriction that the GPR must never be higher than about half the frequency of the lowest formant for satisfactory resynthesis. When GPR is changed while keeping VTL constant, we hear one person repeating an utterance using different pitches, like singing a word on different notes; when VTL is changed keeping GPR constant, we hear something quite different, as though a set of people of different sizes were lined up on a stage, each saying the same word one after another, and all on the same pitch. Utterances recorded from a man can be transformed to sound like women and children. A demonstration of the manipulations possible with STRAIGHT is provided on the website[2] of the Centre for Neural Basis of Hearing. Liu and

---

[2]http://www.pdn.cam.ac.uk/cnbh/

Kewley-Port (2004) have reviewed STRAIGHT and commented favorably on the quality of its production of resynthesized speech. Assmann and Katz (2005) have also shown that a listener's ability to identify vowels is not adversely affected when they are manipulated by STRAIGHT over a reasonable range of GPR and VTL.

### 5.1.1 Speaker Size Discrimination with Vowel Sounds

In Smith et al. 2005, the scaling of VTL was accomplished simply by compressing or expanding the spectral envelope of the speech linearly along a linear frequency axis. On a logarithmic frequency axis, the spectral envelope shifts along the axis as a unit, and this is the form of size change in the frequency domain for information associated with resonance rate. The JND for speaker size was initially measured with single vowels using a two-alternative, forced-choice (2IFC) procedure. One vowel was presented in each interval, and the listener had to choose the interval corresponding to the speaker who sounded smaller. Psychometric functions showing percentage correct as a function of the difference in VTL between the speakers were measured for a variety of test voices with GPR values ranging from 40 to 640 Hz and VTL values from 7 to 24 cm (the average for adult males is about 16 cm). The results showed that detecting a change in speaker size based on a change in VTL is a relatively easy task. The JND was on average about 8%, which compares favorably with the JND for the intensity of a noise (loudness), which is about 10% (Miller 1947). The only exception was for vowels with long VTLs and the highest GPR (640 Hz); in this bottom-right corner of the GPR-VTL plane, the resonance of the vowel becomes long relative to the period of the sound, and the vowel becomes difficult to recognize; the lowest harmonic moves up in frequency beyond the position of the first formant.

By its nature, a change in vocal-tract length produces a predictable shift of the vowel spectrum, as a unit, along a logarithmic frequency axis, and the tonotopic axis along the basilar membrane is quasi-logarithmic. So, it might be possible for a listener to focus on one formant peak and perform the task by noting whether the peak shifted up or down in the second interval. Accordingly, Smith et al. (2005) ran a second version of the discrimination experiment with a more speechlike paradigm, which effectively precluded the possibility of using a simple spectral cue. The paradigm is presented in quasi-musical notation in Figure 3.4. Each interval of the trial contained a sequence of four vowels chosen randomly without replacement from the five used in the experiment, and the vowels were presented with one of four pitch contours, again chosen randomly. The duration of the vowels was shortened to about 400 ms to make the sequences sound more natural. The starting point for each pitch contour was varied randomly over a 9% range, and the level of the vowels in a given interval was roved in intensity over a 6-dB range. The only fixed parameter within an interval was VTL, and the only consistent change between intervals for all of the vowels was VTL. As before, the listener's task was simply to choose the interval with the smaller speaker. In this paradigm, the listener cannot do the task by

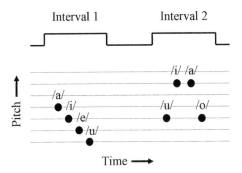

FIGURE 3.4. Schematic of the vowel-phrase paradigm for the VTL discrimination experiment of Smith et al. (2005). The only consistent difference between the vowels in the two intervals is vocal tract length.

listening to a single spectral component and noting whether it shifts up or down in the second interval. It is also the case that this paradigm naturally prompts the listener to think of the sounds in the two intervals as coming from two different speakers; the natural prosody of the sequences discourages listening for spectral peaks.

The experiment with speechlike vowel sequences produced JNDs that were similar to those obtained with single vowels. Together, the experiments with single vowels and vowel sequences show that listeners can make fine judgments about the relative size of two speakers, even when other properties of speech are varying, and that they can make size judgments for vowels scaled well beyond the normal range in both VTL and GPR. The JND for VTL information in vowels was less than 10% over a wide area of the GPR–VTL plane. When the GPR is 160 Hz, there are approximately 10 JNDs in speaker size between VTL values of 7 and 24 cm, so a JND corresponds to a VTL difference of about 2 cm. This supports the hypothesis that acoustic scale functions as a dimension in auditory perception (Irino and Patterson 2002).

### 5.1.2 Speaker Size Discrimination with Syllable Phrases

The experiments of Smith et al. (2005) on speaker size discrimination have since been extended to syllable phrases by Ives et al. (2005) in a study that greatly increased the variability of the stimulus set and made the task much more like that experienced in everyday speech. Ives et al. created a large, balanced database containing 90 consonant–vowel (CV) syllables and 90 vowel–consonant (VC) syllables. Each of the CV and VC categories contained three groups of 30 syllables distinguished by consonant category: sonorants, stops, and fricatives. In each group, six consonants of a specific type were paired with each of five vowels (/a/, /e/, /i/, /o/, and /u/). The full set of 180 syllables is presented in Ives et al. (2005), Table (1). The perceptual center of each syllable was determined (Marcus 1981; Scott 1993) and then used to ensure that the rhythm in the syllable phrases was fairly even, as it is in speech.

The vowels were scaled in GPR and VTL using STRAIGHT to simulate five categories of speaker with GPRs ranging from 80 to 320 Hz and VTLs ranging from 9 to 17 cm. The speaker types included three common categories with typical combinations of GPR and VTL, a large male, a large female, and a small child; and two unusual categories, one with a short vocal tract and a low pitch and one with a long vocal tract and a high pitch. The listeners were presented with two phrases of four syllables in a 2IFC discrimination paradigm very similar to that of Smith et al. (2005). There was a consistent difference in VTL between the two phrases, and the difference was varied over trials to determine the JND for VTL. The syllables in each phrase were selected randomly, with replacement, from one of the six groups within the database (e.g., CV-sonorants, CV-stops, CV-fricatives, VC-sonorants, VC-stops, or VC-fricatives). The level of the syllables in each phrase was roved between phrases over a 6-dB range, and the GPR of each of the syllables within the phrase was varied along one of four pitch contours.

The JNDs for the adult male and female speakers are just over 4% for all six syllable types. The JNDs for the other three categories are slightly larger (5%–6%), due mainly to worse performance on syllables containing stop conso-nants (/b/, /d/, /g/, /p/, /t/, /k/). Thus, the average for all speaker categories and syllable types is about 5%, which is considerably less than the value observed with vowels in a similar paradigm, and the reduction in the JND occurs despite the increased complexity of the stimulus set. Ives et al. (2005) attribute the improvement in performance to the greater naturalness of the speech in the syllable experiment. Although Smith et al. (2005) recorded natural vowels, they extracted the sustained portion in the center of the vowel and applied a cosine onset envelope to all of the vowels, which made them more similar and less natural. In the syllable experiment, the natural onset of each individual syllable was preserved, and the stimuli sounded considerably more natural as a result.

Finally, a note of caution is due with respect to predicting a speaker's height from his or her voice. Although there is a strong correlation between vocal tract length and speaker height over the full range of heights (Fitch and Giedd 1999), and although this makes it easy to distinguish children from adults, it is nevertheless the case that within small groups of adult men or adult women, you cannot expect to predict height differences from the voice differences with great accuracy. There are two reasons for this: First, the standard deviation for height in adult populations is relatively small, only about 4% of mean height, both for adult men and adult women. So the average height difference is relatively small in percentage terms. Second, the correlation between VTL and height is not perfect; on average, in the data of Fitch and Giedd (1999), the standard deviation for VTL, *given height*, is still about 6%. Thus, it is not surprising to find that the correlation between formant frequency and height is weak in small groups of adult men or adult women (González 2004; Owren and Anderson 2005; Rendall et al. 2005). It is also the case that in syllable phrases, the JND for the perception of a change in VTL is about 5% (Ives et al. 2005). So, with your eyes closed, you are not likely to be able to reliably discriminate the height of two men, or two women, drawn randomly from the population, because the difference in VTL will probably be only one or two JNDs.

### 5.1.3 Resonance Rate Discrimination and Profile Analysis

In retrospect, it seems odd that the perception of speaker size has received so little attention in hearing and speech research. In spectral terms, the effect of a change in speaker size is theoretically very simple: If the GPR is fixed and the frequency axis is logarithmic, the profile for a given vowel has a fixed shape, and VTL changes simply shift the profile as a unit, toward the origin as size increases and away from it as size decreases. The analysis of spectral profiles is a well-known topic in psychoacoustics since it was introduced by Spiegel et al. (1981); see Green (1988) for a review. However, in the main, people have elected to follow Green and colleagues and concentrate on profiles constructed from sets of equal-amplitude sinusoids whose frequencies are equally spaced on a *logarithmic* axis. These stimuli are not like the voiced parts of speech; they do not have a regular harmonic structure, the excitation is not pulsive, and they sound nothing like vowels. Moreover, the task in traditional profile analysis (PA) is to detect an increment in one of the sinusoidal components, which is very different from detection of a shift in the spectral location of the profile as a whole.

Drennan (1998) provides an excellent overview of PA research that includes a few PA experiments in which the stimuli are composed of sets of harmonically related components that are intended to simulate vowel sounds to a greater or lesser degree (see, e.g., Leek et al. 1987; Alcántara and Moore 1995). However, there is no attempt to simulate the filtering action of the vocal tract and produce realistic vowel profiles; nor is there any attempt to simulate changes in VTL or measure sensitivity to coherent spectral shifts.

## 5.2 Discriminating Instrument Size from Changes in Resonance Rate

Further support for the hypotheses that acoustic scale is perceived as a dimension of auditory perception is provided by a recent study on the perception of size in musical instrument sounds. The instruments of the orchestra come in families, and within a family, the different instruments have the same shape and construction. The members of a family differ mainly in size. Musical sounds are pulse-resonance sounds, and although the mechanisms they use to produce their notes are sometimes very different from the way humans produce syllables, the notes of music carry size information in the form of a pulse rate and a resonance rate.

van Dinther and Patterson (2006) performed an experiment with scaled musical notes from four instrument families to determine the JND for a change in instrument size over a large range of pulse rates and resonance rates. The experiment focused on the baritone range in four instrument families: strings, woodwind, brass, and voice. Thus for the string family, it was the cello; for the woodwind family, the tenor saxophone; for the brass family, the French horn, and for the human voice, the baritone. The notes were taken from a high-fidelity database of musical sounds (Goto et al. 2003). Each note was extracted with its natural onset to preserve the attack timbre of the instrument, and a cosine-squared envelope was applied to the end of the waveform to produce a smooth 50-ms offset. The total duration of the waveform was 350 ms. These specific instrument families were chosen because they produce sustained notes with similar temporal envelopes, and the notes have the pulse-resonance structure that STRAIGHT is most successful in scaling (Kawahara and Irino 2004).

STRAIGHT was used to modify the pulse rate (PR) and resonance rate (RR) of the notes and produce the small changes required for the discrimination experiment. The JND was measured for five combinations of pulse rate and resonance rate in a pattern similar to that used in Ives et al. (2005). The pulse rates were G1, G2, and G3 (49, 98, and 198 Hz), and the resonance rate was scaled up or down by two-thirds of an octave; the design is illustrated in Figure 3.5. The procedure was similar to that employed in the speechlike version of the discrimination experiment in Smith et al. (2005). Each interval of a trial contained a short melody, instead of a single note, to preclude listeners from performing the task on the basis of a shift in a single spectral peak. The melodies also promote a musical mode of listening (synthetic rather than analytic). Scaled versions of the notes were then used to generate two-sided psychometric functions showing how much the resonance rate of the instrument has to be decreased or increased from that of the standard for reliable discrimination. The psychometric functions

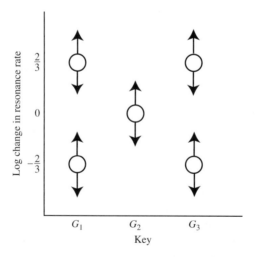

FIGURE 3.5. The combinations of pulse rate and resonance rate used as "standards" in the experiment of van Dinther and Patterson (2006) on discrimination of instrument size. The abscissa shows the pulse rate in musical notation; the ordinate shows the factor by which the resonance rate was modified (in log units). The arrows show the direction in which the JNDs were measured.

were measured for each PR–RR combination of each of the four instruments (cello, saxophone, French horn, and baritone voice).

The results show that listeners are able to discriminate size in instrument sounds, and they can specify which is the smaller of two instruments from short melodies. Within a family, the JND is relatively consistent, typically varying by no more than a factor of two across conditions. The JNDs for the baritone voice are comparable to those observed in the discrimination experiments of Ives et al. (2005), averaging around 5%. The JNDs for the French horn are around 8%, while those for the saxophone are around 12%. So the JNDs for these two instruments are similar to those for other sensory dimensions (Miller 1947), and about double the JND for speech in syllable phrases (Ives et al. 2005). The JNDs in the music study are largest for the cello. The JND is about 10% when the pitch is low and the instrument is small, or the pitch is high and the instrument is large, but they increase to around 20% when the pitch is low and the instrument is large or the pitch is high and the instrument is small. And overall, listeners have slightly more difficulty when the instrument is large and plays a low-pitched melody.

In summary, the psychometric functions associated with discriminating a change in source size as a function of resonance rate are steep and consistent, supporting the argument that resonance rate functions as a dimension of auditory perception. The slope of the psychometric function shows that listeners need a change of 5%–10 % in resonance rate to discriminate a change in the size of the resonators in the vocal tract or the bodies of musical instruments.

## 5.3  Discriminating Pulse Rate in Vowel Sounds and Click Trains

The basic relationship between the acoustic variable "glottal pulse rate" and the perceptual variable "voice pitch" is straightforward: Pitch increases with pulse rate. Indeed, the relationship is so simple that the pitch of the voice is normally expressed in terms of glottal pulse rate in Hertz. Voice pitch is also expressed as the fundamental, F0, of the harmonic series observed in the magnitude spectrum of the short-term Fourier transform of the note. But for most purposes, it is simpler just to think of voice pitch as glottal pulse rate.

As a child grows up into an adult, its vocal cords become longer and heavier (Titze 1989), and the GPR decreases from about 260 Hz for small children of both sexes to about 220 Hz for women and about 120 Hz for men. The change proceeds smoothly with height as girls grow up into women. For boys, however, when they reach puberty there is an increase in testosterone, which accelerates growth in the laryngeal cartilages (Beckford et al. 1985). As a result, there is a sudden drop in GPR by almost an octave at around 13 years of age. In the adult population, once the effect of sex is removed, there is no direct correlation between body size and GPR; that is, the range of heights in the relatively small populations of men, or women, included in studies of GPR (e.g., Lass and Brown 1978; Künzel 1989; Hollien et al. 1994) is not large enough to reveal a correlation with height, given the variability in GPR. Thus, there is size information in GPR, in the sense that one can reliably distinguish small children from large adults, but within a group of men or women, a difference in GPR is not a reliable indicator of a difference in height. As noted earlier, speakers vary GPR by varying the tension of the vocal cords to indicate prosodic distinctions in speech. So, for a given speaker, the long-term average value of his or her voice pitch, over a sequence of utterances, is size information, but the short-term changes in pitch, over the course of an utterance, are speech information rather than size information.

Although the relationship between voice pitch and the perception of size is somewhat complicated, the discrimination of a change in GPR per se is not. Smith et al. (2005) includes an experiment in which they measured the JND for voice pitch using synthetic vowel sounds for a wide range of combinations of GPR and VTL. On each trial, listeners were presented two vowels with the same VTL and that differed a little in GPR, and over the course of the trials, the GPR difference was varied to produce a psychometric function from which the JND was determined. When the GPR was in the normal range for the human voice, the JND was less than 2%. This performance also extended beyond the range of the human voice up to 640 Hz, and it was largely unaffected by the value of the VTL. That is, the discrimination of changes in voice pitch would appear to be largely independent of the properties of the resonance in speech sounds. When the GPR was reduced to 40 Hz, close to the lower limit of human pitch perception, the JND rose to about 9%.

The majority of data on the discrimination of pulse rate, however, come from research that is ostensibly on pitch perception as opposed to size perception. (In the next section, we describe an experiment that shows that changes in pulse rate interact with changes in resonance rate in the perception of changes in source size.) Temporally regular trains of very-short-duration pulses without resonances produce a sound with a strong pitch and a buzzy, mechanistic timbre. These "click trains" and sets of regularly spaced harmonics (which are their spectral equivalent) have been used to study what has been referred to as "residue pitch" (Schouten 1938), "periodicity pitch" (Licklider 1951), "repetition pitch" (Thurlow and Small 1955), "virtual pitch" (Terhardt 1974), and more recently, "melodic pitch" (Krumbholz et al. 2000; Pressnitzer et al. 2001). The discrimination of click rate is similar to the discrimination of the glottal pulse rate in speech sounds, but without any confounding influence from the vocal resonances. In these discrimination studies, matched pairs of click trains with slightly different click rates are compared (typically in a 2IFC paradigm) to determine the JND for a range of click rates. Krumbholz et al. (2000) provide a review of the studies dating back to Ritsma and Hoekstra (1974). They show that "rate discrimination threshold" (RDT), as it is called, is less than 2% for a wide range of click rates, provided that the stimulus contains energy in the region below 1000 Hz.

Krumbholz et al. (2000) extended the research and used RDT to measure the lower limit of "melodic" pitch (LLMP) as a function of the spectral location of the energy in the stimulus. Their results are similar to those of Pressnitzer et al. (2001), who used bandpass-filtered click trains to construct four-note melodies and measure the LLMP in a musical context. The LLMP is about 32 Hz when the stimulus contains low-frequency energy down to 200 Hz. The LLMP increases as the energy moves up in frequency; the rate of increase is initially slow (the LLMP is still below 50 Hz when the lowest component is 800 Hz), but as the lowest frequency in the stimulus moves above about 1000 Hz, the rate of increase accelerates, and when the energy is all above 3200 Hz, the LLMP is about 300 Hz.

# 6. The Interaction of Resonance Rate and Pulse Rate in the Perception of Source Size

Estimating the *absolute* size of a source from a single auditory event is, theoretically, a much more difficult task than making a judgment about the relative size of two similar sources. The listener has to use experience and/or context to interpret the size information in the sound. Nevertheless, when the radio or the telephone presents us with a new, unknown speaker, we can tell whether the speaker is a child or an adult, which suggests that we have the relevant experience. We also know that there is size information in speech sounds. The length of the vocal tract is highly correlated with speaker height (Fitch and Giedd 1999), and the longer the vocal tract, the lower the formant frequencies (Chiba and Kajiyama 1942; Fant 1970). Specifically, as a child grows between the ages of 4 and 12, the formant frequencies of males decrease by about 32% from their values at age 4, while the formant frequencies of females decrease by about 20% over the same age range (Hollien et al. 1994; Huber et al. 1999).

The contrast between the theoretical problem of estimating the absolute size of a sound source and our apparent ability to do it with relative ease for humans prompted Smith and Patterson (2005) to measure listeners' ability to estimate speaker height for isolated vowels with a wide range of GPRs and VTLs. The data are of particular interest because they reveal an interaction between GPR and VTL in the estimation of speaker height.

## 6.1 The Interaction of GPR and VTL in the Estimation of Speaker Size

Listeners were presented isolated vowels scaled over a large range of GPR and VTL values, and requested to make two judgments about each vowel: the height of the speaker (on a seven-point scale from very short to very tall) and the speaker's natural category (man, woman, boy, or girl). The experiment was performed for two ranges of GPR and VTL values. The narrower range was similar to that encountered in the normal population: GPR varied from 80 to 400 Hz in six logarithmic steps, and VTL ranged from 22.2 cm to 7.8 cm in six logarithmic steps. The wider range was chosen to extend the judgments well beyond the values encountered in everyday speech; GPR varied from 61 to 523 Hz in six logarithmic steps, and VTL ranged from 26.8 cm to 6.5 cm in six logarithmic steps. These VTLs simulate speakers ranging from a small child of height 0.6 m (a VTL of 6.5 cm) to a giant of height 3.7 m (a VTL of 26.8 cm). The data showed that the effect of range was small; that is, judgments of size made during the experiment with the extended range, for combinations of VTL and GPR that are commonly encountered, were essentially the same as the judgments made when vowels with similar combinations of VTL and GPR were presented in the experiment with the smaller range.

The results from the two experiments combined are shown in Figure 3.6 as a size surface over the GPR–VTL plane. The figure shows that listeners reliably reported that vowels spoken with a low GPR and a long VTL came from a very tall person (the upper back part of the surface) and that vowels spoken with a high GPR and a short VTL came from a small person (the lower front part of

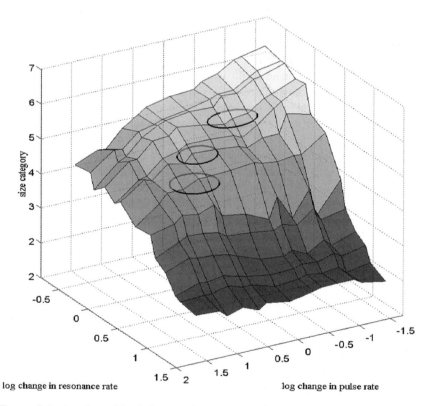

FIGURE 3.6. A surface of size judgments showing the average estimated size for voices with a wide range of combinations of glottal pulse rate and vocal tract length. The surface was constructed from the data of Smith and Patterson (2005). The three ellipses show the normal combinations of GPR and VTL for men, women, and children, estimated from the data of Peterson and Barney (1952).

the surface). However, the surface is not planar, indicating that in at least part of the space, GPR and VTL interact in the determination of the perceived size of the speaker. Broadly speaking, on these log-log coordinates, GPR has a nearly linear effect on perceived size for those VTLs in the normal range for adults and for VTLs longer than those typically encountered in everyday life. The ellipses show the normal range of GPR and VTL values in speech for men, women, and children, derived from the vowels of 76 men, women, boys, and girls speaking ten vowels (Peterson and Barney 1952). The estimates of VTL were calibrated with measurements of the VTL taken from magnetic resonance images (Fitch and Giedd 1999). Each ellipse represents the mean ±2 standard deviations in each dimension for each category of speaker.

In contrast, as VTL decreases through the range associated with children, the effect of GPR decreases *rapidly* for *low* GPRs and *slowly* for *high* GPRs. As a result, for VTLs in the range of most children, and for shorter VTLs, changes in GPR have very little effect on the perceived height of the speaker; all of the vowels are perceived as emanating from very small people. It is also worth noting that the data revealed very little evidence of learning: Listeners could perform at near-asymptotic levels after a few minutes' experience with the task.

## 6.2 The Interaction of GPR and VTL in Instrument Identification

The study by van Dinther and Patterson (2006) of size perception in musical instruments included an experiment to determine the extent to which listeners could recognize instrument sounds when their resonance rates and pulse rates had been increased or decreased with STRAIGHT. They used four families of instruments: strings, woodwinds, brass and voice, and chose four members with different sizes from each family. The specific instruments are listed in their Table 1 with the pitch range, or register. The sixteen starting notes that identify the instruments were scaled up and down by 5, 7, or 12 semitones and up by 7 or 12 semitones *in pulse rate*, and they were scaled up and down by one-third and two-thirds of an octave *in resonance rate*, making a total of 5 × 5, or 25, versions of each note. A 16-alternative forced-choice procedure was used to measure recognition performance using a graphical interface with 16 buttons labeled with the 16 instrument names in the layout shown in their Table 1. On each trial, one of the 25 notes for one of the 16 instruments was selected and played to the listener three times. The listener's task was to identify the instrument from one of the 16 options.

The results showed that listeners could identify the scaled instrument notes reasonably accurately, even for notes scaled well beyond the normal range for that instrument. Performance was above 55% correct for all combinations of pulse rate and resonance rate, and it rose to about 85% correct for the unscaled notes. Chance performance for instrument identification in this task is 6.25% correct. An analysis of the errors showed that listeners were essentially perfect on the identification of instrument family. Moreover, when both the pulse rate and the resonance rate were decreased, if the listener made an error, it was very likely that they would choose a larger instrument from within the same family. Similarly, when both the pulse rate and the resonance rate were increased, if the listener made an error, it was very likely that they would choose a smaller instrument from within the same family.

This prompted van Dinther and Patterson (2006) to summarize the within-family error data in terms of a surface that shows the trading relationship between pulse rate and resonance rate, in order to examine the interaction of pulse rate (PR) and resonance rate (RR) in the perception of instrument size. Specifically, for within-family errors, the percentage of cases in which each listener chose a larger member of a family was calculated as a function of the *difference* in pulse rate and the *difference* in resonance rate between the scaled and unscaled versions of the note. The results were presented as a contour plot (Figure 3.7) in which the dependent variable was the percentage of cases in which the listener chose a larger member of a family given a specific combination of pulse rate and resonance rate.

Consider the 50% correct contour line. It shows that there is a strong trading relation between a change in pulse rate and a change in resonance rate. When the pulse rate is increased on its own, it increases the percentage of cases in which the listener will choose a smaller member of the family. However, this

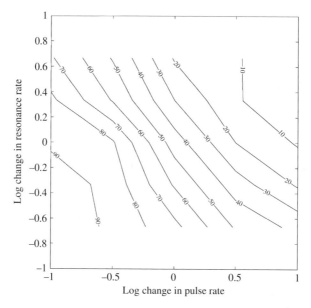

FIGURE 3.7. Contours showing the percentage of within-family errors where the listener chose a larger member of the family, plotted as a function of the difference in pulse rate (abscissa) and resonance rate (ordinate) between the scaled and unscaled versions of the note. The differences are plotted on an octave scale (base-2 logarithmic scale) for both the abscissa and the ordinate. The contours were constructed from the data of van Dinther and Patterson (2006).

can be entirely counteracted by a decrease in resonance rate, which makes the instrument seem larger. Moreover, the contour is essentially a straight line, and on these log-log coordinates, the slope of the line is on the order of −1; that is, in log units, the two variables have roughly the same effect on instrument identification. Very similar, essentially linear, trading relationships are observed for all of the contours between about 20% and 80% correct, and the spacing between the lines is approximately equal. Together, these observations mean that the errors are highly predictable on the basis of just two numbers: the logarithm of the change in pulse rate and the logarithm of the change in resonance rate. The fact that the surface is roughly planar means that the trading relationship can be characterized, except at the extremes, by a plane. For any point in the central range of the plane, a change in pulse rate of PR log units can be counteracted by a change in resonance rate of −1.3 PR log units. This means that when measured in log units, the effect of a change in pulse rate on the perception of size is a little greater than the effect of a change in resonance rate. However, the JND for resonance rate is larger than that for pulse rate, so if we express the relationship in terms of JNDs instead of log units, the relative importance of resonance rate increases. The JND for resonance rate was observed to be about 10%. The JND for pulse rate is more like 2% (Krumbholz et al. 2000; Figure 3.5). Therefore, one JND in resonance rate has about the same effect on the perception of size as four JNDs in pulse rate. The primary point, however, is that there is a very strong interaction between PR and RR in the perception of instrument size. Indeed, van Dinther and Patterson (2006) suggest that much of the difference in timbre between instruments within a family is size information associated with the pulse rate and the resonance rate.

## 6.3 The Interaction of GPR and VTL in Size Discrimination

The size surface of Smith and Patterson (2005) shows that (1) for long VTLs, the slope of the surface is shallow and uniform, (2) for shorter VTLs, in the range of normally sized children, the slope is steep for low GPRs and shallow for high GPRs, and (3) for the shortest VTLs, beyond the normal range, VTL still affects perceived size but GPR does not. The complexity of this surface prompted Gomersall et al. (2004) to develop a method for measuring the slope of the size surface directly using a 2AFC size-discrimination experiment.

It was assumed that the surface in a local region could be approximated by a plane, specifically, that a local region on the surface is reasonably well described by the first-order terms in a two-dimensional Taylor expansion. Listeners were required to discriminate between (1) a four-vowel phrase spoken by a "standard" speaker with a fixed combination of GPR and VTL (that is, a given point on the size surface) and (2) four-vowel phrases from test speakers with combinations of GPR and VTL that differed sufficiently to make their voices discriminable from the test speaker, but not so different as to violate the Taylor expansion criterion (that coefficients above first order in the expansion be small relative to the first-order terms). The vowels for the test and standard speakers were generated using STRAIGHT from recordings of the vowels of one *female* speaker pronounced in /hVd/ format.

The JND for VTL is roughly three times the JND for voice pitch, so on log GPR versus log VTL coordinates, we might expect that the locus of speakers that are equally discriminable from the standard would have combinations of GPR and VTL values that form an ellipse about the standard speaker. The paradigm is illustrated in Figure 3.8, where the open circles show the GPR and VTL combinations for five test speakers, and the filled circles about the open circles show the GPR and VTL combinations for the respective test speakers. On a given trial, a random four-vowel phrase from the standard speaker (with one fixed combination of GPR and VTL values) is presented in one stimulus interval; another random four-vowel phrase from one of the test speakers (with a different, but fixed, combination of GPR and VTL values) is presented in the other interval, and the listener had to choose the interval with the smaller speaker. There were eight test speakers for each standard speaker spaced evenly about the ellipse, as indicated in the figure. The axis of the ellipse was tilted relative to the

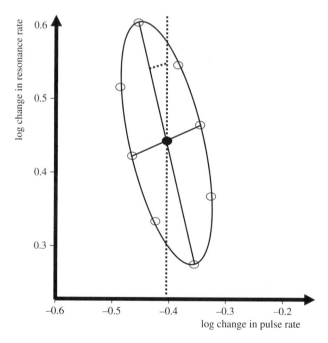

FIGURE 3.8. The ellipse of stimulus conditions used to measure the slope of the size surface for the standard voice representing an adult male. Each of the test voices (open circles) is compared repeatedly with the standard (filled circle) to determine which speaker sounds larger. The data are used to estimate the slope of the plane within the ellipse on the size surface. (Adapted from Gomersall et al. 2004.)

GPR–VTL coordinate system to ensure that both of the experimental variables changed from the first to the second interval on every trial. This helps to prevent listeners from focusing solely on the GPR or the VTL cue.

Test voices with higher GPR values and shorter VTL values tend to be heard as smaller than the standard speaker, and speakers with lower GPR values and longer VTL values tend to be heard as larger than the standard speaker. The eight probabilities, estimated by repeated pairings of a standard voice with each of their respective test voices, can be used to fit a plane to each ellipse of data (Gomersall et al. 2004). The line of steepest descent on the plane provides an estimate of the slope of the size surface at the point of the standard voice, and when the line of steepest descent is projected onto the GPR–VTL plane, the angle of the projected line reveals the tradeoff between VTL and GPR in the determination of perceived size.

The lines of steepest descent for 16 standard voices are presented in Figure 3.9. On this two-dimensional plot, the length of the vector shows the steepness of

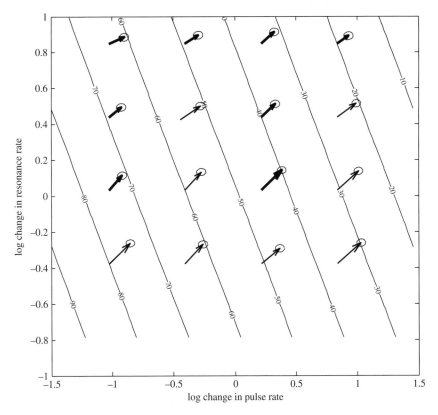

FIGURE 3.9. Size-surface slope vectors showing the angle of steepest descent for the ellipses associated with the 16 standard voices presented in the discrimination experiment of Gomersall et al. (2004).

the gradient (in relative terms), and the angle shows the direction in which the surface goes down fastest. The ellipse at the end of the vector shows the confidence limits for the estimate; that is, there is a 95% probability that the vector ends in the ellipse.

The figure shows that the gradient vectors do not vary substantially, either in terms of their length or their angle, across the GPR–VTL plane as much as might have been expected from the size surface generated with absolute judgments by Smith and Patterson (2005). The results are more like the uniform trading relationship derived from the within-family errors in the musical instrument study (van Dinther and Patterson 2006). The vectors are a little longer and the angles a little larger for the longer VTLs, but the differences are much less than the differences in slope associated with short and long VTLs in the absolute judgments.

## 7. Speaker Normalization

In Chapter 10 of this volume, Lotto and Sullivan argue that the "source" in the case of speech is the message of the communication rather than the pitch of the voice or the shape and length of the speaker's vocal tract. Similarly, at the end of the introduction to this chapter, attention was drawn to the fact that pulse-rate normalization and resonance-rate normalization are not just mechanisms for estimating source size; they also adapt auditory processing to the size of a source and help the auditory system produce a largely size-invariant representation of the message. Indeed, it has been argued, in this chapter and elsewhere (e.g., Irino and Patterson 2002), that pulse-rate adaptation and resonance-rate normalization are general auditory mechanisms that evolved with animal communication to make auditory perception generally robust to variation in source size. It also seems likely that if one could develop an auditory preprocessor that combined pulse-rate adaptation and resonance-rate normalization with source-laterality processing and grouping by common onset, the preprocessor would very likely enhance the robustness of speech-recognition machines considerably.

In speech research, the processes that confer robustness on recognition are collectively referred to as "vowel normalization" (Miller 1989), "vocal tract normalization" (Welling and Ney 2002), or "talker normalization" (Lotto and Sullivan, Chapter 10), depending on the aspect of communication under consideration. Lotto and Sullivan provide a comprehensive review of the adverse effects of the many different forms of speaker variability on the robustness of automatic speech recognition (see their Section 4). They distinguish *intrinsic* normalization techniques "…that rely on information contained solely within the vowel…" from *extrinsic* normalization techniques, which use longer-term aspects of speaker variability, typically at the sentence level rather than the syllable level, for additional normalization (e.g., Ladefoged and Broadbent 1957). Whereas the two classes of normalization techniques receive about equal attention in Chapter 10, the current chapter focuses entirely on just two of the intrinsic techniques: pulse-rate adaptation and resonance-rate normalization. There are several reasons for this: (1) These mechanisms function like mappings that can be applied to the time–frequency representation produced in the cochlea without reference to the context of the communication. (2) They automatically take care of a large portion of the acoustic variability associated with variation in speaker size, and so simplify the process of producing a size-invariant representation of the message at the syllable level. (3) The production of a largely size-invariant version of the message at an early stage in the processing facilitates subsequent extrinsic normalizations involving context; indeed, it may be essential for efficient functioning of extrinsic normalization. (4) There are algorithms for implementing these intrinsic normalizations and integrating them into the mainstream of computational auditory scene analysis; specific algorithms for extrinsic normalization are still in the early stages of development.

In Chapter 10, the primary example of intrinsic normalization is VTL normalization, which is based on formant ratio theory (FRT) (see Miller 1989 for a review). The theory originated with the historic conjecture by Lloyd (1890) that vowels are more readily identified by the ratios of their formant frequencies than by the absolute frequencies of the formants. A physical explanation of how the formant ratios of a vowel might remain largely unchanged as a child grows up and the absolute values of their formant frequencies decrease was provided by the early vocal tract models of Chiba and Kajiyama (1942) and Fant (1970). The basics of FRT from the auditory perspective were presented in Section 3.1 of the current chapter.

In Chapter 10, the importance of intrinsic pulse-rate normalization to speaker normalization and message invariance is largely overlooked, as is often the case in speech and language research. Miller (1989) developed an "auditory-perceptual" approach to speaker normalization, in which FRT was to be augmented by the inclusion of a "sensory reference" (SR). The SR was based on the individual speaker's average pitch, measured relative to the average pitch of the population (Miller 1989, Appendix A), and it was used to adjust formant ratios to improve recognition rates. It is a form of intrinsic GPR normalization, inasmuch as it is used to scale the initial "auditory-perceptual" representation of a sound and to place it within an "auditory perceptual space" that is conceptually similar to the auditory image space described in Section 3. It is technically quite different, however, inasmuch as GPR adaptation precedes the calculation of formant ratios in AIM, whereas it is applied after the calculation of formant ratios in Miller's model. It is also the case that it was developed to accommodate a subset of vowels associated with anomalous pitches, rather than all vowels, so it involves context in a way that is more typical of extrinsic normalization processes. In the end, however, we expect that optimal speech recognition will probably require both of these forms of GPR normalization in either intrinsic or extrinsic forms.

## 8. Summary

The pulse-resonance "syllables" that animals use to communicate, and the pulse-resonance notes that we use to make music, contain information about the size of the source in the pulse rate and the resonance rate. Both decrease as the size of the animal or the instrument increases. Humans perceive changes in resonance rate as changes in source size— either speaker size or instrument size—and they are very sensitive to changes in source size. Resonance rate appears to be a dimension of auditory perception just like musical pitch, and there is a tradeoff between pulse rate and resonance rate in the perception of source size.

The perceptual data support the hypothesis of Irino and Patterson (2002), Smith et al. (2005), and van Dinther and Patterson (2006) that the auditory system adapts to the pulse rate and normalizes for resonance rate as it constructs a largely size-invariant representation of the message of the syllable or the musical note.

*Acknowledgments.* Support for the writing of this paper and much of the research described in it was provided by the UK Medical Research Council (G990369, G0500221) and the German Volkswagen Foundation (VWF 1/79 783).

## *References*

Alcántara JI, Moore BCJ (1995) The identification of vowel-like harmonic complexes: Effect of component phase, level and fundamental frequency. J Acoust Soc Am 97:3813–3824.

Assmann PF, Katz WF (2005) Synthesis fidelity and time-varying spectral change in vowels. J Acoust Soc Am 117:886–895.

Beckford NS, Rood SR, Schaid D (1985) Androgen stimulation and laryngeal development. Ann Otol Rhinol Laryngol 94: 634–640.

Boersma P (2001) Praat, a system for doing phonetics by computer. Glot Int 5(9/10): 341–345.

Chiba T, Kajiyama M (1942) The Vowel: Its Nature and Structure. Tokyo: Tokyo-Kaiseikan.

Cohen L (1993) The scale transform. IEEE Trans Acoust Speech Signal Proc 41:3275–3292.

Cooke M (2006) A glimpsing model of speech perception in noise. J Acoust Soc Am 119:1562–1573.

Drennan W (1998) Sources of variation in profile analysis: Individual differences, extended training, roving level, component spacing and dynamic contour. PhD thesis, Indiana University.

Fant G (1970) Acoustic Theory of Speech Production, 2nd ed. Paris: Mouton.

Fitch WT, Giedd J (1999) Morphology and development of the human vocal tract: A study using magnetic resonance imaging. J Acoust Soc Am 106:1511–1522.

Fitch WT, Reby D (2001) The descended larynx is not uniquely human. Proc R Soc Lond B 268:1669–1675.

Fletcher NH, Rossing TD (1998) The Physics of Musical Instruments. New York: Springer-Verlag.

Gomersall P, Walters T, Turner R, Patterson RD (2004) The relative contribution of glottal pulse rate and vocal tract length in size discrimination judgements. Poster presented at the British Society of Audiology meeting, Sept. London (available on the CNBH Website: http://www.pdn.cam.ac.uk/cnbh/).

González, J (2004) Formant frequencies and body size of speaker: A weak relationship in adult humans. J Phonet 32:277–287.

Goto M, Hashiguchi H, Nishimura T, Oka R (2003) RWC music database: Music genre database and musical instrument sound database. In ISMIR is International Symposium on Music Information Retrieval. pp. 229–230.

Green DM (1988) Profile Analysis. London: Oxford University Press.

Hollien H, Green R, Massey K (1994) Longitudinal research on adolescent voice change in males. J Acoust Soc Am 96:3099–3111.

Huber JE, Stathopoulos ET, Curione GM, Ash T, Johnson K (1999) Formants of children, women and men: The effects of vocal intensity variation. J Acoust Soc Am 106:1532–1542.

Irino T, Patterson RD (2002) Segregating information about the size and shape of the vocal tract using a time-domain auditory model: The stabilized wavelet-Mellin transform. Speech Commun 36:181–203.

Ives DT, Smith, DRR, Patterson RD (2005) Discrimination of speaker size from syllable phrases. J Acoust Soc Am 118:3816–3822.

Kawahara H, Irino T (2004) Underlying principles of a high-quality speech manipulation system STRAIGHT and its application to speech segregation. In: Divenyi P (ed) Speech Segregation by Humans and Machines. Dordrecht: Kluwer Academic, pp. 167–180.

Kawahara H, Masuda-Kasuse I, de Cheveigne A (1999) Restructuring speech represen- tations using pitch-adaptive time-frequency smoothing and instantaneous-frequency- based $F0$ extraction: Possible role of repetitive structure in sounds. Speech Commun 27(3–4):187–207.

Krumbholz K, Patterson RD, Pressnitzer D (2000) The lower limit of pitch as determined by rate discrimination. J Acoust Soc Am 108:1170–1180.

Künzel HJ (1989) How well does average fundamental frequency correlate with speaker height and weight? Phonetica 46:117–125.

Ladefoged P, Broadbent DE (1957) Information conveyed by vowels. J Acoust Soc Am 29:98–104.

Lass NJ, Brown WS (1978) Correlational study of speakers, heights, weights, body surface areas and speaking fundamental frequencies. J Acoust Soc Am 63:1218–1220.

Leek MR, Dorman MF, Summerfield Q (1987) Minimum spectral contrast for vowel identification by normal-hearing and hearing-impaired listeners. J Acoust Soc Am 81:148–154.

Licklider JCR (1951) A duplex theory of pitch perception. Experientia 7:128–133.

Liu C, Kewley-Port D (2004) STRAIGHT: A new speech synthesizer for vowel formant discrimination. ARLO 5:31–36.

Lloyd RJ (1890) Speech sounds: Their nature and causation (I). Phoneticia Studien 3:251–278.

Marcus SM (1981) Acoustic determinants of perceptual centre (P-centre) location. Percept Psychophys 30:247–256.

Meddis R, Hewitt MJ (1991) Virtual pitch and phase sensitivity of a computer model of the auditory periphery. I: Pitch identification. J Acoust Soc Am 89:2866–2882.

Miller GA (1947) Sensitivity to changes in the intensity of white noise and its relation to masking and loudness. J Acoust Soc Am 19:609–619.

Miller JD (1989) Auditory-perceptual interpretation of the vowel. J Acoust Soc Am 85:2114–2133.

Owren MJ, Anderson JD (2005) Voices of athletes reveal only modest acoustic correlates of stature. J Acoust Soc Am 117:2375.

Patterson RD (1994) The sound of a sinusoid: Time-interval models. J Acoust Soc Am 96:1419–1428.

Patterson RD, Holdsworth J (1996) A functional model of neural activity patterns and auditory images. In: Ainsworth WA (ed) Advances in Speech, Hearing and Language Processing, Vol. 3, Part B. London: JAI Press.

Patterson RD, Robinson K, Holdsworth J, McKeown D, Zhang C, Allerhand MH (1992) Complex sounds and auditory images. In: Cazals Y, Demany L, Horner K (eds) Auditory Physiology and Perception, Proceedings of the 9th International Symposium on Hearing. Oxford: Pergamon Press, pp. 429–446.

Patterson RD, Allerhand M, Giguère C (1995) Time domain modeling of peripheral auditory processing: A modular architecture and a software platform. J Acoust Soc Am 98:1890–1894.

Patterson RD, Anderson TR, Francis K (2006) Binaural auditory images for noise-resistant speech recognition. In: Ainsworth W, Greenberg S (eds) Listening to Speech: An Auditory perspective. The Publisher, LEA, is Lawrence Erlbaum Associates City is Mahwah, NJ pp. 257–269.

Peterson GE, Barney HL (1952) Control methods used in the study of vowels. J Acoust Soc Am 24:175–184.

Pressnitzer D, Patterson RD, Krumbholz K (2001) The lower limit of melodic pitch. J Acoust Soc Am 109:2074–2084.

Rendall D, Vokey JR, Nemeth C, Ney C (2005) Reliable but weak voice-formant cues to body size in men but not women. J Acoust Soc Am 117:2372.

Ritsma RJ, Hoekstra A (1974) Frequency selectivity and the tonal residue. In: Zwicker E, Terhardt E (eds) Facts and Models in Hearing. Berlin: Springer, pp. 156–163.

Schouten JF (1938) The perception of subjective tones. Proc Kon Ned Akad Wetensch 41:1086–1093.

Scott SK (1993) P-centres in speech an acoustic analysis. PhD thesis, University College London.

Slaney M, Lyon RF (1990) A perceptual pitch detector. In: Proceedings of the IEEE International Conference on Acoustics, Speech, Signal Processing, Albuquerque, New Mexico.

Smith DRR, Patterson RD (2005) The interaction of glottal-pulse rate and vocal-tract length in judgements of speaker size, sex and age. J Acoust Soc Am 118:3177–3186.

Smith DRR, Patterson RD, Turner R, Kawahara H, Irino T (2005) The processing and perception of size information in speech sounds. J Acoust Soc Am 117:305–318.

Spiegel MF, Picardi MC, Green DM (1981) Signal and masker uncertainty in intensity discrimination. J Acoust Soc Am 70:1015–1019.

Sprague MW (2000) The single sonic twitch model for the sound production mechanism in the weakfish, *Cynoscion regalis*. J Acoust Soc Am 108:2430–2437.

Terhardt E (1974) Pitch, consonance, and harmony. J Acoust Soc Am 55:1061–1069.

Thurlow WR, Small AM Jr (1955) Pitch perception for certain periodic auditory stimuli. J Acoust Soc Am 27:132–137.

Titze IR (1989) Physiologic and acoustic differences between male and female voices. J Acoust Soc Am 85:1699–1707.

Turner RE, Al-Hames MA, Smith DRR, Kawahara H, Irino T, Patterson RD (2006) Vowel normalisation: Time-domain processing of the internal dynamics of speech. In: Divenyi P, Greenberg S, Meyer G. (eds) Dynamics of Speech Production and Perception. Amsterdam: IOS Press, pp. 153–170.

van Dinther R, Patterson RD (2006) Perception of acoustic scale and size in musical instrument sounds. J Acoust Soc Am 120:2158–2176.

Welling L, Ney H (2002) Speaker adaptive modelling by vocal tract normalization. IEEE Trans Speech Audio Process 10:415–426.

Yang C-S, Kasuya H (1995) Dimension differences in the vocal tract shapes measured from MR images across boy, female and male subjects. J Acoust Soc Jpn E 16:41–44.

Yost WA, Patterson RD, Sheft S (1996) A time-domain description for the pitch strength of iterated rippled noise. J Acoust Soc Am 99:1066–1078.

# 4
# The Role of Memory in Auditory Perception

LAURENT DEMANY AND CATHERINE SEMAL

## 1. Introduction

Sound sources produce physical entities that, by definition, are extended in time. Moreover, whereas a visual stimulus lasting only 1 ms can provide very rich information, that is not the case for a 1-ms sound. Humans are indeed used to processing much longer acoustic entities. In view of this, it is natural to think that "memory" (in the broadest sense) must play a crucial role in the processing of information provided by sound sources. However, a stronger point can be made: It is reasonable to state that, at least in the auditory domain, "perception" and "memory" are so deeply interrelated that there is no definite boundary between them. Such a view is supported by numerous empirical facts and simple logical considerations. Consider, as a preliminary example, the perception of loudness. The loudness of a short sound, e.g., a burst of white noise, depends on its duration (Scharf 1978). Successive noise bursts equated in acoustic power and increasing in duration from, say, 5 ms to about 200 ms are perceived not only as longer and longer but also as louder and louder. Loudness is thus determined by a temporal integration of acoustic power. This temporal integration implies that a "percept" of loudness is in fact the content of an auditory memory.

A commonsense notion is that memory is a consequence of perception and cannot be a cause of it. In the case of loudness, however, perception appears to be a consequence of memory. This is not a special case: Many other examples of such a relationship between perception and memory can be given. Consider, once more, the perception of white noise. A long sample of white noise, i.e., a completely random signal, is perceived as a static "shhhhh..." in which no event or feature is discernible. But if a 500-ms or 1-s excerpt of the same noise is taken at random and cyclically repeated, the new sound obtained is rapidly perceived as quite different. What is soon heard is a repeating sound pattern filled with perceptual events such as "clanks" and "rasping" (Guttman and Julesz 1963; Warren 1982, Chapter 3; Kaernbach 1993, 2004). It can be said that the perceptual events in question are a creation of memory, since they do not exist in the absence of repetitions. Kubovy and Howard (1976) provided another thought-provoking example. They constructed sequences of binaural "chords" in which each chord consisted of six simultaneous pure tones with different interaural phase differences (IPDs). The tones had the same frequencies, ranging from 392 to 659 Hz, in all chords. In the first chord, the IPDs were an arbitrary function of frequency. All the subsequent chords were identical to the first chord except for a modification in the IPD of a single tone. The tone with the modified IPD changed from chord to chord, in a sawtooth manner, going gradually from 392 to 659 Hz in some sequences and vice versa in other sequences. Initially, the chords making up such a sequence are perceived as identical stimuli, but after a few iterations an ascending or descending melodic pattern emerges: In each chord, the tone with the modified IPD is perceptually segregated from the other tones, and the listener tracks this tone from chord to chord. The segregation is based on nothing but memory since the segregated tones are, intrinsically, similar to the other tones. The phenomenon is not observable when the chords are separated by very long interstimulus intervals (ISIs), but silent ISIs of 1 s are not too long.

In order to see that perception and memory are deeply interrelated, it is in fact unnecessary to consider specific stimuli. The term "perception" is quite generally used to mean, more precisely, "discrimination" or "identification." In both cases, what is "perceived" is a relation between a stimulus belonging to the present and memory traces of previous stimuli (possibly a single stimulus memory trace in the case of discrimination). When John or Jack says that he perceives the pitch of a newly presented sound as high, he means more precisely that the pitch in question is higher than the average pitch of sounds that he has heard in the past, and memorized. Consider, besides, what psychoacousticians do when they want to assess the "sensation noise" inherent to the perception of some acoustic parameter, that is, the imprecision with which this acoustic parameter is encoded by the auditory system. The only possible method to quantify sensation noise is to measure just-noticeable differences or some other index of discrimination. Thus, one must present successive stimuli and require the listener to compare them. But in such a situation, the internal noise limiting performance may include, in addition to "sensation noise," a "memory noise." Performance will be maximal for some ISI, typically several hundreds of milliseconds if the stimuli are brief. Choosing this optimal ISI does not ensure that the memory noise will be inexistent or even smaller than the sensation noise: For the optimal ISI, the only certainty will be that the memory noise is as small as it can be.

There are multiple forms of auditory memory, and they are certainly based on a variety of neural mechanisms. The present chapter will not consider all of them. For instance, although it has been noted above that one form of auditory memory is involved in the perception of loudness, temporal integrations of this kind (also observable for other auditory attributes) will be ignored in the following. The starting point of the chapter is the general idea that any sound, *once it has ended*, leaves in the brain neural traces that affect the perception of future sounds (and can also, in fact, play a role in the perceptual analysis of the ended sound). The aim of the chapter is to describe a number of interesting psychophysical phenomena illustrating this general idea, and to relate them, as far as possible, to neurophysiological facts.

## 2. Neural Adaptation and Its Possible Perceptual Consequences

A very primitive and short-lived form of auditory memory manifests itself in the phenomenon of *forward masking*. The detection threshold of a brief sound is elevated by the presentation of a previous sound if the two sounds have similar or overlapping power spectra and if the time interval separating their offsets does not exceed about 200 ms (Zwislocki et al. 1959). The amount of masking, or in other words, the size of the threshold elevation, is a decreasing function of the time interval in question. The slope of this function increases with masker intensity, so that masking effects last about 200 ms more or less independently of masker intensity. Because a monaural forward masker has at most a very weak masking effect on a probe sound presented to the other ear, it is believed that the physiological substratum of forward masking is located at a relatively peripheral level of the auditory system. What is this substratum? Two hypotheses can be put forth. According to the first one, forward masking is due to a persistence of the neural excitation produced by the masker beyond its physical offset; the detection threshold of the following probe sound is elevated because, in order to be detectable, the probe must produce a detectable increment in the residual excitation produced by the masker; the just-detectable increment is an increasing function of the residual excitation, as predicted by Weber's law. According to the second hypothesis, in contrast, the trace left by the masker is negative rather than positive: forward masking is due to an "adaptation" phenomenon, that is, a decrease in the sensitivity of the neural units stimulated by the masker following its presentation; in order to be detectable, the following probe must overcome this adaptation and thus be more intense than in the masker's absence. Houtgast and van Veen (1980) and Wojtczak and Viemeister (2005) provided psychophysical evidence in support of the adaptation hypothesis. In their experiments, a 10-ms binaural probe was presented shortly after, during, or shortly before a longer and more intense masker presented to only one ear. On the ear stimulated by the masker, the level of the probe was such that the probe was partially masked but detectable. On the other ear (essentially not subjected to the masker influence), the level of the probe was controlled by the listener, who was required to adjust it to the value producing a mid-plane localization of the probe. For this physical level, one could assume that the "internal level" of the probe was the same at the two ears. When the probe was presented shortly after the masker, it appeared that the physical level adjusted at the unmasked ear was lower than the physical level at the masked ear, as if the masker attenuated the probe. This effect was smaller or absent when the probe was presented during the masker or before it. Such an outcome is consistent with the adaptation hypothesis and was not expected under the persistence hypothesis.

Adaptation effects have been observed by physiologists at the level of the auditory nerve. Is it the neural basis of forward masking? In order to answer that question, Relkin and Turner (1988) and Turner et al. (1994) measured the detection thresholds of probe signals preceded by maskers in psychophysical experiments (on human listeners) as well as physiological experiments (on chinchillas), using one and the same two-interval forced-choice procedure in both cases. In the physiological experiments, the relevant neural information was supposed to be the number of spikes appearing in a single auditory nerve fiber within a temporal window corresponding to the period of the probe presentation. The results suggested that the amount of forward masking resulting from adaptation in the auditory nerve is too small to account for the psychophysical phenomenon, and thus that an additional source of masking exists at a higher level of the auditory system. Meddis and O'Mard (2005) proposed a different scenario. They supposed that the detection of an auditory signal is not simply determined by the quantity of spikes conveyed by the auditory nerve, but requires coincidental firing of a number of nerve fibers. Using this assumption in a computer model of the auditory nerve response to probe signals in a forward masking context, they arrived at the conclusion that the model ingredients were sufficient to predict the forward making effects observable psychophysically. Nonetheless, the temporal rules of forward masking seem to be similar for normal listeners and for cochlear implant patients (Shannon 1990), which suggests that the phenomenon mainly takes place beyond the auditory nerve.

Adaptation effects indeed exist throughout the auditory pathway. Ulanovsky et al. (2003, 2004) recently described an interesting form of adaptation in the primary auditory cortex of cats. The animals were presented with long sequences of pure tones separated by silent ISIs of about 500 ms and having two possible frequencies, with different probabilities of occurrence within each sequence (e.g., f1 f1 f1 $f2$ f1 f1 f1 f1 f1 $f2$ f1 f1 $f2$ ...). Measures of spike count were made in neurons which, initially, were equally responsive to the two frequencies. What the authors found, in the course of various sequences, is a decrease in spike count in response to the more common frequency, but very little or no concomitant decrease in response to the rarer frequency. This stimulus-specific adaptation could be observed even when the two presented frequencies were only a few percent apart. It did not seem to exist subcortically, in the auditory thalamus. It appeared to have a short-term component, reflecting an effect of one tone on the response to the next tone, but also much slower components, revealing a surprisingly long neural memory: the authors uncovered an exponential trend with a time constant of tens of seconds. It was also found that stimulus-specific adaptation could be elicited by sequences of tones differing in intensity rather than frequency.

The cortical adaptation described by Ulanovsky et al. may not be a source of forward masking. However, it is likely to play a role in another perceptual phenomenon, called *enhancement*. An enhancement effect occurs when, for example, a sum of equal-amplitude pure tones forming a "notched" harmonic series (200, 400, 600, 800, 1200, 1400, 1600 Hz; note that 1000 Hz is missing) is followed, immediately or after some ISI, by the complete series (200, 400, 600, 800, *1000*, 1200, 1400, 1600 Hz). The second stimulus is not heard as a single sound with a pitch corresponding to 200 Hz, as would happen in the absence of the first stimulus (the "precursor"). Instead, the second stimulus is heard as a sum of two separate sounds: (1) a complex tone similar to the precursor; (2) the 1000-Hz pure tone that was not included in the precursor; this pure tone "pops out,", as if the other tones were weaker due to an adaptation by the precursor. Amazingly, according to Viemeister (1980), enhancement effects are observable even when the precursor and the following stimulus are separated by several minutes or indeed hours (although stronger enhancement is of course obtained for very short ISIs). However, at least for stimulus configurations such as that considered above, enhancement appears to be a monaural effect: it is not observable when the two successive stimuli are presented to opposite ears. Viemeister and Bacon (1982) wondered whether the enhancement of a tone $T$ increases its forward masking of a subsequent short probe tone with the same frequency. This was indeed verified in their experiment. Using a precursor that was temporally contiguous to the complex including $T$, they found an increase of forward masking by as much as 8 dB, on average. This increase in forward masking, normally requiring an increase of about 16 dB in masker intensity, could not be ascribed to forward masking of the probe by the precursor itself, because the latter effect was too small. Therefore, the experiment showed that an enhanced tone behaves as if it were increased in intensity. To account for that behavior, the authors noted that in the absence of the precursor, the complex including $T$ produced less masking of the probe than $T$ alone. This suggested that somewhere in the auditory system, $T$ was attenuated by the other components of the complex. [The finding in question was in fact consistent with previous psychophysical studies on the mutual interactions of simultaneous pure tones (Houtgast 1972).] Viemeister and Bacon thus interpreted enhancement as a decrease, caused by adaptation, in the ability of a sound to attenuate other, simultaneous sounds. However, Wright et al. (1993) have cast doubts on the validity of this interpretation.

Enhancement effects are not elicited only when a pure tone is added to a previously presented sum of pure tones. In a white noise presented after a band-reject noise, one can hear clearly, as a separate sound, the band of noise that was rejected in the initial stimulus. Similarly, one can hear a given vowel in a stimulus with a flat spectrum if this stimulus is preceded by a precursor consisting of the "negative" of the vowel in question, i.e., a sound in which the vowel formants—corresponding to spectral peaks—are replaced by "antiformants" corresponding to spectral troughs (Summerfield et al. 1987; see also Wilson 1970). In one of the experimental conditions used by Summerfield et al., the precursor consisted of wideband *noise* with a uniform spectrum while the subsequent stimulus was a *complex tone* with very small spectral "bumps" at frequencies corresponding to the formants of a given vowel. This stimulus configuration still produced significant enhancement: the vowel was identified more accurately than in the absence of the precursor. Moreover, it appeared that the benefit of the noise precursor for identification was not smaller than the benefit of a comparable *tonal* precursor with the same pitch as the subsequent stimulus. This is important because it suggests that the occurrence of enhancement does not require from the listener the perception of a similarity between the precursor and part of the subsequent stimulus. In turn, this supports the idea that the effect is caused by "low-level" mechanisms such as adaptation. However, since there are actually various forms of neural adaptation in the auditory system, it remains to be determined which one(s) matter(s) for enhancement. In fact, there may be different forms of enhancement, based on different forms of adaptation. Consider again, in this regard, the perceptual phenomenon discovered by Kubovy and Howard (1976) and described at the beginning of the chapter. Essentially, the authors found that a binaural tone can be made to pop out, in a mixture of other binaural tones, by virtue of its relative novelty. This is apparently an enhancement effect. But interestingly, the novelty involved here is neither a new frequency nor a new intensity; it is only a new interaural delay for a given frequency. If enhancement is mediated, in that case again, by adaptation, the corresponding adaptation may well be different from that underlying the enhancement of new energy in some spectral region.

In their papers about stimulus-specific cortical adaptation, Ulanovsky et al. (2003, 2004) do not relate their physiological observations to the auditory phenomenon of enhancement. However, they do state that this form of adaptation "may underlie auditory novelty detection" (Ulanovsky et al. 2003, p. 394). More specifically, they view it as the neural basis of the *mismatch negativity* or MMN. The MMN, initially identified by Näätänen et al. (1978), is a *change-specific* component of the auditory event-related potential recordable on the human scalp. One can also measure it with a brain-imaging tool such as functional magnetic resonance imaging (fMRI). Näätänen and Winkler (1999) and Schröger (1997, 2005) reviewed the enormous literature (about 1000 articles, up to now) devoted to this brain response. An MMN is typically obtained using a stimulus sequence in which a frequent "standard" sound and a rarer "deviant" sound are randomly interleaved. In such conditions, a subtraction of the average potential evoked by the standard from the average potential evoked by the deviant reveals a negative wave peaking at 100–250 ms following stimulus onset. This negative wave is the MMN. A similar wave is obtained by subtracting the response to the deviant when presented in an "alone" condition from the response to the same stimulus in the context of the sequence including more frequent iterations of the standard. In the latter sequence, the ISI between consecutive sounds may be, for instance, 500 ms, but it is not a very critical parameter: An MMN is still recordable for ISIs as long as 7 or 9 s when the standard and deviant stimuli are two tones differing in frequency by 10% (Czigler et al. 1992; Sams et al. 1993). The main source of MMN is located in the auditory cortex (e.g., Kropotov et al. 2000). It seems that any kind of acoustic change can give rise to MMN: The standard and deviant stimuli can differ in frequency, intensity, spectral profile, temporal envelope, duration, or interaural time delay. An increase in the magnitude of the change produces an increase in the amplitude of the MMN and a decrease in its latency. Giard et al. (1995) reported that the scalp topographies of the MMNs elicited by changes in frequency, intensity, and duration are not identical, which suggests that these three types of change are processed by at least partly distinct neural populations. Analogous results were recently obtained by Molholm et al. (2005) in a study using fMRI rather than electroencephalography. The idea that there are separate and specialized MMN generators is also supported by experiments in which changes occurred on two acoustic dimensions simultaneously: The MMN obtained in response to a two-dimensional change in frequency and interaural relation, or frequency and duration, or duration and intensity, is equal to the sum of the MMNs elicited by its one-dimensional components, exactly as if each of the combined one-dimensional components elicited its own MMN (e.g., Schröger 1995).

A crucial property of the MMN is that it is a largely automatic brain response. It is usually recorded while the subject is required to ignore the stimuli and to read a book or to watch a silent film. The automaticity of the MMN tallies with the suggestion by Ulanovsky et al. (2003) that its neural basis is an adaptation mechanism already taking place in primary auditory cortex. Other authors have also argued that adaptation is the whole explanation (e.g., Jääskeläinen et al. 2004). In this view, the fact that an MMN can be elicited by, for example, a *decrease* in sound intensity, or a change in duration, would mean that certain neurons prone to adaptation are optimally sensitive to particular intensities or durations. However, several experimental results do not fit in with the adaptation hypothesis. For instance, Tervaniemi et al. (1994) recorded a significant MMN in response to occasional *repetitions* of a stimulus in a sequence of "Shepard tones" (sums of pure tones one octave apart) perceived as an endlessly descending melodic line. In the same vein, Paavilainen et al. (2001) report that an MMN can be elicited by the violation of an abstract rule relating the intensity of a pure tone to its frequency ("the higher the frequency, the higher the intensity"). Such findings suggest that even though the MMN is generated pre-attentively, the MMN generator is endowed with some intelligence allowing it to detect novelties more complex than mere modifications of specific sound events. Jacobsen and Schröger (2001) and Opitz et al. (2005) used an ingenious method to identify the respective contributions of adaptation and more "cognitive" operations in the MMN generation process. Consider the two sequences of pure tones displayed in Table 4.1. The "oddball" sequence consists of nine presentations of a 330-Hz standard tone, followed by one presentation of a 300-Hz deviant tone. In the "control" sequence, on the other hand, all tones differ from each other in frequency; however, one tone is matched in both frequency and temporal position to the deviant tone of the oddball sequence, and another tone is matched to the

TABLE 4.1. Frequencies (in Hz) of tones forming an oddball sequence and the corresponding control sequence in the experiment of Opitz et al. (2005)

| Oddball | 330 | 330 | 330 | 330 | 330 | 330(A) | 330 | 330 | 330 | 300(B) |
|---------|-----|-----|-----|-----|-----|--------|-----|-----|-----|--------|
| Control | 585 | 363 | 532 | 399 | 440 | 330(C) | 484 | 707 | 643 | 300(D) |

standard tone of the oddball sequence. A significant difference between brain responses to the tones labeled "A" and "B" may arise from both adaptation or cognitive operations. However, suppose that an MMN is observed when the response to D is subtracted from the response to B. Since both B and D are tones with a novel frequency, one can assume that this MMN is due to cognitive operations rather than to adaptation; a contribution of adaptation is very unlikely, because the frequency difference between B and its predecessors is not larger than the frequency difference between D and any of its predecessors. On the other hand, if the response to A differs from the response to C, the main source of this effect is identifiable as adaptation rather than cognitive operations, because neither A nor C violates a previously established regularity. Following this rationale, Jacobsen and Schröger (2001) and Opitz et al. (2005) obtained evidence that both adaptation and cognitive operations contribute to the "B minus A" MMN. Taking advantage of the fMRI tool, Opitz et al. localized the adaptation component in the primary auditory cortex and the cognitive component in nonprimary auditory areas.

It has been argued that the MMN has a functional value and must be interpreted as a warning signal, drawing the subject's attention toward changes in the acoustic environment. According to Schröger (1997), a change will be detected consciously if the MMN exceeds a variable threshold, the threshold in question being low if the subject pays attention to the relevant auditory stimulation and higher otherwise. Is it clear, however, that the MMN is directly related to the conscious perception of acoustic changes? In support of this idea, Näätänen et al. (1993) and Atienza et al. (2002) found that improvement in the conscious (behavioral) discrimination of two very similar stimuli, following repeated presentations of these stimuli, can be paralleled by the development of an MMN initially not elicited by the same stimuli (see also Tremblay et al. 1997). Besides, according to Tiitinen et al. (1994), an increase in the frequency difference between two tone bursts produces precisely parallel decreases in (1) the subject's behavioral reaction time to the corresponding frequency change and (2) the latency of the MMN recordable with the same stimuli (while the subject is reading a book). The data of Tiitinen et al. suggest in addition that in the vicinity of 1000 Hz, the minimum frequency change able to elicit an MMN (in the absence of attention) is roughly similar to the frequency difference limen measurable behaviorally, under normal conditions. However, Allen et al. (2000) obtained very different results in a study on the discrimination of synthetic syllables. They found that a significant MMN could be measured for stimulus changes that were much too small to be perceived consciously. In itself, this is not inconsistent with Schröger's hypothesis on the relation between the MMN and conscious change detection. But Allen et al. also found essentially identical MMNs in response to inaudible and audible stimulus changes. They were thus led to state that "the neural generators responsible for the MMN are not necessarily linked to conscious perception" (Allen et al. 2000, p. 1389). Another argument against the idea that the MMN-generating mechanism is crucial for the conscious detection of acoustic changes is that according to several authors (see especially Cowan et al. 1993), a given stimulus elicits a detectable MMN only if it is preceded by *several* presentations of a different stimulus. For the conscious perception of an acoustic change, a sequence of only two sounds is of course sufficient.

In summary, it has been pointed out above that neural adaptation in the auditory system—a "negative" form of auditory memory—is likely to be the cause of forward masking and to play a role in the conscious detection of novelties in the acoustic environment. With respect to the detection of novelty, stimulus-specific adaptation is useful because it makes novel sounds more salient: In a noisy jungle, as pointed out by Jääskeläinen et al. (2004), it is a matter of life or death to detect a novel sound such as that of a twig cracking under the paw of a stalking predator. On the other hand, it may be that adaptation does not help a listener to perceive consciously the *relationship* between a novel sound and a previous sound. Indeed, from a certain point of view, adaptation should impair our ability to judge whether two successive sounds are identical or differ in intensity, because if the two sounds are physically identical, their neural representations will nonetheless be systematically different in consequence of adaptation. In a later section of this chapter, it will be seen that people can consciously detect frequency changes on the basis of automatic neural processes that are apparently unrelated to adaptation as well as to other potential sources of the MMN.

## 3. Preperceptual Storage

Vision researchers have firmly established that soon after its termination, an optical stimulus has two types of representation in visual memory. Compelling evidence for this duality was provided, in particular, by Phillips (1974). In his experiments, observers had to make same/different judgments on visual patterns produced by randomly filling cells in a square matrix. On each trial, two successive matrices were displayed, both of them for a time of 1 s. These two matrices always had the same number of cells, but the number in question (an index of complexity) was an independent variable, as well as the ISI. In addition, the two matrices could be displayed either exactly in the same position or in slightly different positions. Finally, an irrelevant matrix acting as a mask could be either presented or not presented during the ISI. When the ISI was short (< 100 ms), discrimination performance was strongly dependent on the "position" factor (displacements impaired performance) and strongly affected by the mask; however, in the absence of displacement or mask, performance was excellent regardless of the number of cells. When the ISI was longer (600 ms), opposite results were obtained: the number of cells had a large effect on performance, which was quite poor for $8 \times 8$ matrices and still not perfect for $5 \times 5$ matrices; however, the position factor had no effect and the mask had only a weak effect. These findings, as well as other results, led to the distinction between: (1) a very short-lived but high-capacity "iconic memory," tied to spatial position and very sensitive to masking; (2) a more enduring but limited-capacity "short-term visual memory," not tied to spatial position and less sensitive to masking.

In the auditory domain, is there a similar duality of memory systems? Cowan (1984) has posited that the answer is yes. In any case, a perceptual phenomenon known as *backward recognition masking* (BRM) seems to imply that one must distinguish a "preperceptual" auditory memory (PPAM) from "postperceptual" short-term auditory memory (STAM). [In the literature, unfortunately, STAM is sometimes referred to as "echoic memory"; this is misleading since STAM is the auditory counterpart of short-term visual memory, not iconic memory.] The phenomenon of BRM was investigated in detail by Massaro (for a short review, see Massaro and Loftus 1996). In his initial experiment, Massaro (1970a) requested listeners to identify as "high" or "low" a 20-ms burst of sinusoidal sound taking two possible frequencies: 870 Hz (correct response: "high") or 770 Hz (correct response: "low"). On each experimental trial, one of the two corresponding test tones was presented and followed by a 500-ms tonal masker of 820 Hz, after an ISI randomly determined among a set of ISI values ranging from 0 to 500 ms. The masker and test tones had the same intensity. Before data collection, the three tested listeners were trained in the task for about 15 hours. The results are displayed in Figure 4.1. It can be seen that identification performance, measured in terms of percent correct, improved markedly and steadily as the ISI increased from about 40 ms to about 250 ms, and then plateaued. Let us stress that for any ISI, the test tones were clearly audible; for small ISIs, the difficulty was not to detect them but only to recognize their pitch. Subsequent experiments indicated that BRM affects, in addition to pitch judgments, judgments of loudness, duration, timbre, spatial position, and speech distinctions.

To account for these results and related ones, Massaro (1972) (see also Massaro and Loftus 1996) essentially argued that "perception takes time." According to his theory, when a short sound $S_1$ is presented to a listener, an image or

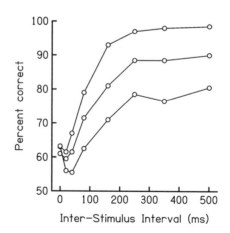

FIGURE 4.1. Results of an experiment on backward recognition masking. Identification performance as a function of the ISI for three different listeners. (Adapted with permission from Massaro DW 1970a; ©American Psychological Association.)

representation of this sound is initially stored in a PPAM system. This storage does not start at the end of the sound but at its onset (or very soon after the onset), and the temporal span of PPAM is about 250 ms, independently of the sound itself. The "perception" (or perceptual analysis) of the sound corresponds to a progressive readout of the information stored in PPAM. If a second sound $S_2$ is presented less than 250 ms after $S_1$, the perceptual analysis of $S_1$ is interrupted because $S_2$ replaces $S_1$ in PPAM. Hence, it is not possible to identify $S_1$ as accurately as in the absence of $S_2$. Otherwise, the perceptual analysis of $S_1$ continues until the disappearance of its image in PPAM. Thus, the time available and needed for an optimal perceptual analysis is the fixed temporal span of PPAM. The product of perceptual analysis is progressively transferred (as soon as perceptual analysis begins) into a different and more enduring memory system, STAM (called "synthesized auditory memory" by Massaro).

This theory met with skepticism in the psychoacoustic community. Several research teams obtained results similar to those of Massaro (1970a) in variants of the experiment described above, but it was also found that the type of psychophysical procedure used to study BRM can have a strong influence on the results (Watson et al. 1976; Yost et al. 1976). Another finding was that no BRM occurs when the mask consists of noise and is thus perceptually dissimilar to the test tones. According to Sparks (1976), even a narrowband noise in the spectral region of the test tones is an ineffective masker. From such a finding, it has been inferred that BRM does not reveal the existence of a *pre*perceptual memory and is instead a *post*perceptual effect—an interference effect in STAM. However, the inference in question is unwarranted. It is based on the erroneous idea that "preperceptual" means "not yet processed by the auditory system." On the contrary, if PPAM does exist, its neural substratum is presumably located in the auditory cortex, very far from the cochlea. The absence of BRM of tones by noise may only mean that (contrary to a hypothesis favored by Massaro) PPAM is not a single-channel memory store, completely filled by any type of sound. In this view, tones and noise have separate representations in PPAM.

Hawkins and Presson (1977) reported evidence that, to some extent, BRM depends on attention. In one of their experiments, a two-alternative pitch identification judgment ("high" or "low") had to be made on a monaural tone burst followed by a tonal masker whose frequency varied unpredictably from trial to trial. In separate blocks of trials, the masker was respectively presented (1) to the same ear as the test tone; (2) to the opposite ear; (3) diotically. Since these three conditions were blocked, it could be expected that in conditions 2 and 3, listeners would be able to filter out the masker attentionally and thus to reduce BRM. This did not happen: the results obtained in the three conditions were exactly the same, and similar to those displayed in Figure 4.1. In a second experiment, however, the authors varied the masker frequency *between* blocks of trials rather than within blocks, and this slight change in procedure had spectacular consequences: BRM was now essentially absent in conditions 2 and 3, whereas in condition 1 results similar to those shown in Figure 4.1 were once more obtained. Overall, therefore, Hawkins and Presson's study suggests that selective attention can largely reduce BRM, but that a purely spatial attentional filter is not sufficient to do so when there is a spatial difference between the masker and test tones. Another suggestion of Hawkins and Presson's study is that a purely spectral attentional filter is also not sufficient to prevent BRM. Bland and Perrott (1978) supported that view: Paradoxically, according to these investigators, a fixed tonal masker has a stronger masking effect when its frequency is far away from the test tones' frequencies than when all stimuli are close in frequency. However, Sparks (1976) found just the opposite.

The fact that BRM seems to be to some extent dependent on attention cannot be taken as an argument against the PPAM concept. On the other hand, as pointed out by Massaro and Idson (1977), it is possible to argue that none of the studies mentioned above provides conclusive evidence for PPAM. In all of them, subjects were required to make absolute judgments: On each trial, the percept evoked by the presented test tone had to be compared with a representation of the two possible test stimuli in a "long-term" memory store. In such conditions, the deleterious effect of the masker may occur while the presented test tone is perceptually analyzed, but also following this perceptual analysis, while the percept (stored in STAM) is compared to the long-term internal representations and a decision is being made. In order to get rid of this ambiguity, Massaro and Idson (1977) replaced the original BRM paradigm by an experimental situation in which listeners simply had to make comparisons between two successive 20-ms tones, differing in frequency and separated by a variable ISI. On each trial, the frequency of the first tone ($S_1$) was selected at random within a frequency range of several semitones, and the frequency of the second tone ($S_2$) was, at random, slightly higher or lower. The task was to judge whether $S_2$ was higher or lower in pitch than $S_1$. In this situation, again, it appeared that performance increased as the ISI increased, up to at least 250 ms. This could not be explained by assuming that, for short ISIs, $S_1$ had a deleterious effect on the processing of $S_2$ because *forward* recognition masking of pitch is nonexistent (Ronken 1972; Turner et al. 1992). A conceivable alternative hypothesis would be that for short ISIs, the main difficulty was not to perceive accurately the frequency of $S_1$ or $S_2$ but to identify the temporal order in which the two tones were presented. However, this hypothesis is ruled out by the fact that the temporal order of two spectrally remote short sounds can be reliably identified as soon as the onset-to-onset interval exceeds about 20 ms (Hirsh 1959). Thus, it seems very hard to account for results such as those of Massaro and Idson without admitting the existence of PPAM and the idea that perception takes time, much more time than the stimulus itself if the stimulus is very short. Kallman and Massaro (1979) provided additional evidence that BRM is at least partly due to interference in PPAM rather than STAM.

Massaro and Idson (1977) were actually not the first to report that an increase in ISI can improve performance in a two-interval auditory discrimination task. This had been previously found by several authors (e.g., Tanner 1961). Recently, the present authors also observed such a trend in a study concerned with frequency discrimination (Demany and Semal 2005). In this respect, our data support the main points of Massaro's theory, in particular the PPAM notion. However, the data are inconsistent with an important detail of Massaro's theory: the idea that the amount of time needed for an optimal perceptual analysis has a fixed value of about 250 ms regardless of the stimulus. In one of our experiments, the two tones presented on each trial ($S_1$ and $S_2$) consisted of either 6 or 30 sinusoidal cycles. They were separated by an ISI that varied between blocks of trials, up to 4 s. The frequency of $S_1$ varied unpredictably within a very wide range: 400–2400 Hz. Performance was assessed in terms of $d'$ (Green and Swets 1974) for relative frequency shifts (from $S_1$ to $S_2$) amounting on average to ±5.8% in the "6 cycles" condition and ±1.3% in the "30 cycles" condition. The results are displayed in Figure 4.2A. For each number of cycles, as the ISI increased from 200 ms to 4 s, $d'$ first increased, rapidly, and then decreased, more slowly. The ISI for which $d'$ was maximal (the optimal ISI) provided an estimation of the duration needed to perceive $S_1$ as accurately as possible. According to Massaro's theory, this optimal ISI should have been nearly the same for the two classes of stimuli. It can be seen, however, that this was not the case: The optimal ISI appeared to be about 400 ms in the "6 cycles" condition and markedly longer, about 1 s, in the "30 cycles" condition. In another experiment, only 30-cycle stimuli were used, but their perceptual uncertainty was manipulated. There were two uncertainty conditions, in which the frequency shifts had the same relative size, on average ±0.8%. In the "high-uncertainty" condition, the frequency of $S_1$ could take any value from 400 Hz to 2400 Hz on each trial. Of course, $S_2$ varied in about the same range. In the "low-uncertainty" condition, on the other hand, $S_2$ could have only three possible frequencies: 400, 980, and 2400 Hz (immediate repetitions of the same $S_2$ from trial to trial being precluded). Figure 4.2B displays the results. It can be seen, firstly, that performance was globally better in the low-uncertainty condition than in the high-uncertainty condition, and secondly that the optimal ISI was longer in the

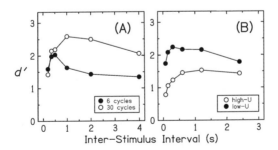

FIGURE 4.2. Results of two experiments by Demany and Semal (2005). In one of them **(A)**, one independent variable was the number of sinusoidal cycles making up each tone: 6 or 30. In the other experiment **(B)**, the number of cycles was fixed (30), but the uncertainty of the stimuli was manipulated; this uncertainty was either high ("high-U") or low ("low-U"). For both experiments, each data point is based on a total of 3000 trials (750 trials for each of the four tested listeners). (Reprinted with permission from Demany L, Semal C 2005; ©Psychonomic Society.)

latter condition. Each of these two experiments shows that frequency information is stored in a PPAM system with a temporal span of at least 1 s. Given its span, the memory store in question is apparently different from that involved in phenomena such as temporal integration of loudness, contrary to a suggestion of Cowan (1984).

## 4. Short-Term Auditory Memory and the Binding of Successive Sounds

Durlach and Braida (1969) made a distinction between two *memory operating modes*: a "sensory-trace" mode and a "context-coding" mode. In the sensory-trace mode, a percept is directly put in relation with the memory trace of a previous percept. In the context-coding mode, a percept is compared with a *set* of memory traces of previous percepts, including possibly quite ancient traces, and the outcome of the comparison is represented by a more or less precise verbal label (e.g., "halfway between /a/ and /o/" in the domain of timbre). Durlach and Braida pointed out that the context-coding mode is necessarily used in identification tasks (absolute judgments on single stimuli) but may also be used in a discrimination task. If, for instance, one has to make a same/different judgment on two stimuli separated by a 24-hour ISI during which many similar stimuli are presented, then a context coding of the two stimuli to be compared will presumably be the most efficient strategy. It will be so because, in contrast to a sensory trace, a verbal label can be perfectly memorized for a very long time, without any deleterious effect of interfering stimuli. However, the sensory-trace mode is undoubtedly the most efficient mode in a discrimination task such as that used by Massaro and Idson (1977) or the high-uncertainty condition of Demany and Semal (2005).

As one would expect, if two successive tones that differ very slightly in frequency or in intensity are separated by a silent ISI for as long as 5 or 10 s, their behavioral discrimination is definitely poorer than if the ISI is shorter, e.g., 1 s. This is especially true if in the test, the two stimuli presented on each trial vary in a wide range from trial to trial (Harris 1952; Berliner and Durlach 1973). Using such a "roving" procedure, Clément et al. (1999) found that the degradation of discrimination performance as the ISI increases is initially slower for frequency discrimination than for intensity discrimination. The degradation observable for frequency discrimination with silent ISIs and a roving procedure is of special interest, because in that case, the possible influence of context coding on performance is probably minimized and negligible. If so, the data reflect a pure sensory-trace decay. How to account for such a decay? The simplest model that can be thought of in the framework of signal detection theory (Green and Swets 1974) was formulated by Kinchla and Smyzer (1967). According to this model, the discrimination of two successive stimuli is limited by a sum of two independent internal noises: a "sensation noise" corresponding to the imperfect perceptual encoding of the two stimuli, and a "memory noise" resulting from a random walk of the trace of the first stimulus during the ISI. The random walk assumption implies that the memory noise is a Gaussian variable whose variance is proportional to the ISI. As noted by Demany et al. (2005), one prediction of the model is that when the ISI increases, the relative decay of discrimination performance ($d'$) will be slower if the sensation noise is large than if the sensation noise is small (because sensation noise and memory noise are supposed to be additive). The veracity of this prediction was actually questioned by Demany et al. (2005) on the basis of the data shown in Figure 4.2A and other data. In visual short-term memory, according to Gold et al. (2005), the fate of a trace is deterministic rather than random, contrary to the principal assumption of Kinchla and Smyzer (1967). With regard to physiology, it has been assumed that the maintenance of a sensory trace during a silent ISI is due to a sustained activity of certain neurons within this time interval. In studies on auditory frequency discrimination by monkeys, such neurons have indeed been found, at the level of the auditory cortex (Gottlieb et al. 1989) but also the dorsolateral prefrontal cortex (Bodner et al. 1996).

One might think that in humans, a frequency discrimination task is not appropriate for the study of "pure" sensory-trace decay because a pitch percept is liable to be rehearsed with profit by humming. However, this hypothesis is wrong: In fact, humming during the ISI is not profitable (Massaro 1970b; Kaernbach and Hahn, unpublished data). More generally—not only in the case of pitch—there seems to be a complete *automaticity of retention* in STAM, at least during silent ISIs. In this respect, STAM appears to be very different from short-term *verbal* memory, which is strongly dependent on attention and rehearsal processes.

One experiment suggesting that STAM is automatic was performed by Hafter et al. (1998). Their stimuli were 33-ms audiovisual signals (tone bursts coupled with colored disks). On each trial, two such signals were successively presented, and the subject had to make, in separate blocks of trials, intensity or luminance comparisons between: (1) their auditory components alone; (2) their visual components alone; (3) their auditory components *and* their visual components (which varied independently). The results obtained with a roving procedure showed that in the dual task of the third condition, the division of attention between auditory and visual signals had no deleterious effect on discrimination performance: For each sensory modality, performance was the same in the dual task and the restricted task. In this experiment, however, the ISI was short (301 ms). One can argue that different results might have been found for longer ISIs.

The present authors used longer ISIs in a purely auditory study (Demany et al. 2001). Our stimuli were 500-ms amplitude-modulated tones with three independent parameters, randomly varying from trial to trial: carrier frequency (2000–3500 Hz), modulation frequency (30–100 Hz), and intensity (48–86 dB SPL). The second of the two stimuli presented within a trial ($S_2$) differed from the first ($S_1$) with respect to only one of the three parameters. The identity of the parameter in question was selected at random, as well as the direction of the shift. The task was to indicate whether the shift was positive or negative (without specifying the identity of the shifted parameter). In the experiment proper, the sizes of the shifts were fixed (in percent) for a given listener and parameter. These sizes had been previously adjusted in order to obtain similar levels of performance for the three parameters. Six of the eight experimental conditions are depicted in Figure 4.3. In all but one of these six conditions (the exception was condition D), a visual cue indicating the identity of the shifted parameter was provided on each trial between $S_1$ and $S_2$. Thanks to this cue, the listener could attend selectively to the relevant parameter and ignore the remaining perceptual information. Crucially, in conditions B and E, the cue was presented immediately after $S_1$, and the ISI was long (4 or 6 s). It could thus be expected that in these two conditions, the cue would have a positive effect on the memorization of the relevant parameter of $S_1$. If so, performance should have been better in condition B than in condition C, since in the latter condition, the cue was presented much later, shortly before $S_2$ rather than just after $S_1$ (this was the only difference between conditions B and C). For the same reason, performance should have been better in condition E than in condition F. However, as indicated in Figure 4.3, the average $d'$ values obtained in conditions B and C, or E and F, were very similar. There was not even a trend in the direction expected under the hypothesis that

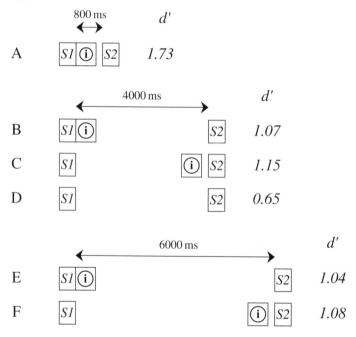

FIGURE 4.3. Results of Demany et al. (2001). "S1" and "S2" are auditory stimuli (amplitude-modulated tones) and each "lock" represents a visual cue. Each $d'$ value is based on at least 4800 trials (at least 1200 trials for each of the four tested listeners). (Adapted with permission from Demany et al. 2001.)

attention aids memory. These data thus support the idea that STAM is automatic. Note that performance was significantly poorer in conditions B and E than in condition A, where the ISI was shorter. This proves that in conditions B and E, performance was limited by memory factors rather than by perceptual factors; a loss of information was taking place during the ISI, and thus one cannot argue that there was no room for a positive effect of attention on memory. It is also important to note that performance was significantly poorer in condition D (no cue) than in conditions B and C. This result, which was predicted by signal detection theory and presumably does *not* reflect an influence of attention on memory (see Demany et al. 2001 for a detailed discussion), rules out a trivial hypothesis according to which the cues were simply ignored.

The just-described study failed to find a benefit of attention for memory while attention was drawn onto one feature of an auditory object among other features of the same auditory object. Is attention more efficacious when, instead, it is drawn onto one auditory object among other auditory objects? This question led Clément (2001) to perform experiments in which, on each trial, the listener was initially presented with a sum of three sinusoidally amplitude-modulated pure tones with distant carrier frequencies (pitches), distant modulation frequencies, and different localizations (one tone was presented to the left ear, another tone to the right ear, and the third tone diotically). These three simultaneous tones, whose carrier frequencies varied from trial to trial, were always perceived as three separate auditory objects. After a silent ISI generally lasting 5 s (sometimes 10 s), their sum was followed by a single tone, identical in every respect to a randomly selected component of the sum except for a slight upward or downward shift in carrier frequency. The task was to identify the direction of this slight frequency shift. In most conditions, a visual cue appearing on the left, middle, or right part of a screen indicated to the listener the relevant component of the tonal complex. As in the study by Demany et al. (2001), this cue occurred either at the very beginning of the ISI or near its end. The outcome was again that the cue's temporal position had no influence on discrimination performance. In one experiment, a cue was always presented at the very beginning of the ISI, but it was invalid on about 20% of trials, unpredictably. On the trials in question, listeners were thus led to rehearse an inappropriate component of the complex. This did not impair performance, to the listeners' own surprise.

The retention of a frequency or pitch trace in STAM is in fact so automatic that paradoxically, it is possible to detect consciously a frequency difference between two tones several seconds apart *in the absence of a conscious perception of the first tone's frequency.* This was shown by Demany and Ramos (2005). In their study, listeners were presented with sums of five synchronous pure tones separated by frequency intervals that varied randomly between 6 and 10 semitones (1 semitone = 1/12 octave). In contrast to the sums of tones used by Clément (2001), these new tonal complexes, or "chords," were perceived as single auditory objects. That was because, among other things, their sinusoidal components did not differ from one another with respect to amplitude envelope or spatial localization. On each trial, a chord was followed, after a silent ISI, by a single pure tone (*T*). Three conditions, illustrated in Figure 4.4A, were run. In the "up/down" condition, *T* was 1 semitone above or below (at random) one of the three intermediate components of the chord (at random again), and the task was to judge whether *T* was higher or lower in frequency than the closest

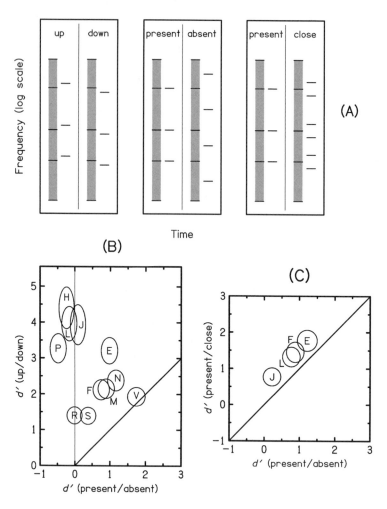

FIGURE 4.4. Experimental conditions and results of Demany and Ramos (2005). (A) Stimulus configurations used in the up/down, present/absent, and present/close conditions. Each horizontal segment represents a pure tone, and the shaded areas represent a possible chord. In the experiments, each chord was followed by a single *T* tone, and the ISI always exceeded the duration of both stimuli. (B) Results of eleven listeners in the present/absent and up/down conditions. Each ellipse (or circle) is centered on the *d'* values measured in the two conditions for a given listener, and its surface represents a 95% confidence area. Oblique lines indicate where the ellipses could be centered if *d'* were identical in the two conditions. (C) Results of four listeners in the present/absent and present/close conditions. (Adapted with permission from Demany and Ramos 2005; ©American Institute of Physics.)

component of the chord. In the "present/absent" condition, $T$ was either identical to one of the three intermediate components or halfway in (log) frequency between two components, and the task was to judge whether $T$ was present in the chord or not. The third condition, "present/close," was identical to the "present/absent" condition except that now, when $T$ was not present in the chord, it was 1.5 semitone above or below (at random) one of the three intermediate components. Figure 4.4B shows the results obtained from 11 listeners in the up/down and present/absent conditions. In the present/absent condition, performance was generally quite poor: the average $d'$ was only 0.46. This reflects the fact that although the chords' components were certainly resolved by the listeners' cochleas, it was essentially impossible to hear them individually. The chords' components were "fused" at a central level of the auditory system, and for this reason they produced on each other an "informational masking" effect (see Chapter 6). In the up/down condition, however, overall performance was very good: the average $d'$ was 2.74. Surprisingly, the judgments required in this condition were relatively easy because the one-semitone frequency shifts elicited percepts of pitch shift even while the component of the chord that was one semitone away from $T$ could not be consciously heard out. Typically, the listeners perceived $T$ as the ending point of a clearly ascending or descending melodic sequence without being able to say anything about the starting point of that sequence. Definite percepts of directional pitch shift were also elicited by the 1.5-semitone frequency shifts occurring on "close" trials in the present/close condition. On "present" or "absent" trials, in contrast, the frequency of $T$ was not so strongly placed in relation with a previous frequency. In the present/close condition, therefore, it was to some extent possible to distinguish the two types of trial on the basis of the audibility of a clear pitch shift. Indeed, as shown in Figure 4.4C, four listeners were more successful in that condition (average $d'$: 1.33) than in the present/absent condition (average $d'$: 0.77). All the data presented in Figure 4.4 were obtained for a chord-$T$ ISI of 0.5 s. However, four listeners were also tested in the up/down and present/absent conditions using longer ISIs. For a 4-s ISI, performance was still markedly better in the up/down condition (average $d'$: 0.94) than in the present/absent condition (average $d'$: 0.33). A subsequent study (Demany and Ramos, in press) showed that performance in the up/down condition was not markedly poorer if the chord was presented to one ear and $T$ to the opposite ear rather than all stimuli being presented to the same ear. In another experiment, the five-tone chords described above were replaced by chords of 10 tones with a constant spacing of 5.5 semitones, and in the up/down condition, $T$ was positioned one semitone above or below any of the chord's 10 components. A 0.5-s ISI was used. Five listeners were still able to perform the up/down task relatively well (average $d'$ : 1.56). In the present/absent condition, in contrast, their performance was at the chance level (average $d'$ : 0.07).

It should be emphasized that these results are not interpretable in terms of adaptation. If listeners' judgments had been based on the relative adaptation of neurons detecting $T$ by the previous chord, then the present/absent condition should have been the easiest condition, since the adaptation of neurons detecting $T$ was respectively maximized and minimized on "present" and "absent" trials.

In reality, the present/absent condition was the most difficult one. The results make sense, however, under the hypothesis that the human auditory system is equipped with automatic "frequency-shift detectors" (FSDs) that can operate on memory traces in STAM. More precisely, it is possible to account qualitatively for the relative difficulties of the three experimental conditions by assuming that such detectors exist and that: (1) some of them are activated only by upward shifts, while others are activated only by downward shifts; (2) within each subset, an FSD responds more strongly to small shifts (such as one-semitone shifts) than to larger shifts; (3) when detectors of upward shifts and downward shifts are simultaneously activated—this was presumably the case in each experimental condition—the dominantly perceived shift is in the direction corresponding to the stronger activation. A similar model had been proposed by Allik et al. (1989) to account for the perception of pitch motion in sequences of chords.

The FSDs that were apparently operating on the sound sequences employed by Demany and Ramos might also play a role in the perceptual detection of very small frequency differences between isolated pure tones. This could explain why frequency discrimination and intensity discrimination are not affected in the same manner by the ISI when the ISI increases from 0.5 s to a few seconds (Clément et al. 1999). However, if evolution provided humans with FSDs, that was probably not specifically to allow them to detect minute frequency changes in an automatic way. A more plausible conjecture is that the FSDs' main function is to *bind successive sounds* and as such to serve as a tool for what Bregman (1990) called "auditory scene analysis" (see also Chapter 11 of this volume). Humans feel that they can *perceive* as a whole a succession of sounds such as a short melody. While the last tone is being heard, the first tone, stored in STAM, still belongs to the same "psychological present" (Fraisse 1967). The perceptual coherence of the whole set of tones may be partly due to the existence of FSDs. Physiologically, the FSD hypothesis is not unrealistic. In the auditory cortex of cats, the response of many neurons to a given tone can be greatly increased by previous presentation of another tone with a different frequency (McKenna et al. 1989; Brosch and Schreiner 2000). These neurons are thus particularly sensitive to tone *sequences*, as expected from FSDs. However, in the just-cited studies, increases in firing rate by a previous tone were observed only for ISIs smaller than 1 s. One must also keep in mind, as pointed out earlier in this chapter, that the mere existence of an interaction between two successive tones in the auditory system does not immediately account for the perception of a relation between them. A problem is to dissociate, in the neural activity concomitant to the presentation of the second sound, what is due to the relation between the two sounds from what is due to the intrinsic characteristics of the second sound.

It has been argued above that retention in STAM does not depend on attention and that the brain automatically puts in relation successive sounds separated by nothing but silence. Nevertheless, attention can have positive effects on auditory perception (Hafter, Chapter 5), and the retention of an auditory trace in STAM may be independent of attention only *after the formation of this trace.*

Demany et al. (2004) showed that the retention of a pitch trace is improved if, *during* the presentation of the tone evoking this pitch, attention is focused on the tone in question rather than on another, simultaneous, tone. Presumably, this occurs because attention improves the formation of the memory trace of the focused pitch. Besides, it should also be pointed out that the perception of relations between sounds is probably more attention-dependent when the sounds are *non*consecutive than when they are separated by nothing but silence. Different memory mechanisms are probably involved in these two cases. In a study by Zatorre and Samson (1991), patients who had undergone focal excisions from the temporal or frontal cortex were required to make same/different judgments (pitch comparisons) on pairs of tones separated by a 1650-ms ISI. The ISI was either silent of filled by six interfering tones, to be ignored. When the ISI was silent, the patients' performance was not significantly different from that of normal control listeners. However, when the ISI was filled by tones, a significant deficit was observed in patients with damage in the right temporal lobe or the right frontal lobe. The memory mechanisms involved in the latter case may be similar to those permitting the detection of repetitions in a cyclically repeated white-noise segment of long duration. Interestingly, whereas for humans repetition detection is possible when the repeated noise segment is as long as 10 s (Kaernbach 2004), cats are apparently unable to discriminate repeated noise from nonrepeated noise as soon as the repeated segment exceeds about 500 ms (Frey et al. 2003). For gerbils, the limit is even lower (Kaernbach and Schulze 2002).

The effect of interfering pure tones on the detectability of a pitch difference between two pure tones has been investigated in detail by Deutsch (for a summary of her work, see Deutsch 1999). She showed, among other things, that performance is much more impaired by an interfering tone that is different from the tone to be remembered ($S_1$) but close to it in pitch than by a remote interfering tone. Deutsch and Feroe (1975) provided evidence that the impairment produced by an interfering tone close in pitch to $S_1$ is not due to a destruction of the trace of $S_1$ but rather to an inhibition of this trace: The deleterious effect of an interfering tone $I_1$ can be reduced by the presentation, following $I_1$, of another interfering tone $I_2$, close in pitch to $I_1$ and less close to $S_1$; a natural interpretation is that $I_2$ inhibits the trace of $I_1$, and in doing so disinhibits the trace of $S_1$. Using again an experimental paradigm in which same/different judgments had to be made on pure tones possibly different in pitch, Semal and Demany (1991, 1993) wondered whether the effect of interfering tones on performance is exclusively determined by the pitch of the interfering tones or is also dependent on their other characteristics. It could be expected, for instance, that the deleterious effect of an interfering tone close in pitch to $S_1$ would be reduced if this tone were much less intense than $S_1$ or consisted of harmonics of a missing fundamental instead of being a pure tone like $S_1$. However, this was not the case: Pitch appeared to be the only perceptual parameter affecting performance. In the same vein, Semal et al. (1996) found that if the two stimuli to be compared are no longer tones but monosyllabic words, identical in every respect or slightly different in pitch, interfering words are not more deleterious than interfering tones with the same pitches. These experiments, and a related study by Krumhansl and Iverson (1992), suggested that humans possess a pitch-specific memory module, deaf to loudness and timbre (spectral composition). There are also experimental data suggesting that at least some aspects of timbre are retained in a specific memory module, deaf to pitch (Starr and Pitt 1997).

The pitch-specific memory module, if it does exist, is not likely to play a major role in the auditory phenomenon reported by Demany and Ramos (2005) and described above. In our view, the FSDs hypothesized by Demany and Ramos must be thought of as detectors of *spectral* shifts rather than shifts in pitch per se (i.e., shifts in periodicity regardless of spectral composition). If, as argued above, the raison d'être of the FSDs is to bind successive sounds, then these detectors should clearly operate in the spectral domain rather than an "abstract" pitch/periodicity domain: Listeners do rely on spectral relationships when they analyze a complex auditory scene and make of it a set of temporal streams within which sounds are perceptually bound (van Noorden 1975; Hartmann and Johnson 1991; Darwin and Hukin 2000).

## 5. Long-Term Traces

For humans, the most important function of the auditory system is to permit *speech* communication. The processing of speech (see in this regard Chapter 10 of this volume) makes use of a long-term auditory memory, at least a rudimentary one: In order to identify an isolated spoken vowel that has just been heard, it is unnecessary to reproduce the sound vocally by trial and error; instead, the percept can be directly matched to an internal auditory template of the vowel, which leads to a rapid identification. Does human auditory long-term memory also include representations of higher-order speech entities, such as words? A majority of authors believe that the answer is negative: It is generally supposed that in the "mental lexicon" used to understand spoken words, meanings are linked with *abstract* representations of words (see, e.g., McClelland and Elman 1986); a given word is supposed to have only one representation, although the actual sound sequence corresponding to this word is not fixed but greatly depends on the speaker and various context factors. Goldinger (1996) has challenged this assumption and argued instead that a given word could be represented as a group of episodic traces retaining "surface details" such as intonation. In one of his experiments, listeners were required to identify monosyllabic words partially masked by noise and produced by several speakers. One week later, the same task was performed again, but the words were repeated either in their original voice or a new voice. The percentage of correct identifications was generally higher in the second session, but the improvement was larger for words repeated in their original voice, which implies that surface details had been kept in memory for one week. In another experiment, however, the retention of surface details appeared to be weaker (significant after one day, but absent after one week). During the second session, the listeners now had to judge *explicitly* whether a given word had also been presented in the first session or was new. Therefore, Goldinger's results suggest that surface details of words can be memorized during long periods of time but that the corresponding traces persist mainly in an implicit form of memory (see also, in this respect, Schacter and Church 1992).

Like the perception of speech, the perception of *music* is strongly dependent on long-term memory traces. Shepard and Jordan (1984) performed one of the studies supporting this view. They played to psychology students a sequence of eight pure tones going from middle C (261.6 Hz) to the C one octave above in equal steps corresponding to a frequency ratio of $2^{1/7}$. The task was to judge the relative sizes of the frequency intervals between consecutive tones, from a strictly "physical" point of view and not on any musical basis. It was found that the third and seventh intervals were judged as larger than the other intervals. This can be understood by assuming that despite the instructions, the sequence played was perceptually compared to an internal template corresponding to the major diatonic scale of Western music (do, re, mi, fa, sol, la, ti, do). In this scale, the third and seventh intervals (mi-fa and ti-do) are physically *smaller* than the other intervals. When people listen to music, their knowledge of the diatonic scale and of other general rules followed in what is called "tonal music" generates expectancies about upcoming events, and these expectancies apparently affect the perception of the events themselves. Bigand et al. (2003) recently emphasized that point. In their experiments, they used sequences of chords in a well-defined musical key (e.g., C major). In half of the sequences, the last chord was made dissonant by the addition of an extra tone close in frequency to one of its components. The listeners' task was to detect these dissonances. In the absence of the extra tone, the last chord was either a "tonic" chord, ending the sequence in a normal way according to the rules of tonal music, or a less-expected "subdominant" chord. The harmonic function (tonic or subdominant) of this last chord was independent of its acoustic structure; it was entirely determined by the context of the previous chords. Yet the notes of tonic chords did not appear more frequently among the previous chords than the notes of subdominant chords. Nonetheless, dissonances were better detected in tonic chords than in subdominant chords. Remarkably, this trend was not weaker for nonmusicians than for musicians. In the same vein, Francès (1958/1988) had previously shown that for nonmusicians as well as for musicians, it is easier to perceive a change in a melodic sequence when this sequence is based on the diatonic scale than when that is not the case, all other things being equal as far as possible. Dewar et al. (1977) also found that a change in one note of a diatonic melody is easier to detect when the new melody is no longer diatonic than when the new melody is still diatonic. Such effects are probably due mainly to long-term learning phenomena rather than to intrinsic properties of the diatonic scale, because the scale in question clearly rests, in part, on cultural conventions (Dowling and Harwood 1986; Lynch et al. 1990).

Tillmann et al. (2000) devised an artificial neural network that becomes sensitive to the statistical regularities of tonal music through repeated exposure to musical material and an unsupervised learning process. This model simulates the acquisition of implicit musical knowledge by individuals without any formal musical education, and can account for their behavior in experiments such as those just mentioned. Note, however, that the knowledge stored in the model is "abstract" in that the model does not explain at all how *specific* pieces of music can be memorized in the long term. Yet it is clear that people possess this kind of memory. Indeed, many persons are able to recognize a tune that they had not heard for years, or even decades. The perception of music is probably affected by episodic traces of this type as well as by internal representations of general, abstract rules.

Although most humans can recognize and identify a large number of tunes, very few can make a precise absolute judgment on the *pitch* of an isolated tone. Even among those who received a substantial musical education, the ability to assign a correct note label to an isolated tone—an ability called "absolute pitch"—is rare (Ward 1999). This ability requires an accurate long-term memory for pitch. It has been claimed, however, that ordinary people exhibit a surprising long-term memory for pitch when they are asked to reproduce vocally a familiar song always heard in the same key (Levitin 1994) or even when they simply speak in a normal way (Braun 2001; Deutsch et al. 2004; see also Deutsch 1991). In his influential theory about pitch perception, Terhardt (1974) has supposed the existence, in every normal adult, of an implicit form of absolute pitch. A periodic complex tone such as a vowel or a violin note is made up of pure tones forming a harmonic series, and some of these pure tones (those having low harmonic ranks) are resolved in the cochlea. When presented in isolation, these resolved pure tones evoke quite different pitch sensations. Yet, when they are summed, one typically hears a single sound and only one pitch, corresponding to the pitch of the fundamental. Why is it the case? According to Terhardt, this is entirely an effect of learning. Because speech sounds play a dominant role in the human acoustic environment, humans learn, according to this theory, to perceive any voiced speech sound, and by extension any sound with a harmonic or quasiharmonic spectrum, as a single sound with one pitch rather than as an aggregate of pure tones with various pitches. More specifically, the pitch evoked by a given pure tone of frequency $f$ is associated in a "learning matrix" with the pitches evoked by its subharmonics ($f/2, f/3$, and so on), due to the co-occurrence of pure tones with these frequency relationships in voiced speech sounds. After the formation of the learning matrix (hypothetically taking place in early infancy, or maybe in utero), the associations stored in it would account for a wide range of perceptual phenomena in the domain of pitch.

A fascinating study supporting Terhardt's hypothesis was performed by Hall and Peters (1982). It deserves to be described here in detail. One class of phenomena that Terhardt intended to explain concerned the pitch of quasiharmonic sounds. Consider a sound $S$ consisting of three pure tones at 1250, 1450, and 1650 Hz. The three components of $S$ are not consecutive harmonics of a common fundamental (which would be the case for 1200, 1400, and 1600 Hz). However, their frequencies are close to consecutive integer multiples of 210

Hz: $6 \times 210 = 1260$, $7 \times 210 = 1470$, and $8 \times 210 = 1680$. When listeners are required to match $S$ in pitch with a pure tone of adjustable frequency, they normally adjust the pure tone to about 210 Hz. But they do so with some difficulty, because the pitch of $S$ is relatively weak. The weakness of its pitch is in part due to the fact that the components of $S$ are *high-rank* quasiharmonics of a common fundamental. Indeed, it is well known that sums of low-rank harmonics elicit more salient pitch sensations than sums of high-rank harmonics. This gave to Hall and Peters the following idea. Suppose that $S$ is repeatedly presented to listeners together with a simultaneous sound $S'$ made up of the first five harmonics of 200 Hz (200, 400, ..., 1000 Hz). If, as hypothesized by Terhardt, one can learn to perceive a given pitch simply by virtue of associative exposures, then the pitch initially perceived in $S$ (about 210 Hz) might progressively change in the direction of the pitch elicited by $S'$ (about 200 Hz). This was indeed found in the study of Hall and Peters. They also observed a trend consistent with Terhardt's hypothesis when the frequencies of the components of $S$ were instead 1150, 1350, and 1550 Hz. In this case, the pitch of $S$ rose instead of falling. However, Hall and Peters noted that they failed to induce significant changes in the pitch of perfectly harmonic stimuli.

Terhardt claimed that his theory also elucidated various phenomena relating to the perception of musical intervals. A sum of two simultaneous pure tones is perceived as more consonant when the frequency ratio of these tones is equal to 2:1 (an octave interval) or 3:2 (a fifth) than when the two tones form slightly smaller or larger intervals. In addition, people perceive a stronger consonance or "affinity" between two *sequentially* presented pure tones when they form approximately an octave or a fifth than when they form other musical intervals. Curiously, however, in order to form an optimally tuned melodic octave, two sequentially presented pure tones must generally have a frequency ratio not exactly equal to 2:1 but slightly larger, by an amount that depends on the absolute frequencies of the tones (Ward 1954); this has been called the "octave enlargement phenomenon." According to Terhardt, all these phenomena originate from the learning process that he hypothesized. The rationale is obvious for the origin of consonance, but needs to be explained with regard to the octave enlargement phenomenon. Terhardt's learning hypothesis is a little more complex than suggested above. He actually argued that the pitch elicited by a given harmonic of a voiced speech sound (before the formation of the learning matrix) is in general slightly different from the pitch of an identical pure tone presented alone, due to mutual interactions of the harmonics at a peripheral stage of the auditory system (Terhardt 1971). Consequently, the pitch intervals stored in the learning matrix should correspond to simple frequency ratios (e.g., 2:1) for pure tones presented simultaneously, but to slightly different frequency ratios for pure tones presented sequentially. This would account for the octave enlargement phenomenon.

Terhardt's conjecture on the origin of the octave enlargement phenomenon has been brought into question by Peters et al. (1983), who found that for adult listeners, the pitch of individual harmonics of complex tones does not differ significantly from the pitch of identical pure tones presented in isolation. Moreover, Demany and Semal (1990) and Demany et al. (1991) reported other results challenging the idea that the internal template of a melodic octave originates from the perception of simultaneous pure tones forming an octave. They found that at relatively high frequencies, mistunings of simultaneous octaves are poorly detected, whereas the internal template of a melodic octave can still be very precise. Burns and Ward (1978) have argued that in musicians, the perception of melodic intervals is "categorical" due to a learning process that is determined by cultural conventions and has nothing to do with Terhardt's hypothetical learning matrix. These authors assessed the discriminability of melodic intervals differing in size by a given amount (e.g., 0.5 semitone) as a function of the size of the standard interval. In musicians, they observed markedly nonmonotonic variations of performance; performance was poorer when the two intervals to be discriminated were identified as members of one and the same interval category (e.g., 3.75 and 4.25 semitones, two intervals identified as major thirds) than when the two intervals were labeled differently (e.g., 4.25 and 4.75 semitones, the latter interval being identified as a fourth). In nonmusicians, on the other hand, performance did not depend on the size of the standard interval. The latter observation suggested that musicians' categorical perception of intervals stems from their musical training itself rather than from some "natural" source. However, the results obtained in nonmusicians by Burns and Ward are at odds with those of Schellenberg and Trehub (1996) on infants. According to Schellenberg and Trehub, 6-month-old infants detect more readily a one-semitone change in a melodic interval formed by pure tones when the first interval corresponds to a simple frequency ratio (3:2 or 4:3) than when the first interval is the medieval *diabolus in musica*, i.e., a tritone (45:32). This makes sense in the framework of Terhardt's theory. So, the roots of our perception of melodic intervals continue to be a subject of controversy (recently nourished by Burns and Houtsma 1999 and Schwartz et al. 2003).

It should be noted that Terhardt's theory is not the only pitch theory assuming the existence, in every normal adult listener, of internal templates of harmonic frequency ratios. In most of the other theories, the question of the templates' origin is avoided. However, Shamma and Klein (2000) have proposed a learning scenario that differs from Terhardt's scenario. In their theory, it is unnecessary to hear, as supposed by Terhardt, voiced speech sounds or periodic complex tones of some other family in order to acquire harmonic templates. Broadband noise is sufficient. The templates emerge due to: (1) the temporal representation of frequency in individual fibers of the auditory nerve; (2) phase dispersion by the basilar membrane; and (3) a hypothetical mechanism detecting temporal coincidences of spikes in remote auditory nerve fibers.

In the domain of *spatial hearing*, the question of the influence of learning on perception began to be asked experimentally very long ago. In a courageous study on himself, Young (1928) wore the apparatus depicted in Figure 4.5 for 18 days. This apparatus inverted the interaural time differences that are used by the binaural system to localize sounds in the horizontal plane, and Young's aim

FIGURE 4.5. The "pseudophone" of Young (1928), after Vurpillot (1975). (Reprinted with permission from Vurpillot 1975.)

was to determine whether his perception of sound localization could be reshaped accordingly. This was not the case: After 18 days, in the absence of visual cues, he was still perceiving on the right a sound coming from the left and vice versa. However, less radical alterations of the correspondence between the azimuthal position of a sound source and the information received by the binaural system are able to induce, in appropriately trained adult listeners, real changes in perceptual localization (Shinn-Cunningham 2000). Localization in the vertical plane, which depends on spectral cues related to the geometry of the external ear (pinna), may be even more malleable. If the usual spectral cues are suddenly modified by the insertion of molds in the two conchae, auditory judgments of elevation are at first seriously disrupted but become accurate again after a few weeks (Hofman et al. 1998). Interestingly, according to Hofman et al., the subject is not aware of these changes in daily life. Moreover, the new spectral code learned does not erase the previous one: When the molds are removed after having been worn permanently for several weeks, the subject immediately localizes sounds accurately and doesn't need a readaptation period; yet, a memory of the spectral code created by the molds appears to persist for several days. Learning-induced modifications of perceptual sound localization have been observed not only in humans but also in barn owls (Knudsen and Knudsen 1985; Linkenhoker and Knudsen 2002). The behavioral changes observed in barn owls were found to be associated with changes in the tuning characteristics of individual auditory neurons (Zheng and Knudsen 1999). It is not clear, however, that psychophysical findings such as those of Hofman et al. (1998) have a similar neural basis. Actually, because their subjects did not need a readaptation period when the molds were removed, Hofman et al. avoided interpreting their findings in terms of "neural plasticity." In their view, the changes that they observed were comparable to the acquisition of a new language and were not based on functional modifications of the auditory system itself.

In humans, nonetheless, influences of auditory experience on auditory perception may well be based in part on functional modifications of the auditory system. This has been suggested to account for data concerning, in particular, *discrimination learning*. When some auditory discrimination threshold is repeatedly measured in an initially naive subject, it is generally found that the subject's performance gets better and better before reaching a plateau: The measured thresholds decrease, at first rapidly and then more and more slowly. Thousands of trials may be necessary to reach the plateau, even though the standard stimulus does not change and the challenge is only to perceive a difference between this stimulus and comparison stimuli varying in a single acoustic dimension. Interestingly, however, the amount of practice needed to reach the plateau appears to depend on the nature of the acoustic difference to be detected: A larger number of trials is required for the detection of interaural intensity differences than for the detection of interaural time differences (Wright and Fitzgerald 2001); also, the number of trials needed to optimally detect differences in (fundamental) frequency depends on the spectrum of the stimuli and is apparently longer for complex tones made up of resolvable harmonics than for pure tones or for complex tones made up of unresolvable harmonics (Demany and Semal 2002; Grimault et al. 2002). Even more interestingly, it has been found in several studies that the perceptual gain due to a long practice period with a restricted set of stimuli is, to some extent, stimulus-specific or dimension-specific. For example, after 10 one-hour sessions of practice in the discrimination of a 100-ms temporal interval (between tone bursts) from slightly different intervals, discrimination of temporal intervals is significantly improved for a 100-ms standard but not for a 50-ms or 200-ms standard (Wright et al. 1997). Analogous phenomena have been reported in the domain of pitch (Demany and Semal 2002; Grimault et al. 2002; Fitzgerald and Wright 2005) and binaural lateralization (Wright and Fitzgerald 2001). This specificity of learning clearly indicates that what is learned is genuinely *perceptual*. The rapid improvements of performance observable at the onset of practice are apparently less stimulus-specific. For this reason, they have been interpreted as due mainly to "procedural" or "task" learning (Robinson and Summerfield 1996). However, Hawkey et al. (2004) recently suggested that their main source is in fact a true perceptual change as well.

One way to account for improvements in perceptual discrimination is to assume that the subject learns to process the neural correlates of the stimuli in a more and more efficient manner. For instance, irrelevant or unreliable neural activity could be used initially in the task and then progressively discarded

(Puroshotaman and Bradley 2005). Alternatively, or in addition, the neural correlates of the stimuli could be by themselves modified during the training process, with beneficial consequences in the task; this would mean that auditory experience can be embodied by changes in the functional properties of the auditory system (Weinberger 1995, 2004; Edeline 1999). Fritz et al. (2003) and Dean et al. (2005), among others, reported results supporting the latter hypothesis. Fritz et al. (2003) trained ferrets to detect a pure tone with a specific frequency among sounds with broadband spectra, and they assessed simultaneously the spectrotemporal response field of neurons in the animals' primary auditory cortex. They found that the behavioral task swiftly modified neural response fields, in such a way as to facilitate perceptual detection of the target tone. Some of the changes in receptive fields persisted for hours after the end of the task. According to Dean et al. (2005), the neurometric function (spike rate as a function of stimulus intensity) of neurons in the inferior colliculus of the guinea pig is also plastic; it depends on the statistical distribution of the stimulus intensities previously presented to the animal. In these neurons, apparently, there is an adjustment of gain that optimizes the accuracy of intensity encoding by the neural population as a whole for the intensity range of the recently heard sounds. Recanzone et al. (1993) provided further support to the idea that auditory experience can change the properties of the auditory system itself. For several weeks, they trained adult owl monkeys in a frequency-discrimination task using pure tones and an almost invariable (but subject-dependent) standard frequency. Behavioral discrimination performance improved, and this improvement was limited to the frequency region of the standard. The authors then found that in the monkeys' primary auditory cortices, neural responses to pure tones had been affected by the training. In particular, the cortical area responding to the standard frequency was abnormally large. Remarkably, this widening was not observed in a control monkey that was passively exposed to the same acoustic stimulation as that received by one of the trained monkeys, without being required to attend to the stimuli. In a similar study on cats rather than owl monkeys, Brown et al. (2004) completely failed to replicate the main results of Recanzone et al.; but cats are poor frequency discriminators in comparison with monkeys and humans. In humans, discrimination learning has been shown to induce modifications in neuromagnetic responses of the brain to the stimuli used in the training sessions (Cansino and Williamson 1997; Menning et al. 2000). Moreover, several recent studies suggest that auditory experience is liable to induce functional changes at a subcortical level of the human auditory system (Krishnan et al. 2005; Philibert et al. 2005; Russo et al. 2005). In the study by Philibert et al., for instance, the tested subjects were hearing-impaired persons who were being fitted with binaural hearing aids, thanks to which they became able to perceive high-frequency sounds at medium or high loudness levels. Before and during the rehabilitation period, electrophysiological recordings of their auditory brainstem responses to clicks were made in the absence of the hearing aids. These recordings revealed a progressive shortening of wave-V latency, hypothetically interpreted by the authors as reflecting a neural reorganization in the inferior colliculus.

In the future, undoubtedly, further investigations of the neural correlates of perceptual changes due to experience will be carried out using more and more refined and powerful techniques.

# 6.  Concluding Remarks

The paramount goal of perception is identification, and in the case of auditory perception it is the identification of *sound sources* and *sound sequences* (because, at least for humans, meanings are typically associated with combinations of successive sounds rather than with static acoustic features). In themselves, the mechanisms of auditory identification are as yet unclear (see McAdams 1993 for a review of models). However, it is clear that identification would be impossible in the absence of a long-term auditory memory. Another form of auditory memory, STAM, is essential for the processing of sound sequences, because without STAM, as emphasized above, the successive components of a sequence could not be connected and therefore the sequence could not be perceived as a single auditory object. These considerations suggest that psychoacousticians should intensively work on the various forms of auditory memory in order to clarify their functioning at the behavioral level. Indeed, basic questions remain unanswered in this field. Yet, during the last two decades, there has been little psychophysical research on auditory memory. Most of the experimental studies intended to provide information on auditory memory have been focused on an evoked electric potential (the MMN) which is only an indirect clue, perhaps not tightly related to psychological reality (Allen et al. 2000). The main cause of psychoacousticians' reluctance may have been the belief that auditory memory is strongly dependent on attentional factors and neural structures that are not part of the auditory system per se. Contrary to such a view, the present authors believe that auditory memory (STAM, at least) is largely automatic and that what people consciously *hear* is to a large extent determined by mnemonic machineries. What people consciously *see* may not be less affected by visual memory. However, STAM is possibly more automatic than its visual counterpart: The perceptual phenomenon reported by Demany and Ramos (2005) and described in Section 4 does not seem to have a visual counterpart. Indeed, the paradoxical sensitivity to change demonstrated by this phenomenon stands in sharp contrast to the surprising "change blindness" of vision in many situations (Rensink et al. 1997; Simons and Levin 1997). Nowadays, a variety of visual phenomena crucially involving visual memory are thoroughly investigated by psychophysicists. It is hoped that the present chapter will encourage some of its readers to undertake research of this type in the auditory domain.

*Acknowledgments.* The authors thank Emmanuel Bigand, Juan Segui, and especially Christophe Micheyl for helpful discussions.

# *References*

Allen J, Kraus N, Bradlow A (2000) Neural representation of consciously imperceptible speech sound differences. Percept Psychophys 62:1383–1393.

Allik J, Dzhafarov EN, Houtsma AJM, Ross J, Versfeld HJ (1989) Pitch motion with random chord sequences. Percept Psychophys 46:513–527.

Atienza M, Cantero JL, Dominguez-Marin E (2002) The time course of neural changes underlying auditory perceptual learning. Learn Mem 9:138–150.

Berliner JE, Durlach NI (1973) Intensity perception. IV. Resolution in roving-level discrimination. J Acoust Soc Am 53:1270–1287.

Bigand E, Poulin B, Tillmann B, Madurell F, d'Adamo DA (2003) Sensory versus cognitive components in harmonic priming. J Exp Psychol [Hum Percept Perform] 29:159–171.

Bland DE, Perrott DR (1978) Backward masking: Detection versus recognition. J Acoust Soc Am 63:1215–1217.

Bodner M, Kroger J, Fuster JM (1996) Auditory memory cells in dorsolateral prefrontal cortex. NeuroReport 7:1905–1908.

Braun M (2001) Speech mirrors norm-tones: Absolute pitch as a normal but precognitive trait. Acoust Res Let Online 2:85–90.

Bregman AS (1990) Auditory Scene Analysis. Cambridge, MA: MIT Press.

Brosch M, Schreiner CE (2000) Sequence sensitivity of neurons in cat primary auditory cortex. Cereb Cortex 10:1155–1167.

Brown M, Irvine DRF, Park VN (2004) Perceptual learning on an auditory frequency discrimination task by cats: Association with changes in primary auditory cortex. Cereb Cortex 14:952–965.

Burns EM, Houtsma AJM (1999) The influence of musical training on the perception of sequentially presented mistuned harmonics. J Acoust Soc Am 106:3564–3570.

Burns EM, Ward WD (1978) Categorical perception—phenomenon or epiphenomenon: Evidence from experiments in the perception of melodic musical intervals. J Acoust Soc Am 63:456–468.

Cansino S, Williamson SJ (1997) Neuromagnetic fields reveal cortical plasticity when learning an auditory discrimination task. Brain Res 764:53–66.

Clément S (2001) La mémoire auditive humaine: Psychophysique et neuroimagerie fonctionnelle. PhD thesis, Université Victor Segalen, Bordeaux, France.

Clément S, Demany L, Semal C (1999) Memory for pitch versus memory for loudness. J Acoust Soc Am 106:2805–2811.

Cowan N (1984) On short and long auditory stores. Psychol Bull 96:341–370.

Cowan N, Winkler I, Teder W, Näätänen R (1993) Memory prerequisites of mismatch negativity in the auditory event-related potential (ERP). J Exp Psychol [Learn Mem Cogn] 19:909–921.

Czigler I, Csibra G, Csontos A (1992) Age and inter-stimulus interval effects on event-related potentials to frequent and infrequent auditory stimuli. Biol Psychol 33:195–206.

Darwin CJ, Hukin RW (2000) Effectiveness of spatial cues, prosody, and talker characteristics in selective attention. J Acoust Soc Am 107:970–977.

Dean I, Harper NS, McAlpine D (2005) Neural population coding of sound level adapts to stimulus statistics. Nat Neurosci 8:1684–1689.

Demany L, Ramos C (2005) On the binding of successive sounds: Perceiving shifts in nonperceived pitches. J Acoust Soc Am 117:833–841.

Demany L, Ramos C (2007) A paradoxical aspect of auditory change detection. In: Kollmeier B, Klump G, Hohmann V, Mauermann M, Uppenkamp S, Verhey J (eds) Hearing: From Basic Research to Applications. Heidelberg: Springer, pp. 313–321.

Demany L, Semal C (1990) Harmonic and melodic octave templates. J Acoust Soc Am 88:2126–2135.

Demany L, Semal C (2002) Learning to perceive pitch differences. J Acoust Soc Am 111:1377–1388.

Demany L, Semal C (2005) The slow formation of a pitch percept beyond the ending time of a short tone burst. Percept Psychophys 67:1376–1383.

Demany L, Semal C, Carlyon RP (1991) On the perceptual limits of octave harmony and their origin. J Acoust Soc Am 90:3019–3027.

Demany L, Clément S, Semal C (2001) Does auditory memory depend on attention? In: Breebart DJ, Houtsma AJM, Kohlrausch A, Prijs VF, Schoonhoven R (eds) Physiological and Psychophysical Bases of Auditory Function. Maastricht, the Netherlands: Shaker, pp. 461–467.

Demany L, Montandon G, Semal C (2004) Pitch perception and retention: Two cumulative benefits of selective attention. Percept Psychophys 66:609–617.

Demany L, Montandon G, Semal C (2005) Internal noise and memory for pitch. In: Pressnitzer D, de Cheveigné A, McAdams S, Collet L (eds) Auditory Signal Processing: Physiology, Psychoacoustics, and Models. New York: Springer, pp. 230–236.

Deutsch D (1991) The tritone paradox: An influence of language on music perception. Music Percept 8:335–347.

Deutsch D (1999) The processing of pitch combinations. In: Deutsch D (ed) The Psychology of Music. New York: Academic Press, pp. 349–411.

Deutsch D, Feroe J (1975) Disinhibition in pitch memory. Percept Psychophys 17:320–324.

Deutsch D, Henthorn T, Dolson M (2004) Absolute pitch, speech, and tone language: Some experiments and a proposed framework. Music Percept 21:339–356.

Dewar K, Cuddy L, Mewhort J (1977) Recognition memory for single tones with and without context. J Exp Psychol [Hum Learn Mem Cogn] 3:60–67.

Dowling WJ, Harwood DL (1986) Music Cognition. Orlando: Academic Press.

Durlach NI, Braida LD (1969) Intensity perception. I. Preliminary theory of intensity resolution. J Acoust Soc Am 46:372–383.

Edeline JM (1999) Learning-induced physiological plasticity in the thalamo-cortical sensory systems: A critical evaluation of receptive field plasticity, map changes and their potential mechanisms. Prog Neurobiol 57:165–224.

Fitzgerald MB, Wright BA (2005) A perceptual learning investigation of the pitch elicited by amplitude-modulated noise. J Acoust Soc Am 118:3794–3803.

Fraisse P (1967) Psychologie du Temps. Paris: Presses Universitaires de France.

Francès R (1988) The Perception of Music (JW Dowling, trans). Hillsdale, NJ: Lawrence Erlbaum. Original work: La Perception de la Musique. Paris: Vrin, 1958.

Frey HP, Kaernbach C, König P (2003) Cats can detect repeated noise stimuli. Neurosci Lett 346:45–48.

Fritz J, Shamma S, Elhilali M, Klein D (2003) Rapid task-related plasticity of spectrotemporal receptive fields in primary auditory cortex. Nat Neurosci 6:1216–1223.

Giard MH, Lavikainen J, Reinikainen K, Perrin F, Bertrand O, Pernier J, Näätänen R (1995) Separate representations of stimulus frequency, intensity, and duration in auditory sensory memory. J Cogn Neurosci 7:133–143.

Gold JM, Murray RF, Sekuler AB, Bennett PJ, Sekuler R (2005) Visual memory decay is deterministic. Psychol Sci 16:769–774.

Goldinger SD (1996) Words and voices: Episodic traces in spoken word identification and recognition memory. J Exp Psychol [Learn Mem Cogn] 22:1166–1183.

Gottlieb Y, Vaadia E, Abeles M (1989) Single unit activity in the auditory cortex of a monkey performing a short term memory task. Exp Brain Res 74:139–148.

Green DM, Swets JA (1974) Signal Detection Theory and Psychophysics. Huntington, NY: Krieger.

Grimault N, Micheyl C, Carlyon RP, Collet L (2002) Evidence for two pitch encoding mechanisms using a selective auditory training paradigm. Percept Psychophys 64:189–197.

Guttman N, Julesz B (1963) Lower limits of auditory periodicity analysis. J Acoust Soc Am 35:610.

Hafter ER, Bonnel AM, Gallun E (1998) A role for memory in divided attention between two independent stimuli. In: Palmer AR, Rees A, Summerfield AQ, Meddis R (eds) Psychophysical and Physiological Advances in Hearing. London: Whurr, pp. 228–236.

Hall JW, Peters RW (1982) Change in the pitch of a complex tone following its association with a second complex tone. J Acoust Soc Am 71:142–146.

Harris JD (1952) The decline of pitch discrimination with time. J Exp Psychol 43:96–99.

Hartmann WM, Johnson D (1991) Stream segregation and peripheral channeling. Music Percept 9:155–184.

Hawkey DJC, Amitay S, Moore DR (2004) Early and rapid perceptual learning. Nat Neurosci 7:1055–1056.

Hawkins HL, Presson JC (1977) Masking and preperceptual selectivity in auditory recognition. In: Dornic S (ed) Attention and Performance VI. Hillsdale: Lawrence Erlbaum, pp. 195–211.

Hirsh IJ (1959) Auditory perception of temporal order. J Acoust Soc Am 31:759–767.

Hofman PM, van Riswick JGA, van Opstal AJ (1998) Relearning sound localization with new ears. Nat Neurosci 1:417–421.

Houtgast T (1972) Psychophysical evidence for lateral inhibition in hearing. J Acoust Soc Am 51:1885–1894.

Houtgast T, van Veen TM (1980) Suppression in the time domain. In: van den Brink G, Bilsen FA (eds) Psychophysical, Physiological and Behavioural Studies in Hearing. Delft: Delft University Press, pp. 183–188.

Jääskeläinen IP, Ahveninen J, Bonmassar G, Dale AM, Ilmoniemi RJ, Levänen S, Lin FH, May P, Melcher J, Stufflebeam S, Tiitinen H, Belliveau JW (2004) Human posterior auditory cortex gates novel sounds to consciousness. Proc Natl Acad Sci USA 101:6809–6814.

Jacobsen T, Schröger E (2001) Is there pre-attentive memory-based comparison of pitch? Psychophysiology 38:723–727.

Kaernbach C (1993) Temporal and spectral basis of the features perceived in repeated noise. J Acoust Soc Am 94:91–97.

Kaernbach C (2004) The memory of noise. Exp Psychol 51:240–248.

Kaernbach C, Schulze H (2002) Auditory sensory memory for random waveforms in the Mongolian gerbil. Neurosci Lett 329:37–40.

Kallman HJ, Massaro DW (1979) Similarity effects in backward recognition masking. J Exp Psychol [Hum Percept Perform] 5:110–128.

Kinchla RA, Smyzer F (1967) A diffusion model of perceptual memory. Percept Psychophys 2:219–229.

Knudsen EI, Knudsen PF (1985) Vision guides the adjustment of auditory localization in young barn owls. Science 230:545–548.

Krishnan A, Xu Y, Gandour J, Cariani P (2005) Encoding of pitch in the human brainstem is sensitive to language experience. Cogn Brain Res 25:161–168.

Kropotov JD, Alho K, Näätänen R, Ponomarev VA, Kropotova OV, Anichkov AD, Nechaev VB (2000) Human auditory-cortex mechanisms of preattentive sound discrimination. Neurosci Lett 280:87–90.

Krumhansl CL, Iverson P (1992) Perceptual interactions between musical pitch and timbre. J Exp Psychol [Hum Percept Perform] 18:739–751.

Kubovy M, Howard FP (1976) Persistence of a pitch–segregating echoic memory. J Exp Psychol [Hum Percept Perform] 2:531–537.

Levitin DJ (1994) Absolute memory for musical pitch: Evidence from the production of learned melodies. Percept Psychophys 56:414–423.

Linkenhoker BA, Knudsen EI (2002) Incremental training increases the plasticity of the auditory space map in adult barn owls. Nature 419:293–296.

Lynch MP, Eilers RE, Oller DK, Urbano RC (1990) Innateness, experience, and music perception. Psychol Sci 1:272–276.

Massaro DW (1970a) Preperceptual auditory images. J Exp Psychol 85:411–417.

Massaro DW (1970b) Retroactive interference in short-term recognition memory for pitch. J Exp Psychol 83:32–39.

Massaro DW (1972) Preperceptual images, processing time, and perceptual units in auditory perception. Psychol Rev 79:124–145.

Massaro DW, Idson WL (1977) Backward recognition masking in relative pitch judgments. Percept Mot Skills 45:87–97.

Massaro DW, Loftus GR (1996) Sensory and perceptual storage. In: Bjork EL, Bjork RA (eds) Memory. San Diego: Academic Press, pp. 67–99.

McAdams S (1993) Recognition of sound sources and events. In McAdams S, Bigand E (eds) Thinking in Sound: The Cognitive Psychology of Human Audition. Oxford: Clarendon Press, pp. 146–198.

McClelland JL, Elman, JL (1986) The TRACE model of speech perception. Cogn Psychol 18:1–86.

McKenna TM, Weinberger NM, Diamond DM (1989) Responses of single auditory cortical neurons to tone sequences. Brain Res 481:142–153.

Meddis R, O'Mard LO (2005) A computer model of the auditory-nerve response to forward-masking stimuli. J Acoust Soc Am 117:3787–3798.

Menning H, Roberts LE, Pantev C (2000) Plastic changes in the auditory cortex induced by intensive frequency discrimination training. NeuroReport 11:817–822.

Molholm S, Martinez A, Ritter W, Javitt DC, Foxe JJ (2005) The neural circuitry of pre-attentive auditory change-detection: An fMRI study of pitch and duration mismatch negativity generators. Cereb Cortex 15:545–551.

Näätänen R, Winkler I (1999) The concept of auditory stimulus representation in cognitive neuroscience. Psychol Rev 125:826–859.

Näätänen R, Gaillard AW, Mäntysalo S (1978) Early selective-attention effect on evoked potential reinterpreted. Acta Psychol 42:313–329.

Näätänen R, Schröger E, Karakas S, Tervaniemi M, Paavilainen P (1993) Development of a memory trace for a complex sound in the human brain. NeuroReport 4:503–506.

Opitz B, Schröger E, von Cramon DY (2005) Sensory and cognitive mechanisms for preattentive change detection in auditory cortex. Eur J Neurosci 21:531–535.

Paavilainen P, Simola J, Jaramillo M, Näätänen R, Winkler I (2001) Preattentive extraction of abstract feature conjunctions from auditory stimulation as reflected by the mismatch negativity (MMN). Psychophysiology 38:359–365.

Peters RW, Moore BCJ, Glasberg BR (1983) Pitch of components of complex tones. J Acoust Soc Am 73:924–929.

Philibert B, Collet L, Vesson JF, Veuillet E (2005) The auditory acclimatization effect in sensorineural hearing-impaired listeners: Evidence for functional plasticity. Hear Res 205:131–142.

Phillips WA (1974) On the distinction between sensory storage and short-term visual memory. Percept Psychophys 16:283–290.

Purushotaman G, Bradley DC (2005) Neural population code for fine perceptual decisions in area MT. Nat Neurosci 8:99–106.

Recanzone GH, Schreiner CE, Merzenich MM (1993) Plasticity in the frequency representation of primary auditory cortex following discrimination training in adult owl monkeys. J Neurosci 13:87–103.

Relkin EM, Turner CW (1988) A reexamination of forward masking in the auditory nerve. J Acoust Soc Am 84:584–591.

Rensink RA, O'Regan JK, Clark JJ (1997) To see or not to see: The need for attention to perceive changes in scenes. Psychol Sci 8:368–373.

Robinson K, Summerfield AQ (1996) Adult auditory learning and training. Ear Hear 17: 51S–65S.

Ronken DA (1972) Changes in frequency discrimination caused by leading and trailing tones. J Acoust Soc Am 51:1947–1950.

Russo NM, Nicol TG, Zecker SG, Hayes EA, Kraus N (2005) Auditory training improves neural timing in the human brainstem. Behav Brain Res 156:95–103.

Sams M, Hari R, Rif J, Knuutila J (1993) The human auditory sensory memory trace persists about 10 sec: Neuromagnetic evidence. J Cogn Neurosci 5:363–370.

Schacter DL, Church B (1992) Auditory priming: Implicit and explicit memory for words and voices. J Exp Psychol [Learn Mem Cogn] 18:915–930.

Scharf B (1978) Loudness. In: Carterette EC, Friedman MP (eds) Handbook of Perception, IV: Hearing. New York: Academic Press, pp. 187–242.

Schellenberg EG, Trehub SE (1996) Natural musical intervals: Evidence from infant listeners. Psychol Sci 7:272–277.

Schröger E (1995) Processing of auditory deviants with changes in one vs. two stimulus dimensions. Psychophysiology 32:55–65.

Schröger E (1997) On the detection of auditory deviations: A pre-attentive activation model. Psychophysiology 34:245–257.

Schröger E (2005) The mismatch negativity as a tool to study auditory processing. Acta Acust 91:490–501.

Schwartz DA, Howe CQ, Purves D (2003) The statistical structure of human speech sounds predicts musical universals. J Neurosci 23:7160–7168.

Semal C, Demany L (1991) Dissociation of pitch from timbre in auditory short-term memory. J Acoust Soc Am 89:2404–2410.

Semal C, Demany L (1993) Further evidence for an autonomous processing of pitch in auditory short-term memory. J Acoust Soc Am 94:1315–1322.

Semal C, Demany L, Ueda K, Hallé PA (1996) Speech versus nonspeech in pitch memory. J Acoust Soc Am 100:1132–1140.

Shamma S, Klein D (2000) The case of the missing pitch templates: How harmonic templates emerge in the early auditory system. J Acoust Soc Am 107:2631–2644.

Shannon RV (1990) Forward masking in patients with cochlear implants. J Acoust Soc Am 88:741–744.

Shepard RN, Jordan DS (1984) Auditory illusions demonstrating that tones are assimilated to an internalized musical scale. Science 226:1333–1334.

Shinn-Cunningham BG (2000) Adapting to remapped auditory localization cues: A decision-theory model. Percept Psychophys 62:33–47.

Simons DJ, Levin DT (1997) Change blindness. Trends Cogn Sci 1:261–267.

Sparks DW (1976) Temporal recognition masking—or interference? J Acoust Soc Am 60:1347–1353.

Starr GE, Pitt MA (1997) Interference effects in short-term memory for timbre. J Acoust Soc Am 102:486–494.

Summerfield Q, Sidwell A, Nelson T (1987) Auditory enhancement of changes in spectral amplitude. J Acoust Soc Am 81:700–708.

Tanner WP (1961) Physiological implications of psychophysical data. Ann NY Acad Sci 89:752–765.

Terhardt E (1971) Pitch shifts of harmonics, an explanation of the octave enlargement phenomenon. In: Proceedings of the 7th International Congress on Acoustics (Budapest) 3:621–624.

Terhardt E (1974) Pitch, consonance, and harmony. J Acoust Soc Am 55:1061–1069.

Tervaniemi M, Maury S, Näätänen R (1994) Neural representations of abstract stimulus features in the human brain as reflected by the mismatch negativity. NeuroReport 5:844–846.

Tiitinen H, May P, Reinikainen K, Näätänen R (1994) Attentive novelty detection in humans is governed by pre-attentive sensory memory. Nature 372:90–92.

Tillmann B, Bharucha JJ, Bigand E (2000) Implicit learning of tonality: A self-organizing approach. Psychol Rev 107:885–913.

Tremblay K, Kraus N, Carrell TD, McGee T (1997) Central auditory system plasticity: Generalization to novel stimuli following listening training. J Acoust Soc Am 102:3762–3773.

Turner CW, Zeng FG, Relkin EM, Horwitz AR (1992) Frequency discrimination in forward and backward masking. J Acoust Soc Am 92:3102–3108.

Turner CW, Relkin EM, Doucet J (1994) Psychophysical and physiological forward masking studies: Probe duration and rise-time effects. J Acoust Soc Am 96:795–800.

Ulanovsky N, Las L, Nelken I (2003) Processing of low-probability sounds by cortical neurons. Nat Neurosci 6:391–398.

Ulanovsky N, Las L, Farkas D, Nelken I (2004) Multiple time scales of adaptation in auditory cortex neurons. J Neurosci 24:10440–10453.

van Noorden LPAS (1975) Temporal coherence in the perception of tone sequences. PhD Thesis, Technische Hogeschool Eindhoven, the Netherlands.

Viemeister NF (1980) Adaptation of masking. In: van den Brink G, Bilsen FA (eds) Psychophysical, Physiological and Behavioural Studies in Hearing. Delft: Delft University Press, pp. 190–198.

Viemeister NF, Bacon SP (1982) Forward masking by enhanced components in harmonic complexes. J Acoust Soc Am 71:1502–1507.

Vurpillot E (1975) La perception de l'espace. In: Fraisse P, Piaget J (eds) Traité de Psychologie Expérimentale, vol. VI: La Perception. Paris: Presses Universitaires de France, pp. 113–198.

Ward WD (1954) Subjective musical pitch. J Acoust Soc Am 26:369–380.

Ward WD (1999) Absolute pitch. In: Deutsch D (ed) The Psychology of Music. San Diego: Academic Press, pp. 265–298.

Warren RM (1982) Auditory Perception: A New Synthesis. New York: Pergamon.

Watson CS, Kelly WJ, Wroton HW (1976) Factors in the discrimination of tonal patterns. II. Selective attention and learning under various levels of stimulus uncertainty. J Acoust Soc Am 60:1176–1186.

Weinberger NM (1995) Dynamic regulation of receptive fields and maps in the adult sensory cortex. Annu Rev Neurosci 18:129–158.

Weinberger NM (2004) Specific long-term memory traces in primary auditory cortex. Nat Rev Neurosci 5:279–290.

Wilson JP (1970) An auditory after-image. In: Plomp R, Smoorenburg GF (eds) Frequency Analysis and Periodicity Detection in Hearing. Leiden, the Netherlands: Sijthoff, pp. 303–318.

Wojtczak M, Viemeister NF (2005) Mechanisms of forward masking. J Acoust Soc Am 115: 2599.

Wright BA, Fitzgerald MB (2001) Different patterns of human discrimination learning for two interaural cues to sound-source location. Proc Natl Acad Sci USA 98:2307–12312.

Wright BA, McFadden D, Champlin CA (1993) Adaptation of suppression as an explanation of enhancement effects. J Acoust Soc Am 94:72–82.

Wright BA, Buonomano DV, Mahncke HW, Merzenich MM (1997) Learning and generalization of auditory temporal-interval discrimination in humans. J Neurosci 17:3956–3963.

Yost WA, Berg K, Thomas GB (1976) Frequency recognition in temporal interference tasks: A comparison among four psychophysical procedures. Percept Psychophys 20:353–359.

Young PT (1928) Auditory localization with acoustical transposition of the ears. J Exp Psychol 11:399–429.

Zatorre RJ, Samson S (1991) Role of the right temporal neocortex in retention of pitch in auditory short-term memory. Brain 114:2403–2417.

Zheng W, Knudsen EI (1999) Functional selection of adaptive auditory space map by $GABA_A$–mediated inhibition. Science 284:962–965.

Zwislocki J, Pirodda E, Rubin H (1959) On some poststimulatory effects at the threshold of audibility. J Acoust Soc Am 31:9–14.

# 5
# Auditory Attention and Filters

ERVIN R. HAFTER, ANASTASIOS SARAMPALIS, AND PSYCHE LOUI

## 1. Introduction

The traditional approach to the study of sound sources emphasized bottom-up analysis of acoustic stimuli, whether they were more primitive features such as frequency, sound level, and source direction, or complexes made by combinations of primitives. However, a simple scan of the table of contents of this volume shows that the field has evolved considerably toward a realization of important top-down processes that modulate the perception of sounds as well as control how we derive and interpret the natural acoustical events of everyday life. Probably the most commonly used word in this regard is *attention*, a term whose meaning is "understood" by everyone, but whose scientific description encompasses a variety of operational definitions. The current chapter does not attempt to address all of these approaches. Rather, it concentrates on the listener's ability to extract relevant features of the auditory scene and seeks to understand the seeming ability to focus on some parts of the auditory stream at the expense of others. While perceptual attention is typically defined in terms of internal processes that help us extract relevant information from a complex environment, such a broad view does not tell us about the nature or locus of the selection process. The focus of attentional theories ranges from stimulus cohesion, whereby attention binds sensory features into higher-order percepts (Treisman and Gelade 1980), to segregation, which breaks stimuli into parts so that selective processing can be allotted to the attended element. The concentration here is on the latter, though issues of feature binding become important when segregation is based on a combination of auditory features such as harmonicity and timbre, or on the collective effects of a string of tones in a melody.

Intensive study of attention to specific portions of auditory messages began in the 1950s through seminal experiments and commentary by a group of individuals (see Cherry 1953; Broadbent 1958; Deutsch and Deutsch 1963; Norman 1969; Treisman 1969; Moray 1970) whose focus was on what happens when we are confronted with sounds from multiple auditory sources. Cherry's colorful description of the "cocktail party effect" was seen as an example of the listener's ability to derive information from one stream of speech in the presence of others, and from this grew a theoretical discussion of the presence of specific filters (or channels) that select information that matches their pass bands for interpretation by more central processes. A common methodology in this early work on spatially defined attentional channels was dichotic listening, in which two different streams of spoken material were presented simultaneously, one to each ear, and the subject was instructed to "shadow" the attended speech at one ear by repeating it as it was heard. Inability to recall information from the unattended ear fitted well with Broadbent's (1958) filter theory. In order to model the perceptual bottleneck caused by two messages reaching a sensory buffer at the same time, he proposed a selective filtering process that would allow one message to go forward while leaving the other to decay in short-term memory. In this way, a more robust feature of the unattended stimulus such as its fundamental frequency might persist until after analysis of the selected message was complete. Based on subsequent evidence that some higher-level information, such as the listener's name, might break through to awareness from the unattended source (e.g., Moray 1960), Treisman (1964, 1969) abandoned the notion of all-or-none filters in favor of attentional attenuators that selectively reduce the effectiveness of a stimulus without totally blocking it. From a more cognitive perspective, Deutsch and Deutsch (1963) eschewed the idea of peripheral filtering in favor of late selection that could include semantic factors as well. In the years since, there have been no clear winners in the debate about early and late sites of selection, with evidence for attentional filtering ranging from hard-wired frequency analyzers to segregation of simultaneous auditory streams. One potential reason for this ambiguity is described by Broadbent's (1958) notion that information blocked by an early bottleneck may be later processed in short-term memory. In this regard, Norman (1969) demonstrated that when subjects are asked about what they just heard on the unattended side when a shadowing task is suddenly terminated, they can produce up to 30 seconds of recall.

The more specific a definition of attention, the greater its reliance on the experimental operations used to define it. Traditionally, work on visual attention has relied heavily on measures of reaction times (RTs). Conversely, studies of auditory attention have concentrated on paradigms that measure the ability to extract signals from a noisy background for detection and/or discrimination. In this case, attention is often thought of in terms of the listener's focus on the expected location of a signal along a monitored dimension. While this is the major concern of the current chapter, also noted are cases in which the use of RT in audition has been useful for highlighting the distinction between endogenous cues, whose information directs attention to the predicted locations of meaningful stimuli, and exogenous cues, which provoke a reflexive attraction to a location, regardless of its relation to the task. Finally, it is obvious that auditory attention affects more than just simple acoustical features, having profound influences on such complex processes as informational masking, spatial hearing, and speech understanding. However, because those topics appear elsewhere in this volume (Chapters 6, 8, and 10), the stress here is on the effects of cueing in activating attentional filters based on expected features of the auditory signal.

## 2. Signal Detection

### 2.1 Detecting a Single Signal in Noise

Sensory coding generally begins with the breakdown of complex stimuli into more narrowly defined regions along fundamental dimensions. In audition, this is done by processes in the cochlea that separate the acoustic input into the separate frequency channels leading to frequency-specific activity in the auditory nerve. In the history of psychoacoustics, these frequency channels have taken on various names including "critical bands" and "auditory filters" (see Moore 2003 for a review). Division of this kind provides a distinct advantage for signal detection by increasing the signal-to-noise ratio (S/N) for narrowband signals through the rejection of interference from more distant regions of the spectrum.

An early demonstration of exclusive processing of a single band was evident from the effects of reducing the width of a noise masker on detection of a tonal signal. At first, limiting the masker had no effect on signal detection, but performance rose when the bandwidth was reduced to the point that its edges fell into the so-called critical band surrounding the signal (Fletcher 1940). The ability to respond to the stimulus within a selected region of the spectrum is reminiscent of one of the oldest theories of attention, in which it is pictured as focusing of a searchlight on the relevant information. A demonstration is seen in measures of the "critical ratio" (Fletcher 1940). Here, the bandwidth of the masker remains wide, but the subject is informed about the frequency to be detected by presentation of the signal before the experiment begins. The primary assumption of this method is that the band level of the masker (BL) of the noise within the listening band is proportional to the level of the signal at threshold. Given that the total wideband noise has a spectrum level of $N_0$ (level/Hz), the width of a rectangular equivalent of the listening band is computed by dividing BL by $N_0$. Hence, the term *critical ratio* (see Hartmann 1997). The high correlation between critical ratios and other measures of the bandwidths (Scharf 1970; Houtsma 2004) lends credence to the view that attention can focus on a specific region of an auditory dimension, even when the stimulus covers a much wider range. Throughout this chapter, we will use the term *filter* to describe this kind of attentional selection in a variety of auditory dimensions.

The important attentional assumption of the critical ratio is that the listener is accurately informed about where to listen. As noted above, this is typically done by playing a sample of the signal before data collection begins, and it is often enhanced by feedback presented after each trial. A potential weakness of feedback is that the subject must hold a representation of the stimulus in the signal interval in memory for comparison to the feedback, which may explain improvements in performance at the beginning of tests with weak signals (Gundy 1961). Some experiments have attempted to focus the subject with simultaneous cueing such as by adding the signal to a continuous tonal pedestal set to the same frequency as the signal (see Green 1960) or, for detection of a monaural signal in noise, by a sample of the signal presented to the contralateral ear (Taylor and Forbes 1969; Yost et al. 1972). However, one must be cautious about the possibility that these cues introduce detectable changes in dimensions other than energy. In the former case, this might due to introduction of transients when signals are added to the pedestal (e.g., Macmillan 1971; Leshowitz and Wightman 1972; Bonnel and Hafter 1999), while in the latter it might reflect introduction of binaural effects of the kind responsible for binaural masking level differences (BMLDs) (Jeffress et al. 1956). Perhaps the most efficient way to inform the subject about the signal is to begin each trial with an iconic cue, i.e., one that matches the signal in every respect except level. Typically, the level of this cue is set high enough to make it clearly audible but not so much as to prevent the qualitative impression that both cue and signal are heard as tones in noise. A special advantage of this kind of cueing is that it can provide control data for quantification of the effects of uncertainty when signals are drawn at random from a range of frequencies on a trial-by-trial basis.

## 2.2 Detection with Frequency Uncertainty

Obviously, focusing attention on a single filter is a weak strategy for detecting a tonal signal when there is uncertainty about its frequency. Traditionally, this has been studied by choosing the signal on each trial at random from a set of $M$ frequencies. An assumption of independence between maskers at the outputs of the "auditory filters" centered on the $M$ tones has led to the term M-orthogonal bands (MOB) to describe models couched in signal detection theory (SDT) (Green and Swets 1966). When $M$ is relatively small, it is assumed that the subjects are able to monitor the appropriate bands through repeated testing with feedback or through trial-by-trial iconic cues. In a more complex situation from the subject's perspective, effects of uncertainty have been tested by presenting signals drawn completely at random from a wide range of frequencies. In this case, an MOB model predicts that detection must fall with increasing $M$ due to the increased probability of a false alarm produced by noise alone in one of the nonsignal filters, while views based on issues of shared attention point to such factors as inaccuracy in choosing which filters to monitor as well as higher demands on a limited attentional resource. These will be discussed later, in Section 3.1.

Green (1960) used SDT to examine the MOB model through comparisons between the effects of uncertainty on human behavior and that of a hypothetical ideal observer whose knowledge of the signal is exact. For the "ideal," increasing $M$ has three effects on the psychometric functions. These functions, which relate performance—in units of the percentage of correct decisions in a two-alternative forced-choice (2AFC) task—to signal level are depicted in Figure 5.1. First is a shift to the right, indicative of the need for higher signal levels to maintain constant performance with increased uncertainty. The second is an increase in the slopes of the functions, making the rise in performance from chance to perfect happen over a smaller change of level. The third is a deceleration of the other two effects, with each successive increment in $M$ having less of an effect than the one before. In comparison to the ideal observer with $M = 1$, the

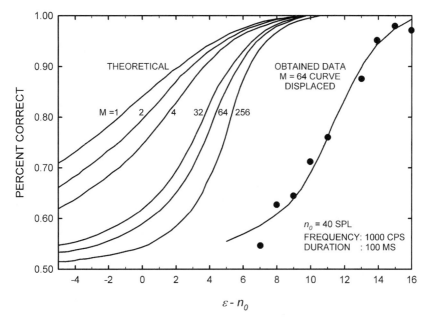

FIGURE 5.1. Predictions of an M-orthogonal band model of the effects of stimulus uncertainty on signal detection for an ideal energy detector with signal known exactly. (Reprinted with permission from Figure 6 in Green 1960).

empirically measured function from human subjects listening for only a single signal, that is, without frequency uncertainty, showed both the rightward shift indicative of reduced performance and a steeper slope. Rather, the human data closely resemble the ideal observer with an $M$ of 64. Green (1961) argued that this steeper slope explains why increasing $M$ from 1 to 51 (with 51 possible signal frequencies spaced evenly over a range of 3000 Hz) produced a rise in threshold of only 3 dB compared to the 6 dB change expected for the ideal observer. From these results, Green (1960, 1961) concluded that listeners, even in the simplest of cases, have a high degree of signal uncertainty about such basic parameters as frequency, phase, duration, and time of occurrence. Lumping uncertainty along multiple parameters produces a single variable that we call "unavoidable uncertainty" (UU). From this perspective, when the experimenter varies one signal parameter from 1 to $M$ and holds all other parameters constant, uncertainty for the human observer is best described as a rise in uncertainty from UU to $M$ + UU. Green (1960) applies this idea to a feature other than frequency, noting that variation of the moment of the signal's presentation over a range of 8 s produced a rise in threshold of less than 2 dB (Egan et al. 1961). Still more support for the idea that human performance is plagued by uncertainty along multiple dimensions is beautifully seen in classic papers by Jeffress (1964, 1970), who showed that receiver operating characteristics (ROCs) based on human performance in a tone-detection task fitted well to comparable ROCs for an ideal observer who has no knowledge of the signal's phase. In summary, this view, based on the MOB model of SDT, sees the human subject as an optimal detector who must rely on noisy data for making judgments. While this seems true, it leaves open questions about top-down processes and questions about how shared attention might interact with stimulus uncertainty. This will be discussed more fully in Section 3.2.

## 2.3 Use of a Probe-Signal Method to Measure the Shapes of the Listening Bands

While the critical ratio offers a way for getting at the width of a listening band employed in wideband noise, it is limited to a single feature of that band, revealing nothing about the structure of the filter. In order to address that problem, Greenberg and Larkin (1968) described a probe-signal method designed to provide a direct description of both the width and shape of the listening band. Their subjects were trained to detect 1000-Hz signals presented in wideband noise. During probe-signal conditions, the subject's expectancy of 1000 Hz was maintained by using that frequency on the majority of trials, while for the remaining trials, frequencies of the "probes" were chosen from a set symmetrically spaced around 1000 Hz. Detection probabilities were lower with probes than with expected signals by an amount that grew with their distances from 1000 Hz. The explanation given for this result was that subjects had responded to stimuli within an internal filter centered on 1000 Hz, thus reducing the levels of probes via attenuation by the skirts of that filter. In support of this proposal, the estimated width of the internal filter was similar to that of the "critical band" measured by other means, a result later confirmed by Dai et al. (1991), who showed that bandwidths obtained with probe signals from 250 to 4000 Hz resembled those found with notched-noise masking (Patterson and Moore 1986). Obviously, listening for an expected frequency does not make the rest of the spectrum inaudible. In a single-interval (yes/no) task, Dai et al. (1991) inserted some probes well outside the "auditory filter" of the expected frequency. Presenting these tones at several levels allowed them to plot psychometric functions that then were used to evaluate the salience of each probe. Results showed a maximum attenuation of only 7 dB for all probes, regardless of their distance from expectation.

Because the most intense probes were higher in level than the expected signals, the authors suggested that subjects might have changed their response strategies, accepting even widely divergent frequencies as valid signals. This addresses a potential criticism of the probe-signal method first raised in Greenberg and Larkin (1968) and referred to in Scharf et al. (1987) as a "heard but not heeded" strategy, which posits that filter-like results could also appear in the normal probe method if subjects chose not to respond to sounds that were not like the expected signal. In order to address this, Scharf et al. (1987) included a condition in which trial-by-trial feedback was provided so as to encourage the subject to respond to probes that differed in perceived quality from the expected signal. When this, as well as other conditions intended to test the idea, had little effect, Scharf et al. (1987) concluded that although the "heard but not heeded" hypothesis might have had some effect on results from the probe-signal method, the width of the listening band was based primarily on sensory filtering. This does not, however, preclude individual differences among listeners based on where they place their attentional filters or how they interpret sounds that differ from the expected signals in qualitative as well as quantitative ways. In this regard, Penner (1972) varied payoffs in a probe-signal method and showed that different subjects used different subjective strategies. This led some to act as if listening through narrower, auditory-filter-like bandwidths and others to show wider filters that were more inclusive of distant probes.

## 2.4 Probe-Signal Measures of Selectivity in Domains Other Than Frequency

The probe-signal method has been used to study other situations in which dimensions might be represented by their own internal filters. For example, Wright and Dai (1994) studied listening bands found with off-frequency probes using signals that could have one of two durations: 5 or 295 ms. In mixed conditions, where the durations of the expected signals and probes did not always match, probes were more poorly detected if the durations of the probe and signal were different, a result interpreted as an indication of attention to distinct locations in the time–frequency plane. In a follow-up experiment, Dai and Wright (1995) looked for evidence of attention to specific filters in the signal-duration domain by measuring performance in a probe task in which the expected signals and probes differed only in their durations. In a condition in which the duration was 4, 7, 24, 86, 161, or 299 ms chosen at random, performance was only slightly less than when each was tested alone, suggesting little effect of duration uncertainty. Then, in separate blocks, with expected signals whose durations were either 4 or 299 ms, the durations of occasional probes were 7, 24, 86, 161 ms. There, seemingly in keeping with the idea of tuning in the duration domain, detection of probes declined to chance as their durations differed from expectation. While the first result shows that subjects could monitor multiple durations with ease, the second seems to demonstrate focus on a specific listening band defined by duration. Some element of the seeming focus on an expected duration may stem from a "heard-but-not-heeded" strategy (Scharf et al. 1987), where all signals are heard but some are ignored because they do not fit the description of an expected signal.

## 3. Attention and Effort

The National Aeronautics and Space Agency (NASA) once asked this lab why trained airline pilots using one of their new flight simulators were crashing on landing at an uncharacteristically high rate. In response, we proposed that landing an airplane requires repeated answers to the yes/no question, "Is it safe to continue?" Based on a traditional SDT perspective, we cited Norman and Bobrow's (1975) argument that $d'$ in basic signal detection is "data-limited," that is, affected by the S/N but not by attention, and argued it was probably a more relaxed response criterion ($\beta$) in the simulator that led to more false alarms, i.e., crashes. When asked by NASA to prove this, we proposed a task in which subjects would be paid either by the hour or by a pay-for-performance scheme meant to be more like real flying, with a false alarm on a rare percentage (2%) of the noise-only trials producing a "crash," sending the subjects home without pay. Subjects in this condition reported heightened attention and, not surprising in the light of Kahneman's (1973) discussion of the relation between attention and effort, a higher level of stress. In order to increase the overall cognitive load of the task, the signal's frequency was varied from trial to trial, drawn at random from a wide range of possibilities. For maximum uncertainty, these tones were presented without cues; for minimum uncertainty; each was preceded by an iconic cue.

Findings reported to NASA (Hafter and Kaplan 1976) showed that the nonstressful (hourly-pay) condition produced a typical uncertainty effect, that is, a difference between thresholds, with and without cues, of about 3 dB. Furthermore, with uncertainty held to a minimum, varying the payoff from easy to stressful had no effect, in accord with Norman and Bobrow (1975). The most interesting result was an interaction between uncertainty and payoff, with the risky scheme reducing the effect of uncertainty by half. Evidence that payoff could improve performance led us to postulate that the widths of the effective listening bands were subject to cognitive processes, with a high cost of shared attention produced associated with uncertainty being somewhat relieved by subjects using fewer, albeit wider, filters. From this perspective, the threat of a potential crash led observers to attend more closely, responding to more, albeit narrower, effective listening bands.

## 3.1 Effects of Signal Uncertainty on the Bandwidths of the Effective Filters

The idea that frequency uncertainty might produce changes in the listening bands is not new, with suggestions varying from a single auditory filter switched to each possible frequency in accord with its probability to the proposition of a single wideband filter that encompasses all possibilities (Swets 1984). However, all such models predict large losses in detectability when the signal is drawn from a large range of possible frequencies, and that is simply not the case. What is more, the idea of locally wider bands seems antithetical to the traditional view that auditory filters are immutably constrained by the mechanics of the cochlea. Scharf and his colleagues (e.g., Scharf et al. 1997) have argued for top-down control of the peripheral filters through efferent innervation of the cochlea by the olivocochlear bundle (OCB). This was based on observations of wider bands found using a probe-signal method in patients whose OCBs had been severed during surgery. In rebuttal to the view of top-down control, however, Ison et al.

(2002) cite reduced OCB function due to aging when noting that probe-signal measures showed only small differences in width between young and old patients. In order to address the idea in Kaplan and Hafter (1976) that uncertainty and, presumably, the increased cost of shared attention had widened the effective bandwidths in Hafter and Kaplan (1976), Schlauch and Hafter (1991) devised a means for use of the probe-signal method to examine the filters as a function of a controlled amount of uncertainty in a way that would be unbiased by the unavoidable uncertainty discussed earlier in conjunction with the M-orthogonal band model (Green 1960).

The probe-signal method is based on a primary assumption that the subject responds only to sounds within the filter centered on the expected frequency. In this case, signals to be detected in wideband noise would be drawn at random from the range 600–3750 Hz, but uncertainty would be erased by beginning each trial with a clearly audible cue that told the subject what frequency to expect. Within-subject comparisons showed that performance with these iconic cues was as good as that in a long block that presented only a single frequency. Expectancy was established by presenting iconic cues (with cue-to-signal ratios $f_c/f_s$ of 1.00) on 76% of the trials. However, in the remaining 24% of the trials, probe signals differed from expectancy values by small distances. In line with the quasi-logarithmic distribution of frequency in the cochlea (e.g., Greenwood 1961; Moore and Glasberg 1983), it was assumed that auditory filters are well characterized by a single quality factor, Q (the ratio of a filter's center frequency to its bandwidth). For analysis of the hypothetical constant-Q filter, probes were set to one of four log-distances from expected frequencies, with $f_c/f_s$ of 0.95, 0.975, 1.025, or 1.05, and data were averaged for each value of $f_c/f_s$, regardless of $f_c$. Preliminary measures of performance taken at several points across the range of possibilities were used to derive a function describing the signal levels needed for equal detectability. Once the experiment began, all signals were set accordingly. Performance was measured as a percentage of correct responses [P(C)] in a 2AFC task.

Ordinarily, filters are plotted as decibels of loss relative to a value of zero dB assigned to their center frequencies. In order to convert P(C) into dB, psychometric functions were derived from additional tests conducted with several overall signal levels. A filter constructed in this way is illustrated in the left panel of Figure 5.2. Data points are averages from three subjects, and the fitted line is a rounded exponential (ROEX) model of the auditory filter (Patterson and Nimmo-Smith 1980) plotted in terms of a single bandwidth parameter, $p$, as in Patterson and Moore (1986). The close fit between this derived filter and that in Patterson and Moore (1986) in both shape and bandwidth suggests that subjects here attended to single auditory filters at the cued locations, and were unaffected when the signal was roved from trial to trial.

In order to study the effects of uncertainty on bandwidth, Schlauch and Hafter (1991) planned to use the same method while increasing the number of bands that the subject must monitor on each trial. For this, they eschewed comparisons to the ideal observer of SDT as in Green 1960, and chose instead to define the

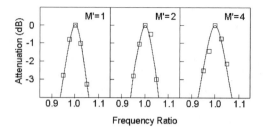

FIGURE 5.2. Listening bands derived for using a probe-signal method for detection of signal frequencies that varied the range 600–3750 Hz (adapted from Schlauch and Hafter 1991). Each trial began with a cue containing $M'=1$, 2, or 4 tones, one of which defined the expected frequency of the signal. The abscissa is plotted in terms of the ratio of cued frequency to probe frequency ($f_c/f_s$). The ordinate shows the loss in effective level of probe signals ($f_c/f_s \neq 1.00$) in dB relative to that found with the signals at the expected value ($f_c/f_s = 1$). These were found using separately obtained psychometric functions to convert performance in P(C) effective signal levels in dB. The fitted curves are ROEX filters (Patterson and Moore 1986) (see text).

case of minimal uncertainty in terms of the subject's own performance. This is described in the leftmost panel of the figure by the label $M' = 1$ to indicate that these data represent the best that a subject could do when required to monitor only a single band on each trial. Uncertainty was defined in terms of the number of tones in each cue, be they one, two or four tones ($M' = 1$, 2, or 4). In all cases, the $M'$ tones were chosen at random from the range of possible frequencies. While only one of the cue tones on a trial matched the signal, the subject would have to monitor filters at all of their frequencies. Results from $M' = 2$ and 4 are plotted in the center and rightmost panels of Figure 5.2, again normalized in order to set performance with a ratio $f_c/f_s = 1.00$ to zero dB. While the fitted curves are ROEX filters (Patterson and Moore 1986), there is a small but consistent increase in bandwidth parameter with increasing $M'$, lending support for the view that listening bands, as measured in auditory masking, are labile in ways that allow them to be affected by attentional factors.

## 3.2 Relative Cueing for Detection at Emergent Levels of Processing

Just as iconic cues can alleviate the loss of detectability due to frequency uncertainty, so too can a variety of other cues that bear a more distant relation to the signal. Although these are typically less effective than iconic cues, we have seen improvement in the detection of randomly chosen tones cued by tones related to the signal by a musical interval (Hafter et al. 1993), by a five-tone harmonic sequence for which the signal is the missing fundamental (Hafter and Schlauch 1989), and with musically trained subjects who have absolute pitch and are cued with a visual description of the signal on a musical score (Plamondon and Hafter 1990). From this, it would seem that successful cueing requires only

that the cue and signal sound alike in the "mind's ear." However, an interesting result brought this into question. Whereas a harmonic sequence helped a
subject listen for its missing fundamental, the reverse was not true. That is, when
the signal to be detected was a randomly chosen harmonic sequence, cueing
with its missing fundamental did not improve detection. A potential explanation
for this begins with the suggestion that the harmonic sequences were detected
on the basis of their emergent property: their complex pitch. The hierarchy of
processing in the auditory nervous system means that signals are represented in
multiple sites along the auditory pathway. If one assumes that attention can be
focused on any level of processing, it follows that while detection of a five-
tone complex might be based on activity in five distinct locations in a neural
representation of frequency, it could also reflect activity in a single location in
a representation of complex pitch. As discussed in Hafter and Saberi (2001),
for optimal performance, attention should be directed toward the level with the
best signal-to-noise ratio (S/N) and, in the case of a complex signal, this is
on the complex feature rather than the primitives. This can be understood in
terms of the probability of false positives in detecting a five-tone complex. If
done on the basis of individual frequencies, a false alarm should happen in
response to multiple peaks in the spectrum of noise-alone trials, regardless of
their frequencies. Conversely, for detections based on complex pitch, a false
alarm should happen only if peaks in the noise spectrum happen to be related
by a common fundamental.

Why, then, the asymmetry in cueing with complexes and their fundamentals?
An alternative to the simple sounds-like hypothesis says that a successful cue
specifies a unique location at the level of processing where detection takes place.
From this perspective, a complex pitch would be able to specify the location of
its fundamental in a representation organized by frequency, but because a pure
tone does not belong to a single harmonic complex, a single frequency cannot
specify a unique location in a representation organized by complex pitch.

In order to test this hypothesis, Hafter and Saberi (2001) compared performance from five conditions in which stimuli might be cued and detected at more
than one level of processing. In all cases, signals and cues were made up of
three tones and presented in continuous background noise. Individual tones were
preset to be equally detectable in accord with the relation between thresholds
and frequency. Specifics of the five conditions as well as results from the 2AFC
detection task are described in Figure 5.3. In Condition 1, signals drawn at
random from the frequency range 400–4725 Hz were presented without cues.
The average level of these tones was set to produce extremely low performance
[P(C) = 0.60] in order to leave room for improvement in other conditions. This
level was then used throughout the experiment. Signals in Condition 2 were
missing-fundamental harmonic complexes. Each was created by first selecting
a frequency designated as a fundamental ($f_0$) from the range 200–675 Hz. Its
next six harmonics ($f_1$ to $f_6$) were computed, and three of them were chosen at
random to be a signal. For example, if the randomly chosen $f_0$ was 310 Hz and
the harmonics chosen for the signal were $f_1$, $f_4$, and $f_6$, the signal would consist

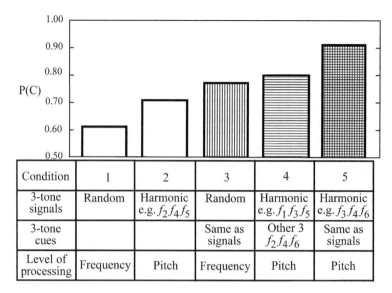

| Condition | 1 | 2 | 3 | 4 | 5 |
|---|---|---|---|---|---|
| 3-tone signals | Random | Harmonic e.g. $f_2 f_4 f_5$ | Random | Harmonic e.g. $f_1 f_3 f_5$ | Harmonic e.g. $f_3 f_4 f_6$ |
| 3-tone cues | | | Same as signals | Other 3 $f_2 f_4 f_6$ | Same as signals |
| Level of processing | Frequency | Pitch | Frequency | Pitch | Pitch |

FIGURE 5.3. Results (adapted from Hafter and Saberi 2001) demonstrating cueing and signal detection based on two different levels of processing, one tied to analysis of individual frequencies and one tied to the emergent property complex pitch (see text). Signals and cues in all five conditions were three-tone complexes.

of 620, 1550, and 2170 Hz. These were also without cues. Although levels here were the same as those in Condition 1, improved performance [P(C) = 0.70)] confirms the prediction that detection should be better if based on a complex pitch than if on a comparable set of independent frequencies. In cued conditions, 3, 4, and 5, cues were set 7 dB above the signals. Signals in Condition 3 were random tones selected as in Condition 1. Cues matched the signals in frequency, thus informing the subject about where to listen in the frequency domain. Comparing results [P(C) = 0.76] to those in Condition 1 shows that cues ameliorated uncertainty about frequencies in the signals. Signals in Condition 4 were selected in the same way as in Condition 2. However, each cue was chosen to match the signal in the complex pitch domain without highlighting its frequencies. For this, after three of the six harmonics had been chosen to be a signal, the remaining three were used as the cue. In terms of the example above for Condition 2, if $f_0$ was 310 Hz and $f_1$, $f_4$, and $f_6$ made up the signal, the cue would be $f_2$, $f_3$, and $f_5$, or 930, 1240, and 1860 Hz. Comparing results [P(C) = 0.79] to those in Condition 2 shows that the cues ameliorated uncertainty about the complex pitch of the signal. Finally, Condition 5 used cues to inform the subject both about frequencies in a signal and its complex pitch. For this, signals were harmonic complexes chosen as in Conditions 2 and 4, but frequencies in each cue were identical to those in the signal. Comparisons of these results [P(C) = 0.91] to those in Conditions 3 and 4 shows the added effectiveness of cueing both signal features frequency and complex pitch. A prediction for the optimal summation of information in the two domains obtained by summing the two $(d')^2$ values from Conditions 3 and 4 finds that performance in Condition 5 was slightly higher than the predicted value, perhaps indicating a form of useful crosstalk whereby cues in one dimension enhanced the efficacy of cues in the other. In a more general sense, these data strongly support the idea that detection of a complex signal can be based on multiple dimensions and that attention can be focused on these dimensions through cues that share a unique level of processing with the signals.

## 4. Expectancy and the Analysis of Clearly Audible Signals

To this point, we have talked about how cueing can improve detection by indicating to the listener where to expect a weak signal along some dimension. Related effects with suprathreshold stimuli show that cues can also change the way that an audible target is perceived. This is especially obvious with running speech, where syntax and semantics affect how speechlike sounds are heard. The focus here will be on nonspeech cueing of audible signals based on connections between cues and signals that range from simple relations such as harmony and interstimulus intervals to more-complex patterns established by presenting stimuli in an auditory stream.

### 4.1 Attention Focused by Musical Expectancy

Attention to sequential information is especially important in music, where the basic nature of the stimulus is represented in the relation between auditory events over time. From this perspective, a musical context can act as a cue to establish expectations of future events in the stream. Many have considered expectancy to be a major feature of music, particularly Meyer (1956), who has postulated that the systematic violation of expectation is a primary factor in the elicitation of emotion by music. Much of the experimental work on musical expectation has used a variant of the probe-tone paradigm of Krumhansl (e.g., Krumhansl and Kessler 1982), which presents a melodic scale followed by a probe tone that is rated by the listener for its goodness of fit to the preceding notes. Ratings are highest for probes that match the melody in tonal and harmonic contexts. For instance, if the melody is in C major, the most highly rated pitch class is C, followed by G, E, and F. Similarly, probe-tone profiles have been shown to reflect the statistics underlying a musical composition, as expectancies are established by melodies composed in a single key (Krumhansl 1990). Reaction time (RT) has also been employed in the study of musical expectation. As an example, Bharucha and Stoeckig (1986) presented pairs of chords and asked subjects to say whether the second chord was consonant or dissonant. Results showed that if the second chord was expected (based on harmonic relations to the first), RTs were faster for consonance, but if the second chord was unexpected, RTs were faster for dissonance. Interestingly, while musically trained subjects were faster on average, the main effect held for those without musical training. It is tempting to speculate on whether musical similarity ratings, especially those tied to harmonicity, derive from fundamental auditory processes as might be predicted by models of frequency discrimination and pitch, or are reflective of musical experience. The learning hypothesis gains credence from the observation that the modern musical interval of the fifth is different from that used in sixteenth-century music but alas, we have no experimental data on similarity ratings from the sixteenth century. As always, the likely answer is that both nature and nurture are probably involved.

## 4.2 Attention Focused Through Internal Oscillations Entrained to Temporal Sequences

Another auditory factor implicated in sequential cueing is the timing of events or rhythm. As with other factors such as pitch contour, regularity of timing can cause the listener to look ahead, prescribing the appropriate moment for evaluation of an upcoming target. Evidence for this kind of entrainment to rhythm is found in the work of Jones and her colleagues (e.g., Jones and Boltz 1989; Large and Jones 1999; Jones et al. 2002), who postulate that consistent timing in a sequence of events produces an anticipatory attentional focus based on the temporal structure of the sequence. Support for this has come from a paradigm in which the subject hears a sequence of tones of different frequencies played at regular intervals. The sequence begins with a reference tone and ends with a target tone that follows the penultimate tone by a variable inter-onset interval (IOI). The listener's task is to judge whether the pitch of the target is the same or different from that of the reference. For IOIs of up to 1200 ms, Jones et al. (2002) found that performance on the discrimination task was maximal when the IOI matched expectations established by the rhythm, but fell off as a function of the difference between the actual IOI and expectancy. As shown in Figure 5.4 (from Jones et al. 2002), this fits with the notion of attentional filters discussed above, implying that regularity in a sequence can be used to select a temporal filter that focuses attention at a specific time. In keeping with that interpretation, the width of the filter was at a minimum when tones in the sequence were presented with a regular rhythm, but grew wider for cases in which the context was less regular. Comparing these results with experiments discussed above, in which subjects listened for a tone at a cued frequency, one might say that just as the earlier study showed sensitivity to a change in level at a specific place in acoustic frequency, Jones et al. (2002) showed sensitivity to a change in frequency at a specific instant in time. These kinds of multifilter interactions, often with separate dimensions examined in tandem, possibly represent a significant part of analyses making up complex perception. A model for how we focus attention in time is proposed by Large and Jones (1999), who posit that the allocation of attention is controlled by a set of nonlinear internal oscillators that can entrain to events in the acoustic stream while tracking complex rhythms.

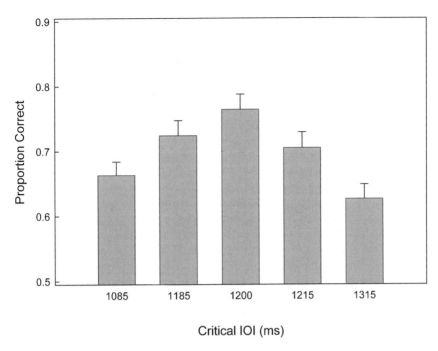

FIGURE 5.4. Results adapted from Jones et al. (2002) in which subjects compared the frequency of the first tone in a rhythmic sequence to the last. They suggest that stimulus regularity can act as a cue, focusing attention on the moment when it is most needed. In this way, the peaked function can be thought of as a kind of filtering in the time domain that diminishes processing before and after the expected time (see text).

## 4.3 Segregation into Multiple Auditory Streams

Most auditory communication relies on information carried in acoustic sequences. It is impossible to reference all of the important work on streaming by Bregman and his colleagues, but for a remarkable compendium of knowledge about streaming, how it works and how it interacts with other features of audition, the book by Bregman (1990) is highly recommended. From the perspective of auditory attention and selective filtering, one can argue that the pattern of acoustical features in a sequence establishes expectancies for higher-order structured relations such as those found in melodies.

A major approach to these issues has been through examination of stream segregation, whereby a sequence may be perceived as two streams that are essentially coexistent in time. An example from music in which a sequence is not parsed into separate streams is called hocketing. It occurs when a melodic line carried by interleaved sequences from different instruments or voices is heard as a single stream. However, the long history of polyphonic music shows that with a greater separation of the notes in a sequence, the percept can be one of two separate streams. Composers from Bach to Moby have utilized this to play separate melodies on alternating notes, even when both are produced by a single-voiced instrument such as the recorder. In the laboratory, one way to see whether a mixture of alternating sound sequences is segregated is simply to ask subjects whether they hear one stream or two. When the sequence is a simple alternation between tones, A and B, segregation depends on the physical difference between A and B as well as the speed of presentation, with tones that are more similar typically reported as a single stream (Bregman 1990). A popular approach, suggested by van Noorden (1975), is to present a sequence of alternating tones whose perceived rhythm is ambiguous, depending upon how the elements are grouped. An example of this paradigm is illustrated in Figure 5.5, where two frequencies, A and B, are presented in ABA triplets. When the frequency separation between alternating tones is small, or when the tempo is not too fast, the listener reports hearing a single stream that resembles the "galloping" rhythm of a horse. Conversely, when the separation is large (typically three semitones or more) (Bregman 1990) or the sequence is played at a brisk tempo, the percept changes to that of two simultaneously occurring rhythms corresponding to the separate A and B patterns in a way that has been likened to Morse code. Thus, Carlyon and his colleagues (e.g., Carlyon et al. 2003) have found it convenient to instruct subjects to refer to the two kinds of percepts with the terms "Horse" and "Morse." A key point in stream segregation

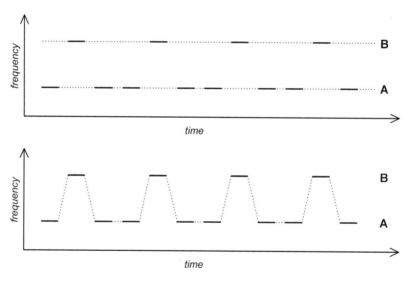

FIGURE 5.5. Schematic description of a common paradigm in which two auditory components alternate in triads. The ordinate plots frequency, and the abscissa, time. Two frequencies labeled A and B are represented by the darker bars (see text). Lighter lines in the two panels portray two different ways in which the stimuli might be perceptually grouped by the listener. The lower panel is heard as a galloping sound produced by grouping the individual triads; the upper panel is heard as two pulsing streams at the two frequencies.

is that while listeners can switch the focus of attention to either percept at will, they generally do not report hearing both at the same time.

A longstanding discussion about the role of attention in auditory grouping concerns whether it is preattentive or is representative of a top-down, more schematic analysis (Bregman 1990). An interesting feature of stream segregation is that it is not instantaneous, but rather builds up over time, sometimes not reaching a peak until many seconds after the onset of the sequence. On the grounds that segregation is a form of grouping, Carlyon and his associates have used the buildup to examine the role of attention when the subject hears the ambiguous stream while responding in a second, independent task. A study by Carlyon et al. (2001) presented 21 s of ABA repetitions, like those shown in Figure 5.5, to the left ear of a subject who used a pair of buttons to report whether the perception was of one stream ("Horse") or two streams ("Morse"). This generally took a few seconds. In a potentially competing task, the 21-s period also began with a presentation to subject's right ear of 10 s of 400-ms bursts of noise whose levels rose quickly to a base amplitude and then, more slowly, either increased or decreased. When the subject was told to ignore the noises, the switch from "Horse" to "Morse" was at about the same time as when sounds were in the left ear alone. However, when the instructions were to spend the first 10 s responding to noises in the right ear as either "approaching" or "withdrawing" and then switch to applying streaming instructions to the tones in the left ear, a buildup of segregation like that seen with no distraction began at the end of the tenth second. From this, the authors concluded that attending to the sequence of tones is important for streaming to build up.

In a later study, Carlyon et al. (2003) looked more deeply into the effects of different kinds of secondary tasks. Here, 13 s of ABA sequences were presented for streaming, but during the first 10 s, subjects were instructed to perform one of three additional tasks that were auditory, visual, or purely cognitive. For auditory distraction, subjects counted the number (3, 4, or 5) of tones in the sequence that had been chosen at random to be perturbed by addition of clearly audible 16-Hz amplitude modulation. Simultaneously, they saw a movie displaying a sequence of 125-ms ovals, most of which were filled by solid lines drawn on an angle from side to side, and for visual distraction, counted the number (3, 4, or 5) of ovals with short line segments. In the nonperceptual cognitive task, subjects were instructed to count backward from a random number presented before each the trial. At the end of the 10 s, the display told subjects to respond to the ABA sequence as "Horse" or "Morse." Results showed considerable stream segregation in all conditions, suggesting that segregation could, to some extent, take place while one was attending to another task. However, support for the idea of special demands required by perceptual attention to another sensory domain or to a competing cognitive task was seen in the fact that the number of "Morse" (stream-segregated) responses was least for the condition with mental arithmetic and most when the tokens to be monitored were superimposed on the auditory stream.

As discussed above, the importance of simple auditory dimensions such as frequency in stream segregation has led to discussion of whether it is a process that relies primarily on lower-level attentional or even preattentional processes. However, it is clear that segregation can also occur with streams defined by differences in emergent perceptual properties such as timbre. In this regard, studies of multidimensional scaling in timbre have identified two axes of sound quality defined by spectral and temporal envelope distributions. Using a streaming paradigm like the one described in Figure 5.6, Wessel (1979) showed streaming based on differences of timbre. For the creation of this "Wessel illusion," sequences were made up of repeated sets of three tones, shown in Figure 5.6, played with a constant intertone interval. If the timbres of the tones are identical, the listener hears repeated three-tone sequences that rise in pitch. However, if the alternate tones are set to one of two timbres, say flutes and trumpets, as described by solid and open symbols in the figure, and if the timbres are sufficiently different along a relevant feature such as the spectral centroid or attack time, the stream breaks into two timbrally defined melodies, or *Klangfarbenmelodien*. Now the percept is of two three-tone sequences, with one characterized by the falling sequence of solid symbols and one by the sequence of open symbols.

Another demonstration of higher-order attentional processes in stream segregation can be found in Dowling (1973), who interleaved familiar melodies with sequences of randomly chosen tones. He found the usual result that segregation was easier when the frequency range of the melody and interfering tones did not overlap. However, stream segregation was strengthened by cues that specified in advance the melody to be heard. Dowling et al. (1987) later showed that when the two streams overlapped in frequency, segregation was better if the "on-beat" stream (the one beginning the stimulus) was the target melody. However, when the streams did not overlap in frequency, it made no difference which stream came first.

It is, perhaps, stretching things to say that cueing the positions along an auditory dimension for detection or identification of a weak signal is the same as preparing the listener to respond to qualitative features of clearly audible tones and streams, but a common thread that leads us to think in terms of an umbrella of auditory attention is that detection, identification, and interpretation are all affected by expectancies of the signal to come, be they from preceding stimuli or from long-term memory. Thought of in this way, streaming suggests

FIGURE 5.6. The "Wessel illusion" (Wessel 1979) in which triplets are presented that when of the same timbre, sound like repeated three-tone melodies that rise in pitch. When alternate tones are set to different timbres as described by the solid and open symbols, and the spectral centroids of the two sounds are sufficiently different, the percept is of two three-tone melodies that fall in pitch (see text).

that features in an ongoing auditory sequence can cue the listener to expect what is next in a way that allows the successive stimulus to be accepted into the stream. This is especially interesting with simultaneous or interleaved sequences, for it describes a role for focused attention in separating auditory objects or events on the basis of shared commonalities. Clearly, Wessel's (1979) use of timbre as the dimension on which the streaming illusion is based goes beyond the suggestion (Hartmann and Johnson 1991) that stream segregation is based simply on channeling in the auditory periphery through differences between tones in fundamental dimensions such as frequency, spatial separation, and duration. In this regard, Moore and Gockel (2002) have examined evidence that suggests that any "sufficiently salient perceptual difference may lead to stream segregation."

## 5. Reflexive Attraction of Attention

To this point we have discussed informational cues that tell the listener about what or where a signal may be. A distinction has been made between these and another kind of cue, such as an unexpected shout, that seems to pull attention to a place in space, regardless of its importance.

### 5.1 Comparisons Between Endogenous and Exogenous Cueing

Posner (1980) posited a distinction between two kinds of attentional orienting mechanisms, endogenous and exogenous. Up to this point, we have discussed primarily the former. Endogenous cue essentially "push" (Jonides 1981) attention to a place in an auditory dimension where a signal is likely to be, thus telling the subject where to listen. As implied by the word endogenous, the subject must extract relevant information in the cue and make the connection to the expected location of the signal. An endogenous cue need not be identical in form or place to the signal, the requirement being only that it supply useful information that can be deciphered by the listener. That is demonstrated in the case discussed above in which a successful cue tone was related to the signal by a known frequency distance, the musical fifth. For an endogenous cue to be meaningful, the location to which it points must be significantly correlated with the actual position of the signal. Conversely, exogenous cues carry no information about where the signal will be. Rather, they "pull" (Jonides 1981) attention to a location in a reflexive manner without regard for where the signal will actually appear. Such cues can affect performance, positively if they happen to match the signal and negatively if they do not. When a cue of either type correctly defines the location of a signal, it is called "valid"; when it does not, it is called "invalid."

The most common experimental demonstration of the difference between these two kinds of cues is through measurement of reaction times (RTs). They are easily described in the classic visual spatial-cueing paradigm developed by

Posner (1980), whereby the subject fixates on a point between potential signal locations placed, say, on either side of fixation, and the instruction is to respond as quickly as possible to a signal presented at either of these locations. Since an endogenous cue must, by definition, match the signal with a probability greater than chance, subjects who use the cue will show a faster RT on valid trials and a slower RT on invalid trials. In this situation, purely endogenous cues might be arrows at the point of fixation pointing to the left or right, or a written script on the screen with the words "left" or "right," or even an auditory cue with the subject trained to expect a signal on the left after a high pitch and on the right after a low pitch. Exogenous cueing needs no training, acting as it does at a more primitive level in much the way that an unexpected shout might draw one's attention without consideration for what was said. In the visual task described above, a successful exogenous cue might be a presignal flash presented at random at one of the two potential signal locations. Although totally unrelated to where the signal will be, a flash that happens, by chance, to be valid will speed RT, and one that is invalid will slow it. Because an endogenous cue relies on the use of information, it is thought to elicit top-down control, while the more primitive response to an exogenous cue is thought to be based on bottom-up processes. The time between the cue and signal, often called the stimulus-onset asynchrony (SOA), is critical for RT, in that the reflexive effect of an exogenous cue is gone after a relatively short SOA, i.e., generally less than a second, while the effect of an endogenous cue can last much longer, covering SOAs on the order of seconds.

In the auditory modality, the first conclusive evidence of a difference between endogenous and exogenous cueing was observed by Spence and Driver (1994), who used RT to measure auditory discrimination in a spatial cueing task. For these experiments, several speakers were arranged spatially around the listener. Each trial consisted of two sounds—a cue and a target—separated by a variable SOA. In the endogenous condition, subjects were told that the location of the cue predicted the location of the target in 75% of the trials, while in the exogenous condition, subjects were told to ignore the cue because if offered no information about the target. The cue was a pure tone played from one of the speakers; the target, a tone or burst of noise. The task was either to identify the spatial location of the target (front/back or up/down) or, with tonal signals, to identify its pitch (high/low). Effective cueing was defined as RTs that were faster with valid cues than with invalid cues. Effects with exogenous cues were small, short-lived, and observed only for spatial localization, while those with endogenous cues were larger, persisted over longer SOAs and occurred both in frequency discrimination and localization. In subsequent work (Spence and Driver 1998, 2000), in which the focus was on visual and tactile as well as auditory cues, valid exogenous cues again produced faster RTs at short SOAs, unlike endogenous cues, for which the RT advantage lasted over longer delays. A difference, however, was that the effects of both kinds of cueing were observed in discriminations of differences in frequency as well as in spatial location.

Differential roles for endogenous and exogenous cueing in detection have been discussed by Green and McKeown (2001). They note that the use of the probe-signal method to measure the shapes of listening bands probably reflects elements of both kinds of cueing. The idea is that the cue is endogenous in the sense that it predicts where, in frequency, the majority of signals will be. It is exogenous because of a possible reflexive pull of attention to the frequency of the cue, regardless of its predictability. Arguing that filters should be wider without the exogenous component, they note that data from Hafter et al. (1993), where bandwidths found when the frequency of the expected signal was a musical fifth above that of the cue, were wider than those seen when the cue and signal had the same frequency.

Another difference between endogenous and exogenous cueing in masking is also seen in Johnson and Hafter 1980, where the two are put into opposition. In a yes/no detection task, the signal on any trial could be one of two widely spaced frequencies (500 and 1200 Hz). Tonal cues were presented in alternation, with 500 Hz on odd trials and a 1200-Hz cue on even trials. Given that endogenous cueing acts to reduce inaccuracy in the filters to be monitored, one would expect these cues to provide cumulative information over the course of a session, constantly reminding the subject about where to listen. Not surprisingly, on valid trials, this accumulated knowledge plus any exogenous effects of the cues produces a 1.5-dB increase in gain relative to an uncued control condition. Conversely, on invalid trials, one might expect the endogenous benefit, but it would be in conflict with the reflexive pull of the exogenous cue to the wrong filter. This is seen in a loss in performance relative to the control conditions of 1.0 dB.

## 5.2 Inhibition of Return in Exogenous Cueing

Mondor and Lacey (2001) compared the effects of exogenous auditory cueing on RT in tasks based on discrimination of auditory properties of targets other than their spatial direction. In three separately tested conditions, the subject was asked to make a rapid discrimination between two stimulus values in one of three dimensions: level, frequency, or timbre. Cues in each condition presented one of the same two values used as targets, but this was done randomly in such a way that the cue matched the target in one-half of the trials and not in the other half. Thus, the cues were exogenous, offering no information about the correct response. Targets followed cues by SOAs of 150, 450, or 750 ms. All sounds came from a single speaker, so stimulus direction was not a factor. Of interest was the difference in RT between invalid and valid cues, where exogenous cueing generally works only for short SOAs. The RT difference was positive, with an SOA of 150 ms, indicating of a reflexive pull to the appropriate place in the stimulus dimension being tested. The RT difference was zero when the SOA was 450 ms, indicative of no effect of cueing, and was negative for the SOA of 750 ms. The latter effect, in which responses were actually faster to an invalid cue, has been likened to a kind of late rebound. Generally called inhibition of return (IOR) (Posner and Cohen 1984), it is thought to represent an internal inhibition of responses to the cued location.

In summary, comparisons between endogenous and exogenous cueing in audition show similar results to those found in vision. With endogenous cueing, valid (informative) cues can speed responses for SOAs of several seconds, while invalid (counterinformative) cues tend to slow them down. With exogenous cueing, results are more temporally dependent; valid cues produce faster RTs for short SOAs but slower RTs with longer SOAs, as observed in IOR.

## 6. Auditory Attention and Cognitive Neuroscience

The surging interest in cognitive neuroscience has led to numerous methods for observing neural activity while the brain is engaged in auditory attention. Two of the most common neuroimaging techniques are functional magnetic resonance imaging (fMRI) and positron-emission tomography (PET). Measures of fMRI estimate neural activity through measures of the blood oxygenation level dependent response (BOLD), which represents a coupling between hemodynamic and neural activity. Neural activity in PET relies on radioactive tracers taken up by the most active neurons. In an example of the use of fMRI for the study of auditory streaming, Janata et al. (2002) used polyphonic music in which the subject listened to a duet whose melodies differed in their respective timbres. In one condition, a subject was asked to attend to and track one of the melodies. In separate comparison tasks, subjects either listened to the duets passively or rested in silence. Results showed that active listening produced a greater BOLD response in the superior temporal gyrus (STG) and some frontal and parietal areas including precentral gyrus, supplementary and presupplementary motor areas (SMA), and intraparietal sulcus (IPS). Because the STG is primarily involved in auditory processing (e.g., Zatorre et al. 2002), while the IPS, precentral gyri, SMA, and pre-SMA are implicated in more general working-memory and attentional tasks, results from this experiment are in agreement with other studies using fMRI (Petkov et al. 2004) and PET (e.g., Zatorre et al. 1999) in implicating a frontoparietal network that couples with domain-specific sensory cortices during sustained attention to auditory and musical stimuli.

Perhaps the most developed approach to the study of human neuroscience and auditory attention has been in the domain of electrophysiology using event-related potentials (ERPs), a technique that allows detailed study of the time course of attention. Several components of the ERP have been specifically linked to auditory attention. The earliest known sound-evoked potential is the auditory brainstem response, or ABR, which is a complex of seven individual waveform components starting at around 10 ms after the onset of a brief sound. The ABR complex is thought to be independent of attentional modulation, and is therefore described by Hackley (1993) and Woldorff et al. (1998) as being strongly automatic, differentiating neural processes in the brainstem from the brain's attentional network. The first known cortical ERP shown to be related to hearing is the N1, a negative waveform that is largest at around 100 ms after the onset of any auditory stimulus. Hillyard et al. (1973) found the N1 component to be sensitive to modulation by attention in a study using dichotic listening, where a rapid stream of tone pips was sent randomly to each ear. Listeners were required to perform an attention-demanding fine pitch discrimination task on tones coming into one of their ears. Comparisons between ERPs evoked by pips to the left and right ears showed a larger N1 for stimuli in the attended ear than for those in the unattended ear. Thus, the neural generators of the N1 component were said to be partially automatic, because while sensory stimulation without attention was sufficient for its elicitation, it was sensitive to the enhancement or modulation of attention. Woldorff and Hillyard (1991) later localized the N1 to the primary auditory cortex through the use of magnetoencephalography (MEG).

Another interesting component of the ERP that has been implicated in attention is called the mismatch negativity (MMN). It is elicited in the so-called oddball paradigm, whereby a standard stimulus is presented repeatedly with high probability, but an occasional deviant stimulus is introduced at random locations in the stimulus stream. When comparing ERPs evoked by the standard and deviant stimuli, the MMN is a negative waveform in the deviant response at about 150–200 ms poststimulus. Because this occurs in all senses that have been tested and reflects any stimulus that differs from its surrounding context, Näätänen (1988) claimed that the MMN is independent of attention. However, in a subsequent study using the dichotic oddball paradigm with stimuli presented at fast rates, Woldorff and Hillyard (1991) found that the MMN was larger on the attended side than on the unattended side, suggesting that while mismatch detection, like sensory processing, is partially automatic in its elicitation, it can be modulated by attention. Using MEG, Woldorff et al. (1998) also found attention modulation of a magnetic analogue of the MMN localized in the primary auditory cortex. Based on these findings, Hackley (1993) and Woldorff and Hillyard (1991) proposed that, unlike the case with other evoked potentials such as the auditory brainstem reflexes, the N1 and MMN in early sensory cortices are "partially automatic," such that attention is not required for cortical activity but can strongly enhance or modulate cortical processing. This is supported by Sussman et al. (1998, 1999), who linked the MMN to auditory attention and auditory streaming through tasks in which subjects heard a sequence of tones that alternated between high and low frequencies. If segregated, each of the pitch streams was organized into its own simple melody. On a small number of trials, the order of tones was changed in the low-frequency range in order to test the hypothesis that an MMN would appear only if the two melodies were segregated. As expected, this was true when the tempo was fast, both when the subjects were instructed to attend to the tones and when they were asked to perform the secondary task of reading a book. However, when the tempo was slow, in which case typically there was less streaming, an MMN was seen only when the subjects were instructed to attend to the auditory stimuli.

## 7. Summary

There are many manifestations of top-down control of sensory processing that fall under the great heading of attention. Internal states, emotions, social constraints, deep knowledge, all can affect what we hear and how we interpret it. In this chapter we have concentrated on one specific role of attention, the ability to extract signals from a background. From this perspective, we have looked at some of the putative internal filters that constrain the processing of a signal, especially when there is subjective uncertainty about where the signal may fall along a relevant dimension. This has been done through examination of informational cues that tell the listener where to listen, be it in frequency, at a higher level of processing such as complex pitch, at a specific moment in time marked by temporal rhythms, or at places in pitch prescribed by musical melodies. In addition, we have discussed cases in which the relation between sequential sounds can affect whether the sequence is heard as perceptual whole or divided into separate streams. We have avoided the excellent literature on attention with speech and speechlike sounds as well as "informational masking," because those topics appear elsewhere in this volume. Also absent is mention of a proposed mechanism for attention, though as briefly noted in the final section, there is a rapidly growing movement in human neuroscience to look for the neural networks that produce results seen in the behavior. From our perspective, the proposal is that the kinds of psychoacoustic measures discussed here are of value for understanding the role of attention in the perception of more-complex arrays of sound in the natural world.

*Acknowledgments.* Great thanks are in order for Anne-Marie Bonnel, whose many valuable ideas added immeasurably to this chapter. Also, we thank Brian Moore, whose tough and insightful comments of an earlier draft are greatly appreciated.

## References

Bharucha JJ, Stoeckig K (1986) Reaction-time and musical expectancy—priming of chords. J Exp Psychol [Hum Percept Perform] 12:403–410.

Bonnel AM, and Hafter ER (1998) Divided attention between simultaneous auditory and visual signals. Perception and Psychophysics 60(2), 179–190.

Bregman AS (1990) Auditory Scene Analysis: The Perceptual Organization of Sound. Cambridge, MA: Bradford Books, MIT Press.

Broadbent DE (1958) Perception and Communication. New York: Pergamon Press.

Carlyon RP, Cusack R, Foxton JM, Robertson IH (2001) Effects of attention and unilateral neglect on auditory stream segregation. J Exp Psychol [Hum Percept Perform] 27: 115–127.

Carlyon RP, Plack CJ, Fantini DA, Cusack R (2003) Cross-modal and non-sensory influences on auditory streaming. Perception 32:1393–1402.

Cherry EC (1953) Some experiments on the recognition of speech with one and with two ears. J Acoust Soc Am 24:975–979.

Dai H, Wright BA (1995) Detecting signals of unexpected or uncertain durations. J Acoust Soc Am 98:798–806.

Dai HP, Scharf B, Buus S (1991) Effective attenuation of signals in noise under focused attention. J Acoust Soc Am 89:2837–2842.

Deutsch JA, Deutsch D (1963) Attention: Some theoretical considerations. Psychol Rev 70:80–90.

Dowling WJ (1973) The perception of interleaved melodies. Cogn Psych 5:322–337.

Dowling WJ, Lung KM-T, Herrbold S (1987) Aiming attention in pitch and time in the perception of interleaved melodies. Percept Psychophys 41:642–656.

Egan JP, Greenberg GZ, Schulman AI (1961) Intervals of time uncertainty in auditory detection. J Acoust Soc Am 33:771–778.

Fletcher H (1940) Auditory patterns. Rev Mod Phys 12:47–65.

Green DM (1960) Psychoacoustics and detection theory. J Acoust Soc Am 32:1189–1203.

Green DM (1961) Detection of auditory sinusoids of uncertainty frequency. J Acoust Soc Am 33:897–903.

Green DM, Swets JA (1966) Signal Detection Theory and Psychophysics. New York: John Wiley & Sons.

Green TJ, McKeown JD (2001) Capture of attention in selective frequency listening. J Exp Psychol [Hum Percept Perform] 27:1197–1210.

Greenberg GZ, Larkin WD (1968) Frequency-response characteristic of auditory observers detecting signals of a single frequency in noise: The probe-signal method. J Acoust Soc Am 44:1513–1523.

Greenwood DD (1961) Critical bandwidth and the frequency coordinates of the basilar membrane. J Acoust Soc Am 33:1344–1356.

Gundy RF (1961) Auditory detection of an unspecified signal. J Acoust Soc Am 33: 1008–1012.

Hackley SA (1993) An evaluation of the automaticity of sensory processing using event-related potentials and brainstem reflexes. Psychophysiology 30:415–428.

Hafter ER, Kaplan R (1976) The interaction between motivation and uncertainty as a factor in detection. NASA project report, Ames Research Center, Moffit Field, CA.

Hafter ER, Saberi K (2001) A level of stimulus representation model for auditory detection and attention. J Acoust Soc Am 110:1489–1497.

Hafter ER, Schlauch RS (1989) Factors in detection under uncertainty. J Acoust Soc Am 86:S112.

Hafter ER, Schlauch RS, Tang J (1993) Attending to auditory filters that were not stimulated directly. J Acoust Soc Am 94:743–747.

Hartmann WM (1997) Signals, Sounds, and Sensation. New York: AIP Press.

Hartmann WM, Johnson D (1991) Stream segregation and peripheral channeling. Music Percept 9:155–184.

Hillyard SA, Hink RF, Schwent VL, Picton TW (1973) Electrical signs of selective attention in the human brain. Science 182:177–80.

Houtsma AJM (2004) Hawkins and Stevens revisited with insert earphones (L). J Acoust Soc Am 115:967–970.

Ison JR, Virag TM, Allen PD, Hammond GR (2002) The attention filter for tones in noise has the same shape and effective bandwidth in the elderly as it has in young listeners. J Acoust Soc Am 112:238–246.

Janata P, Tillmann B, Bharucha JJ (2002) Listening to polyphonic music recruits domain-general attention and working memory circuits. Cogn Affect Behav Neurosci 2: 121–140.

Jeffress LA (1964) Stimulus-oriented approach to detection. J Acoust Soc Am 36: 766–774.

Jeffress LA (1970) Masking. In: Tobias J (ed) Foundations of Modern Auditory Theory 1. New York: Academic Press, pp. 87–114.

Jeffress LA, Blodgett HC, Sandel TT, Wood CL III (1956) Masking of tonal signals. J Acoust Soc Am 3:416–426.

Johnson DM, Hafter ER (1980) Uncertain-frequency detection——Cueing and condition of observation. Percept Psychophys 28:143–149.

Jones MR, Boltz M (1989) Dynamic attending and responses to time. Psychol Rev 96: 459–491.

Jones MR, Moynihan H, MacKenzie N, Puente J (2002) Temporal aspects of stimulus-driven attending in dynamic arrays. Psychol Sci 13:313–319.

Jonides J (1981) Voluntary versus automatic control over the mind's eye. In: Long J, Baddeley AD (eds) Attention and Performance IX. Hillsdale, NJ: Lawrence Erlbaum.

Kahneman D (1973) Attention and Effort. Upper Saddle River, NJ: Prentice-Hall.

Krumhansl CL (1990) Tonal hierarchies and rare intervals in music cognition. Music Percept 7:309–324.

Krumhansl CL, Kessler EJ (1982) Tracing the dynamic changes in perceived tonal organization in a spatial representation of musical keys. Psychol Rev 89:334–368.

Large EW, Jones MR (1999) The dynamics of attending: How people track time-varying events. Psychol Rev 106:119–159.

Leshowitz B, Wightman FL (1972) On the importance of considering the signal's frequency spectrum: Some comments on Macmillan's Detection and recognition of increments and decrements in auditory intensity experiment. Percept Psychophys 12:209–212.

Macmillan NA (1971) Detection and recognition of increments and decrements in auditory intensity. Perception and Psychophysics, 10:233–238.

Meyer L (1956) Emotion and Meaning in Music. Chicago: University of Chicago Press.

Mondor TA and Lacey TE (2001) Facilitative and inhibitory effects of cuing, sound duration, intensity, and timbre. Percept Psychophys 63:726–736.

Moore BCJ (2003) An Introduction to the Psychology of Hearing, 5th ed. San Diego: Academic Press.

Moore BCJ, Glasberg BR (1983) Suggested formulas for calculating auditory-filter bandwidths and excitation patterns. J Acoust Soc Am 74:750–753.

Moore BCJ, Gockel H (2002) Factors influencing sequential stream segregation. Acta Acust 88:320–333.

Moray N (1960) Broadbent's filter theory—postulate H and the problem of switching time. Q J Exp Psychol 12:214–220.

Moray N (1970) Attention: Selective Processes in Vision and Hearing. New York: Academic Press,.

Näätänen R. 1988. Implications of ERP data for psychological theories of attention. Biological Psychology, 26(1–3):117–63

Norman DA (1969) Memory while shadowing. Q J Exp Psychol 21:85–93.

Norman DA, Bobrow DG (1975) On data-limited and resource-limited processes. Cogn Psych 7:44–64.

Patterson RD, Moore BCJ (1986) Auditory filters and excitation patterns as representations of frequency resolution. In: Moore BCJ (ed) Frequency Selectivity in Hearing. New York: Academic Press, pp. 123–177.

Patterson RD, Nimmo-Smith I (1980) Off-frequency listening and auditory-filter asymmetry. J Acoust Soc Am 67:229–245.

Penner MJ (1972) Effects of payoffs and cue tones on detection of sinusoids of uncertain frequency. Percept Psychophys 11:198–202.

Petkov CI, Kang X, Alho K, Bertrand O, Yund EW, and Woods DL (2004) Attentional modulation of human auditory cortex. Nat Neurosci 7:658–663.

Plamondon L, Hafter ER (1990) Selective attention in absolute pitch listeners. J Acoust Soc Am Suppl 1 88:S49.

Posner MI (1980) Orienting of attention. Q J Exp Psychol 32:3–25.

Posner MI, Cohen Y (1984) Components of visual orienting. In: Bouma H, Bouwhuis DG (eds) Attention & Performance X. Cambridge, MA: MIT Press, pp. 531–555.

Scharf B (1970) Critical Bands. In: Tobias J (ed) Foundations of Modern Auditory Theory 1. New York: Academic Press, pp. 159–202.

Scharf B, Quigly S, Aoki C, Peachy N, Reeves A (1987) Focused auditory attention and frequency selectivity. Percept Psychophys 42:215–223.

Scharf B, Magnan J, Chays A (1997) On the role of the olivocochlear bundle in hearing: 16 case studies. Hear Res 103:101–122.

Schlauch RS, Hafter ER (1991) Listening bandwidths and frequency uncertainty in pure-tone signal detection. J Acoust Soc Am 90:1332–1339.

Spence CJ, Driver J (1994) Covert spatial orienting in audition: Exogenous and endogenous mechanisms. J Exp Psychol [Hum Percept Perform] 20:555–574.

Spence C, Driver J (1998) Auditory and audiovisual inhibition of return. Percept Psychophys 60:125–139.

Spence C, Driver J (2000) Attracting attention to the illusory location of a sound: Reflexive crossmodal orienting and ventriloquism. NeuroReport 11:2057–2061.

Sussman E, Ritter W, Vaughan HG Jr (1998) Attention affects the organization of auditory input associated with the mismatch negativity system. Brain Res 789:130–138

Sussman E, Ritter W, Vaughan HG Jr (1999) An investigation of the auditory streaming effect using event-related potentials. Psychophysiology 36:22–34

Swets JA (1984) Mathematical models of attention. In: Parasuraman R, Davies DR (eds) Varieties of Attention. London: Academic Press, pp. 183–242.

Taylor MM, Forbes SM (1969) Monaural detection with contralateral cue (MDCC). I. Better than energy detector performance by human observers. J Acoust Soc Am 46:1519–1526.

Treisman AM (1964) Monitoring and storage of irrelevant messages in selective attention. J Verb Learn Verb Behav 3:449–459.

Treisman AM (1969) Strategies and models of selective attention. Psychol Rev 76:282–299.

Treisman AM, Gelade G (1980) Feature-integration theory of attention. Cogn Psycho 12:97–136.

van Noorden LPAS (1975) Temporal coherence in the perception of tone sequences. PhD thesis, Eindhoven University of Technology.

Wessel DL (1979) Timbre space as a musical control structure. Comp Music J 3:45–52.

Woldorff MG, Hillyard SA (1991) Modulation of early auditory processing during selective listening to rapidly presented tones. Electroencephalogr Clin Neurophysiol 79:170–191.

Woldorff MG, Hillyard SA, Gallen CC, Hampson SR, Bloom FE (1998) Magnetoencephalographic recordings demonstrate attentional modulation of mismatch-related neural activity in human auditory cortex. Psychophysiology 35:283– 292.

Wright BA, Dai H (1994) Detection of unexpected tones with short and long durations. J Acoust Soc Am 95:931–938.

Yost WA, Penner MJ, Feth LL (1972) Signal detection as a function of contralateral signal-to-noise ratio. J Acoust Soc Am 51:1966–1970.

Zatorre RJ, Mondor TA, Evans AC (1999) Auditory attention to space and frequency activates similar cerebral systems. NeuroImage 10:544–554.

Zatorre RJ, Belin P, Penhune VB (2002) Structure and function of auditory cortex: Music and speech. Trends Cogn Sci 6:37–46.

# 6
# Informational Masking

GERALD KIDD, JR., CHRISTINE R. MASON, VIRGINIA M. RICHARDS,
FREDERICK J. GALLUN, AND NATHANIEL I. DURLACH

## 1. Introduction

The problem of perceiving the sounds emanating from a particular sound source becomes much more difficult when sounds from other independent sources occur at the same time. These unwanted or "masking" sounds compete with the desired or "target" sound at a variety of levels within the auditory system. The study of masking has a long history in the auditory literature, a portion of which is reviewed below. What is emphasized here is the psychoacoustic approach to the study of masking which typically applies rigorous empirical methods in an attempt to quantify the degree of interference that results from the competition among sources.

This chapter reviews the rapidly growing literature concerning *informational* masking. The term "informational masking," and the complementary term "energetic masking" with which it is usually associated and contrasted, first appear in the auditory literature (to our knowledge) in an oft-cited abstract of a presentation by Irwin Pollack at the spring meeting of the Acoustical Society of America in 1975 (Pollack 1975). Informational masking subsequently attracted considerable attention through the work of Charles Watson and his colleagues using a novel experimental technique in which the discriminability of an alteration in some aspect of an element of a sequence of tones was measured as a function of uncertainty. Very large differences in the ability of listeners to discern changes in the target element (e.g., an alteration in frequency or intensity) were observed as the degree of uncertainty of the surrounding context tones was manipulated. The series of articles by Watson and colleagues has recently been reviewed (Watson 2005), and the reader is referred to that summary (see also Watson and Kelly 1981; Watson 1987) for an in-depth discussion of that work. Most of the literature considered in this chapter dates from an article by Donna Neff and David Green (1987) in which uncertainty was created by the simultaneous presentation of a set of random-frequency masker tones. The procedure they used will be referred to as the "multitone masking experiment." However, as will become apparent, contemporary studies of informational masking span a wide range of experimental stimuli and tasks. As should also be obvious from this chapter, informational masking is not a single phenomenon but rather may be the result of the actions of any of several stages of processing beyond the auditory periphery and is intimately connected to perceptual grouping and source segregation, attention, memory, and general cognitive processing abilities.

The chapter is organized in the following manner: The first section (after this introduction) attempts to provide historical context for the concepts of energetic and informational masking. In subsequent sections, findings from the multitone masking experiment are reviewed and many of the factors that influence the results of that experiment are discussed. Included is a section describing quantitative approaches to accounting for the basic results from these studies. The multitone masking experiment is emphasized because there are more data available using that procedure than other procedures, there have been attempts to model the results, and it is useful for illustrating the influences of such factors as a priori knowledge and perceptual grouping and segregation that affect a wide range of tasks. Subsequent sections cover topics including informational masking in both nonspeech and speech discrimination, spatial factors, consideration of age-related effects, and informational masking in listeners with hearing loss.

## 2. Definitions and Historical Perspectives on Masking

Pollack (1975) is credited with coining the term "informational masking." However, what is often overlooked is that in that same abstract, he also apparently was the first to use the term "energetic masking." While there has been considerable discussion recently about how best to define informational masking, there has hardly been any corresponding discussion of how to define energetic masking. Operationally, many auditory researchers define informational masking as masking that occurs beyond that which can be attributed to energetic masking (always assuming that the observer is generally attentive and motivated to solve the task, is fully informed about the experiment, etc.). Although appealingly simple and unifying across a wide range of tasks, such a definition says more about what the phenomenon is not than what it is and does not go very far toward providing a deep understanding. However, energetic and informational masking remain linked in the literature, and an important first question to ask is, what exactly is meant by energetic masking and how well is it understood?

"Masking" is one of the fundamental concepts of hearing. The modern study of masking is often considered to have originated in a series of experiments by researchers from Bell Labs, including the seminal studies of Wegel and Lane (1924) and summary by Fletcher (1929). The idea that masking could occur at different physiological sites was considered even in these early studies. For example, Wegel and Lane (1924) state that

These [results] may be explained by assuming that there are two kinds of masking, central and peripheral, the former being generally relatively small and resulting from the conflict of sensations in the brain, and the latter originating from overlapping of stimuli in the end organ. Central masking is probably always present to a certain extent, whereas peripheral masking can only occur when the two tones excite the same region on the basilar membrane. All large amounts of masking may be attributed to peripheral masking... (p. 273).

Energetic masking usually means peripheral masking, although Durlach et al. (2003a) have suggested that energetic masking could occur at higher physiological sites even if it was not present or dominant in the periphery. The essence of this idea is that the auditory neurons at a particular physiological site are so engaged by one stimulus that they cannot adequately represent another stimulus competing for the same neurons. Although it is not clear from the literature, it is reasonable to speculate that the term "energetic masking" was meant to refer to masking predicted by the critical-band energy-detector model that was widely accepted and very influential at the time of Pollack's abstract. A thorough explanation of that model may be found in Green and Swets 1974, Chapter 8. Essentially, the model consists of a "critical band" filter (Fletcher 1940) followed by a physiologically inspired rectifier, an integrator (based on psychophysical estimates of temporal integration), and a mechanism that based decisions on stimulus energy. Some variability in the process was assumed always to be present due to "internal noise" independent of the properties of the acoustic input. This model was used to successfully account for a number of studies of tone-in-noise detection. Later work, however, revealed several conditions in which such a single-channel energy model could not account for key masking results (e.g., Pfafflin and Mathews 1966). For example, equating the masker and target (or "signal"; these terms will be used interchangeably) plus masker energy (e.g., Hall and Grose 1988) or introducing a random rove in level (Gilkey 1987; Kidd et al. 1989) defeats such a model, yet human performance is not greatly affected. These findings have led to the view that decision variables based on broadband spectral shape (e.g., Green 1983, 1988; Durlach et al. 1986) or within-band temporal fluctuations (e.g., Richards 1992; Kidd et al. 1993; Viemeister and Plack 1993; Dau et al. 1996) appear more plausible.

Nonetheless, it is obvious that energetic masking may dominate certain listening situations. As mentioned above, one way of defining informational masking uses the amount of energetic masking as a reference. Thus, at the very least it would be helpful in attempting to quantify informational masking if there were a precise model of energetic masking that could accurately predict performance for a wide range of stimuli or a measurement procedure in which one could be certain that only energetic masking was present. To date, most models proposed to account for energetic masking are psychoacoustic models (e.g., Fletcher 1940; Bos and de Boer 1966; Zwicker 1970; Patterson et al. 1982; Glasberg and Moore 1990) based on observations of masking in various conditions, such as for the detection of a tone in filtered Gaussian noise. There have been some efforts to characterize energetic masking by the application of an Ideal Observer analysis (e.g., Heinz 2000; Heinz et al. 2002) at a defined physiological level (e.g., the auditory nerve). As suggested by Durlach et al. (2003a), a thorough understanding of energetic masking would be based on such an analysis at various physiological sites within the auditory system. Unfortunately, we currently do not have models or measurement techniques that are adequate to determine the necessary quantities with sufficient precision for the types of sounds used in informational masking experiments. One difficulty lies in the several types of possible physiological codes (e.g., rate, synchrony, across-frequency coincidence detection, envelope fluctuations) that could be used as decision variables in detection, discrimination, or identification tasks (e.g., Erell 1988; Delgutte 1990, 1996; Carney et al. 2002; Colburn et al. 2003). Furthermore, as discussed more fully below, there are some indications that nonenergetic masking (whatever label is attached to it) is present even in conditions under which uncertainty is minimal, thus complicating efforts to separate energetic and informational factors. If energetic masking cannot be well described empirically or through modeling, then defining informational masking in reference to energetic masking is problematic.

In a chapter of Stevens's *Handbook of Experimental Psychology*, Licklider (1951) offered the following definition of masking: "Masking is thus the opposite of analysis; it represents the inability of the auditory mechanisms to separate the tonal stimulation into components and to discriminate between the presence and the absence of one of them. The degree to which one component of a sound is masked by the remainder of the sound is usually determined by measuring two thresholds" (p. 1005). This definition appears to have formed the basis for the ANSI standard (ANSI 1994).[1] A few years after the publication of Stevens's *Handbook*, Tanner (1958) wrote an article succinctly titled "What is masking?" in which he gave examples of phenomena causing a difference between detection thresholds (and thus presumably falling under the definition proposed by Licklider) but questioned whether the term "masking" was an appropriate descriptor. The primary problem that appears to have motivated Tanner's interest in the definition of masking was that of the effect of uncertainty regarding the frequency of a target on detectability. During the mid-1950s through the early 1960s, the Electronic Defense Group at the University of Michigan was engaged in applying the Theory of Signal Detectability to psychophysical research. The task of detecting a signal of uncertain frequency was an early problem examined by several investigators from that group (e.g., Tanner 1958 citing earlier work; Veniar 1958a,b; Creelman 1960; Green 1961). Although the magnitude of the effect varied with the specific condition tested (e.g., range of possible frequencies), the gist of the empirical findings was that uncertainty about the frequency of a target tone presented in noise on a trial-by-trial basis elevated detection thresholds a maximum of about 3 dB relative to the case in which

---

[1]The word "masking" has, historically, been combined with a variety of modifiers: peripheral, central, backward, forward, perceptual, recognition, remote, contralateral remote, etc. Although these various combined forms—including energetic and informational masking—often reflect the actions of overlapping or even entirely separate mechanisms, they all have in common the basic idea that one sound interferes with the reception or processing of another sound.

the same target(s) were held constant in frequency across trials (Green 1961). This small elevation in threshold was interesting in its own right and led to consideration of a number of models of processing based on a single tunable filter or simultaneous monitoring of multiple filters (e.g., Creelman 1960; Green 1961; Buus et al. 1986). Tanner's (1958) primary concern, however, was whether the elevation in threshold due to uncertainty ought to be called "masking." Although Tanner did not use the term energetic masking, his description of the expected performance in the known-frequency tone-in-noise experiment was consistent with what is now termed energetic masking. What to call masking that was *not* energetic masking—caused by "distraction," "memory loss," and "other factors"—seems to be fundamental to the problem he raised.

Extending Licklider's (1951) characterization of masking as a failure of analysis, Green and Swets (1974, p. 276) state,

frequency analysis is evident in the ability to distinguish among different parts of a complex sound stimulus, and in the ability to ignore unwanted, interfering sounds and hence to listen to only certain aspects of the total stimulus configuration. What are still unknown are the specifics of this process—what is the extent of this ability to ignore unwanted sounds, and how good is this ability in a variety of stimulus situations?

Here the terms "complex sound stimulus" and "total stimulus configuration" seem to imply multiple sounds, although the former could perhaps still be taken to mean—similar to Licklider—hearing out a partial in a single complex sound. However, clearly, the authors are addressing the case in which there are multiple audible sounds and the listener must choose among them. This obviously implies selectivity at a higher level than peripheral filtering: Sounds or portions of sounds are audible, and the listener chooses to attend to or to ignore the sounds.

To summarize, then, the early theoretical view of "masking" was dominated by consideration of detecting tones masked by noise or by other tones, and overlapping patterns of excitation in the cochlea were generally regarded as the physiological basis for masking. However, it was also acknowledged that other factors—clearly not related to overlapping patterns of excitation —could influence the amount of masking observed even in fairly simple experiments. Although acknowledged, it also seems that there was some discomfort among auditory scientists about how such influences should be described both on practical and theoretical grounds.

There were also other, seemingly unrelated, indications of the distinction between peripheral and central masking from very different types of studies. In the 1960s, Carhart and colleagues were engaged in a series of speech recognition experiments, many of which had to do with how using two ears can improve performance relative to one ear or how multiple signals combined to produce masking of target speech. They noted (e.g., Carhart et al. 1969) that listeners sometimes demonstrated significantly more masking from the combination of a speech masker and a modulated noise, or two separate speech maskers, than would be expected based on a simple summation of their separate effects. They state that "…one is carried to the conclusion that the task of abstracting a primary message from multiple competition containing meaningful speech is more difficult than is accounted for by the simple algebraic spectrum of the combined maskers" (p. 695). They termed this additional interference "perceptual masking" but also allowed that the term "cognitive interference" could be applied as a descriptor as well. At the time, they believed that the mechanisms supporting perceptual masking were unique to speech, but speculated about other examples of "excess additivity" (e.g., Bilger 1959; Green 1967) that were also considered (as with their speech findings) to be "...the product of cumulative interference involving two or more independent mechanisms..." and stress that "...excess additivity is demonstrable in various ways" (p. 701). Of particular interest here was the idea that the masking of speech could be influenced by masker semantic or linguistic factors that were clearly not explainable based on peripheral overlap of excitation. Carhart et al. (1969) not only provide a compelling example in support of that idea, but also review several earlier studies that presented data or conclusions concordant with their own.

Around the time of Pollack's (1975) published abstract, the work he was engaged in concerned (among other things) the identification of random auditory waveforms (pseudorandom pulse sequences) and determining the factors that interfered with identification (Pollack 2001, Personal communication). In one particularly relevant study, Pollack (1976) concluded that interference in the identification of specific random auditory waveforms was more closely related to the interruption of auditory processing due to "informational interference" (from irrelevant pulses added before, during, or after the pattern to be identified, in either the ipsilateral or the contralateral ear) than to "traditional" masking. Although the terminology varied somewhat the distinction between peripheral/energetic masking and central/informational masking is clear from Pollack's writings.

At about the same time, Watson and his colleagues (Watson et al. 1975, 1976) were studying conditions under which masker frequency uncertainty adversely affected frequency or intensity discrimination performance for target tones embedded in sequences of "context" tones. They applied Pollack's informational masking concept initially to experiments with varying degrees of context-tone uncertainty with low-uncertainty conditions producing small amounts of informational masking and high-uncertainty conditions producing large amounts (and sometimes *very* large amounts) of informational masking. Apparently because of concerns that the varying context tones affected the audibility of the target tones, Watson and Kelly (1981) reported decreases in detectability/detectatrility for target tones in sequences of context tones as uncertainty was increased. They also called this decline in performance with increasing uncertainty informational masking. Watson's work in particular drew considerable attention because the effects were often so large but also because they could be produced using standard discrimination procedures while varying uncertainty in a controlled manner.

In a fundamental shift in paradigm from that used by Watson and colleagues, Spiegel et al. (1981) examined signal and masker uncertainty for simultaneously—rather than sequentially—presented context tones. They found that masker uncertainty was generally more detrimental than signal uncertainty and, in an important technical and conceptual development instrumental in the series of "profile analysis" studies (Green 1988), demonstrated that randomly roving the level of stimuli on every presentation had much smaller effects on performance than the critical-band energy-detector model would predict (see also Mason et al. 1984; Richards and Neff 2004). Part of their interest in this topic was due to the earlier findings by Tanner and colleagues that signal frequency uncertainty generally produced small elevations in thresholds. Although the task used by Spiegel et al. was to discriminate an intensity increment to a single tone of a simultaneous multitone complex, the connection to the classical tone-in-noise experiment was closer than the sequential discrimination procedure used by Watson and colleagues, and the "masking" that was produced by uncertainty raised again some of the earlier issues discussed above (see also Spiegel and Green 1982). In those studies, though, as in Watson's work, the masker samples were always the same on both intervals of a given trial, so that the only alteration in the stimulus was that caused by the target (unlike most procedures for detecting a tone in Gaussian noise in which each noise burst is a different sample). Later, Kidd et al. (1986) examined how masker uncertainty affected profile analysis, but in their study, the maskers were different samples on the two intervals of every trial as well as on different trials.

In a study that was seminal in the informational masking literature, Neff and Green (1987) generated maskers composed of $N$ randomly selected tonal components, where $N$ was varied over a wide range. The task was to detect a pure-tone target of fixed and known frequency located within a distribution of potential masker components, for masker samples drawn at random on every presentation. These multitone maskers were essentially sparse samples of components from Gaussian noise, with the density of components the parameter of interest. They found large amounts of masking for small numbers of components that, according to classical masking theory, should produce very little energetic masking. In subsequent years, this simultaneous multitone masking procedure would be used by many investigators to study informational masking (e.g., Neff et al. 1993; Kidd et al. 1994; Oh and Lutfi 1998; Wright and Saberi 1999; Richards et al. 2002; Durlach et al. 2003b, 2005; Richards et al. 2004).

In the time since Watson's early work with tone sequences, and the original simultaneous multitone masking study by Neff and Green, a rapidly increasing body of work has appeared that addresses the issue of whether the masking observed in a given task—regardless of whether it was simply based on a difference in two threshold measurements—consists of energetic masking or informational masking, or both. Several authors have attempted to clarify and distinguish between energetic and informational masking. For example, Leek et al. (1991) state that "Informational masking is broadly defined as a degradation of auditory detection or discrimination of a signal embedded in a context of similar sounds; it is not related to energetic masking caused by physical interactions between signal and masker" (p. 205). Here, the important concept of signal-masker similarity in informational masking is raised. Also, from Neff (1995), "Informational masking is…the elevation in threshold produced by stimulus uncertainty, and can be contrasted with energy-based masking, in which signal detection is determined by the ratio of signal-to-masker energy within a presumed auditory filter…" (p.1909).

The concepts of energetic and informational masking now appear to be firmly entrenched in the auditory literature independent of their original meanings or whether each is sufficiently descriptive. Defining informational masking as masking beyond energetic masking is only useful to a certain degree. Limitations on many perceptual and cognitive processes, such as a failure to segregate a sound source, focusing attention on the wrong source or feature of a sound, insufficient processing capacity to comprehend rapid information from a source, limitations on memory, all of these factors qualify as interference that cannot be attributed to energetic masking but that may be manifested in masking experiments.

## 3. The Multitone Masking Paradigm

The original article by Neff and Green (1987) describing the simultaneous "multitone masking" paradigm was discussed briefly in a historical context in the section above. Here, more detail is provided about the basic paradigm and initial findings. In subsequent sections a number of the major findings and phenomena reported in the literature using this procedure or variations of it are discussed.

Figure 6.1 shows a schematic example of the stimuli typically used in the multitone masking experiment with the two panels indicating two possible masker draws, one in each observation interval. The target and "protected region" a range of frequencies surrounding the target in which masker components are excluded, not used in Neff and Green (1987) are also shown. The purpose of the protected region is to limit energetic masking (cf. Neff and Callaghan 1988). Originally, Neff and Green (1987) were interested in the question of how many frequency components were required to create "noise," at least with respect to

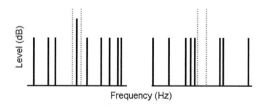

FIGURE 6.1. A schematic illustration of two possible random-frequency masker samples in the multitone masker experiment plotted as line spectra. The left panel shows target-plus-masker and the right panel shows masker only. The target is the taller solid line in the center of the spectrum, while the dashed lines indicate the boundaries of the "protected region" where masker components are not allowed to fall.

its masking properties (cf. Schafer et al. 1950; also Hartmann et al. 1986). The frequencies that constitute a Gaussian noise are continuous within the bandwidth, and the amplitudes of the components are Rayleigh distributed (cf. Hartmann 1997, Chapter 23). Neff and Green drew $N$ random samples of these components on each interval of every trial, where $N$ was the variable of interest. The signal frequency was fixed and known throughout a block of trials. Results for three different signal frequencies, 0.25, 1, and 4 kHz, were presented. For maskers composed of small numbers of components, very little masking was expected based on intuitions drawn from the critical-band energy-detector model because it would rarely happen that the masker components would fall near enough to the signal to energetically mask it. Surprisingly, large amounts of masking (more than 50 dB for 1 and 4 kHz) were found for maskers composed of as few as ten components, and significant amounts of masking were observed for only two components. Furthermore, the maximum amount of masking did not occur for the masker consisting of the most components (i.e., true Gaussian noise) but instead for a masker made up of about 10–20 frequency components (also Oh and Lutfi 1998). When the masker sample was randomized only between trials, but was the same on both intervals within a trial, the amount of masking was substantially reduced.

Typical findings from the multitone masking experiment are shown in Figure 6.2 as filled circles (Oh and Lutfi 1998). The figure shows group mean thresholds measured as a function of the number of components in the masker. The sound pressure level of the maskers was held constant. Note the initial

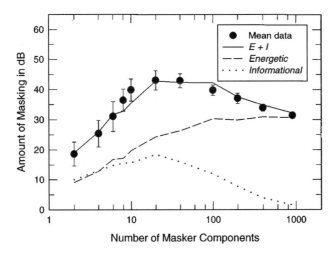

FIGURE 6.2. Typical results from the multitone masker experiment. The abscissa is the number of frequency components in the masker, while the ordinate is amount of masking. Group mean data are plotted as filled symbols. The solid line shows the predicted thresholds based on the CoRE model, while the dashed and dotted lines are estimates of the amounts of energetic and informational masking, respectively. (Reprinted with permission from Oh and Lutfi 1998.)

increase in thresholds followed by a plateau followed by a decrease in thresholds. The interpretation is that when there are very few components, masking is dominated by informational masking, while at the opposite extreme, when the masker is true Gaussian noise, the masking observed is almost entirely energetic in nature. Varying the number of masker components thus changes the ratio of energetic to informational masking. The plateau region from roughly 10 to 100 components indicates component densities for which the successive samples are sufficiently different perceptually to create significant uncertainty while also producing significant amounts of energetic masking.

## 4. The Component Relative Entropy (CoRE) Model and Other Quantitative Approaches Applied to the Multitone Masking Experiment

Any comprehensive model of masking must be able to account for both energetic and informational factors. The component-relative entropy (CoRE) model, first proposed by Lutfi (1993), is the only model to date that has been used to explain the results from a wide range of studies by taking into account both energetic and informational masking. In the original article describing the CoRE model, Lutfi (1993) demonstrated that it could predict the results from a variety of multitone masking experiments, such as those reported by Neff and Green (1987). However, it also accurately predicted the findings from other types of studies such as the profile analysis experiment described by Kidd et al. (1986) in which sensitivity to differences in spectral shape was measured for randomly perturbed reference spectra. The model relies on the statistical summation over trials of the outputs of a set of auditory filters spanning the audible frequency range. The amount of masking that is predicted is related to both the target-to-masker ratio (T/M) in the band containing the target and the variability of the outputs of the attended-to nontarget bands. For instance, during detection of a pure-tone target in Gaussian noise, the variability in the outputs of the nontarget bands, computed across trials, would be relatively low (depending on bandwidth, duration, etc.). The threshold for the target would primarily be determined, therefore, by the target-to-masker energy in the target's band, with very little contribution to the overall amount of masking from nontarget bands. In contrast, for a random-frequency multitone masker comprising a few components and with a "protected region" surrounding the target (refer to Figure 6.1), the variability of the outputs of the nontarget bands across trials may be quite large, while the T/M at masked threshold in the target's band may be very high. This situation consists of a small amount of energetic masking with a relatively large amount of informational masking. The predictions of the CoRE model are also illustrated in Figure 6.2. The data points are group-mean masked thresholds, and the solid line is the total amount of masking predicted by the CoRE model. The two lower curves illustrate the model predictions for energetic and informational masking. The sum of the two types of masking equals the overall masked threshold curve. The change in the proportion of energetic to informational masking as the density of the multitone masker increases is obvious.

Although the CoRE model provides an excellent account of the data shown in Figure 6.2, as well as a number of other informational masking conditions, there are some findings that it cannot explain. For example, Oh and Lutfi (2000) report that the CoRE model does not provide a satisfactory explanation for the large decrease in masking found by causing a target tone to be mistuned slightly in a multitone masker having masker components drawn at random from a set of harmonically related tones. Also, Kidd et al. (2003a) note that the CoRE model does not capture the trend in masking apparent for the multiple-bursts different masker (discussed below) as the number of masker bursts and interburst interval are varied. Nonetheless, the CoRE model represents an important conceptual tool that can help explain many of the findings from a large subset of informational masking studies.

One aspect of informational masking that is clear for most studies and a diverse set of experimental procedures is that the listener attends to frequency regions that provide no useful information for solving the task. Neff et al. (1993) sought to determine whether listeners who were very susceptible to informational masking ("high-threshold" listeners) exhibited wider listening bandwidths than less-susceptible ("low-threshold") listeners (see also Richards and Tang 2006). In order to obtain listening bandwidth estimates, they adapted the techniques normally used to measure "auditory filter" characteristics (e.g., Patterson et al. 1982) and applied them to the multitone masking experiment. In the more common procedure, a set of threshold estimates is obtained for pure-tone targets masked by notched-filtered noise as the bandwidth of the notch is varied. Based on these threshold estimates, a best-fitting set of filter parameters (making an assumption about the type of filter) is then computed. Neff et al. performed a similar analysis using data obtained from the multitone masking experiment in which the variation in the width of the "notch" was accomplished by changing the size of the "protected region" around the target frequency (refer to Figure 6.1). They found that the estimated "attentional" filter bandwidths and processing efficiency (related to the target-to-masker ratio in the filter at masked threshold) were both lower and poorer, respectively, in the high-threshold group than in the low-threshold group with large differences found between subjects. Furthermore, the more susceptible high-threshold group generally exhibited large amounts of informational masking even for extremely broad protected regions.

The reliance on the outputs of various frequency channels—both those containing target energy and those that contain masker energy—can be estimated by deriving weights based on listener responses. Richards et al. (2002) obtained channel weights in a multitone masking experiment and proposed that the weights were combined linearly to form the decision variable (see also Tang and Richards 2003; Richards and Tang 2006). The weighting functions they observed are illustrated in Figure 6.3 for four different observers.

There are several aspects of weighting functions that reveal the underlying strategies used by an observer. First, the weights may or may not be ideal. The

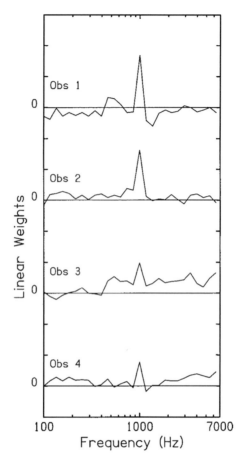

FIGURE 6.3. Weighting functions for four individual listeners. The target frequency was fixed at 1000 Hz. (Reprinted with permission from Richards et al. 2002.)

extent to which listeners adopt nonideal strategies can help to explain some of the large intersubject differences often found in informational masking studies. Second, listeners may weight certain frequency regions differently than others, again presumably providing some insight into the underlying decision processes. In the weighting functions illustrated in Figure 6.3, two of the listeners show positive weights (roughly corresponding to a tendency to respond that the signal was present when that masker component was relatively high in level) in the higher regions of the nonsignal frequency bands. It is possible that such weights could indicate that those listeners were more prone to experience interference in target detection for high-frequency masker tones than for low-frequency masker tones. The issue of which types of sounds, and their composition, are most effective in producing informational masking occurs frequently in the literature.

Durlach et al. (2005) compared performance for ten randomly chosen multitone masker samples in conditions in which a given masker sample was held constant on every trial throughout a block of trials versus conditions under which that same masker was randomly mixed with the others. The procedure they used was single-interval YES-NO, allowing separate estimates of sensitivity and bias. Although they also employed the usual protected region around the target, they noted that masker energy still could directly affect the target by leaking into the filter containing the target. In conditions in which masker samples are chosen at random on every presentation, the leakage from masker components into the target filter would act like a random rove in level (cf. Green 1988, pp. 19–21). Based on the thresholds obtained in the fixed condition of the experiment, and taking into account the "rove" in level of the output of the target's filter caused by mixing masker samples, they computed the filter widths necessary to account for masked thresholds. Surprisingly, this version of the "energy detector model" did a reasonably good job of predicting the thresholds found in the mixed condition for most of the subjects tested. However, an analysis of the pattern of bias observed in the experiment indicated that such a simple model could not produce a completely plausible account of the data.

## 5. The Effects of A Priori Knowledge: Training, Cuing, and Stimulus Set Size

A classical view of the auditory system is that the frequency spectrum is divided into a set of contiguous bandpass filters and the observer can select a particular filter or filters to attend to and ignore the outputs of irrelevant filters. If that were true, and the listener were able to do so perfectly, then the masking observed in the multitone masking experiment would be solely energetic masking and presumably would be much less than has often been observed empirically. Is it possible that listeners can be trained to accomplish this task or be provided with enough a priori information so that informational masking is completely eliminated? The differences between individuals in terms of the "susceptibility" to informational masking appear to be enormous, much greater than for energetic masking (cf. Leek and Watson 1984; Neff and Dethlefs 1995; Durlach et al. 2003b; Kidd et al. 2003a; Durlach et al. 2005). Does that imply that informational masking is due simply to too little training or an incomplete understanding of what to listen for?

### 5.1 Training

One perspective on the issue of training is that it is irrelevant. The observation that listeners demonstrate large amounts of informational masking in certain circumstances is the important finding, not whether individuals are able to overcome informational masking with practice. However, for many individual subjects, many hundreds of trials fail to reduce masked thresholds in the multitone masking experiment after the initial learning that typically occurs due to familiarization with the task.

Several studies have addressed this issue directly, although to our knowledge, systematic examination of the amount of training in the multitone masking experiment over very long periods of time has not been reported in the literature. However, the amount of training prior to data collection is nearly always reported, and it is possible to draw some conclusions based on those reports.

Neff and Callaghan (1988) present learning curves from four listeners for 2- and 10-component multitone maskers. The plots they present indicate thresholds as a function of the number of trials out to 1800 trials. One subject (L1) demonstrated a large decrease in masked thresholds in the 2-component masker during the first six 100-trial blocks. The same subject showed somewhat less of a decrease in threshold in the 10-component masker. Very little evidence of improvement with practice was apparent for the other three listeners. The authors conclude that genuine individual differences in susceptibility to multitone masking exist that cannot be overcome by simple repetitive training.

Similar findings were reported by Neff and Dethlefs (1995) for five subjects who were chosen from a larger group because they exhibited a wide range of susceptibility to informational masking. No relationship between susceptibility and amount of improvement over time was observed. Some subjects did in fact exhibit significant improvements after extensive training. Neff and Dethlefs conclude, "…the potential for long-term training clearly exists for some listeners and conditions, particularly for maskers with small numbers of components randomized within trials. For the majority of listeners, however, performance is quite stable over time, regardless of whether they are high- or low-threshold listeners…" (p. 133).

With respect to (much) longer-term training and experience, Oxenham et al. (2003) compared performance in a task very similar to the multitone masking experiment (called "multiple bursts same," described below) for highly trained musicians versus nonmusicians. The basic idea was that because musicians normally engage in auditory tasks that require them to "hear out" specific elements (e.g., the melodic line played by a particular instrument embedded in the sounds created by other musical instruments), they might be less susceptible to informational masking. Presumably, because of training (and/or perhaps natural ability) they may be better than nonmusicians at hearing out the pitch of one specific tone embedded in a set of other tones. Oxenham et al. (2003) found that indeed, as a group, highly trained musicians were less susceptible to informational masking than a matched group of nonmusicians. No significant differences were found between groups with respect to auditory filter characteristics, processing efficiency, or performance on another complex masking task that is thought to produce about the same amount of energetic masking as the multiple-bursts same task but much less informational masking (Kidd et al. 1994). It should be noted, however, that the musically trained group still demonstrated about 10 dB of informational masking for the conditions tested compared to about 25 dB for the nonmusician group.

## 5.2 Cuing

Perhaps an even more effective means for directing the attention of the listener to the correct location in frequency, compared with extensive practice, is to provide an exact copy of the target immediately prior to stimulus presentation. Cuing, in this manner, presumably reduces any variation or "noise" in the stored pitch reference (i.e., the pitch of the target to be detected) and provides a strong sensory trace for comparison with the test stimulus. An extensive, explicit examination of the effect of cuing was reported by Richards and Neff (2004). For a fixed and known target frequency in a random-frequency (on every presentation) multitone masker, they found that providing a pretrial cue (an exact copy of the target tone) reduced masking by about 5 dB averaged over subjects and conditions. When both the signal and masker frequencies were randomized, even greater group mean effects—as large as 20 dB in some cases—were found.

Richards et al. (2004) compared the effectiveness of target, masker, and target-plus-masker cues both when the cues were presented before the trial and when they were presented after the trial. Somewhat surprisingly, perhaps, the most effective cue was not the target but the masker alone presented immediately before the stimulus. Although it may be counterintuitive that cuing the masker confers a greater benefit than cuing the target, it is the random variation in the masker that creates uncertainty in the listener. This cuing effect, which appears to be robust, was not attributed to peripheral processes (e.g., adaptation of the auditory nerve) but instead seems to be central in origin.

One possible interpretation of this masker-first advantage is that the listener is able to construct a "rejection filter" corresponding to the masker more readily than an acceptance filter centered on the target. Such a rejection mechanism has been proposed by Durlach et al. (2003a) and referred to as the listener strategy "Listener Min" (contrasted with the acceptance filter strategy referred to as "Listener Max"). This putative strategy is more general than the application to the cuing results discussed here, and could provide an account of performance in a variety of masking experiments. However, the findings reported by Richards et al. (2004) could be explained by several alternative mechanisms and at this point are not thoroughly understood.

## 5.3 Stimulus Set Size

How does uncertainty vary with the size of the set of stimuli encountered in a randomized presentation design? By definition, uncertainty means that the listener is not certain about which stimulus will be presented on a given trial or interval within a trial. The two extreme cases, then, are that the same masker is presented on every interval of every trial and that a different sample is presented on every presentation. The question is, how does informational masking vary as the size of the set of potential maskers increases from one to many?

In the informational masking literature, the first study to address this issue directly was that of Wright and Saberi (1999). They varied the number of masker tokens presented in a given block of trials from 2 to 10. When the randomization of masker samples occurred on every interval of the two-interval trial, thresholds were elevated by about 7–18 dB relative to the fixed condition. When the tokens were randomized across trials, but were the same within a trial, much less of an increase in thresholds with respect to the fixed condition was found. Some of their key results are shown in Figure 6.4. The greater masking when samples are randomized on every presentation vs. when they are the same in the two (or more) intervals of a trial but mixed between trials was also supported by the findings of Tang and Richards (2003), Richards and Neff (2004), and the earlier report by Neff and Dethlefs (1995).

Richards et al. (2002) also measured multitone masking (6 masker components) as masker set size was varied. The number of masker tokens in their study ranged from 3 to 24 plus a completely random condition. However, they also tested multiple sets of tokens. Although their randomized-sample results supported those of Wright and Saberi (1999) discussed above with respect to the increase in masking with increasing masker set size, they also found large differences between subjects and between the specific sets of maskers tested. When masker set size was 3, large differences in thresholds were observed between sets. However, for masker set sizes of 12 or greater, much smaller differences were found. Their findings suggest that masker set sizes of 12 or more create uncertainty that is comparable to that found from truly random masker sampling. A cautionary note to that conclusion, though, is that even for set sizes of 24, listeners were able to perform above chance in a memory test that was designed to determine whether the listeners could remember whether specific masker tokens had been presented in the experiment. Thus, some learning of specific samples appeared to have occurred, a result that could affect the uncertainty in the experiment.

Durlach et al. (2005) compared the masking produced by a set of ten random-frequency multitone complexes when the masker tokens were fixed across a block of trials to when the same ten masker tokens were randomly drawn on each stimulus presentation within a block of trials. Their data take the form of psychometric functions obtained in a YES-NO detection task. The results from five listeners (rows) are shown in Figure 6.5. In both columns, the obtained psychometric function for the "mixed" condition (R) is shown as the heavy

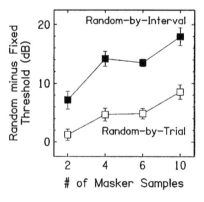

FIGURE 6.4. The increase in masking due to between-trial (open symbols) and within-trial (filled symbols) random selection of masker samples using fixed-sample thresholds as a reference. The abscissa is the number of samples in the set. Data are group means. (Reprinted with permission from Wright and Saberi 1999.)

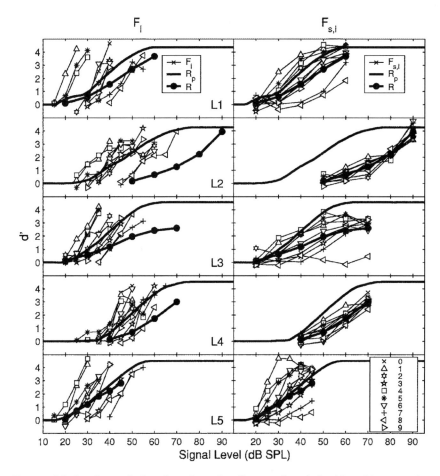

FIGURE 6.5. Psychometric functions from five listeners (rows) for 10 multitone masker samples. In the left column, the psychometric functions are displayed for each masker sample for a condition in which the sample was fixed (the only masker presented) throughout a block of trials ($F_i$). The right column also shows psychometric functions for all ten masker samples sorted from blocks of trials when the samples were mixed and chosen randomly on every presentation ($F_{s,i}$). The heavy solid line (no data points) represents the pooling of the ten psychometric functions computed from the fixed condition ($R_p$), while the heavy dark line with data points (filled circles) shows the psychometric function measured in the mixed (all 10 samples) condition (R). (Reprinted with permission from Durlach et al. 2005.)

black line with filled symbols, and the predicted mixed function (from averaging psychometrics from individual fixed tokens, left column) is shown as the heavy smooth psychometric function ($R_p$). The psychometric functions for individual tokens (key lower right panel) extracted in the mixed condition are also shown in the right column. There are two points of interest, in the context of the current topic, to be made regarding these findings. First, large differences in the amount of masking were observed among fixed masker tokens. This is indicated by the lateral spread of the functions in the left column of the figure. The effect of mixing the tokens within a block of trials was generally to decrease the slope of the composite psychometric function (R) relative to the individual functions obtained in the fixed condition. Furthermore, for three of the five subjects, the composite mixed function was shifted substantially to the right (more masking) relative to the average of the fixed functions (right panel; compare solid curve $R_p$ to data points on R). This lateral shift, which was almost 40 dB for one subject, is interpreted as informational masking due to masker sample uncertainty. Second, not only are there large differences between fixed masker tokens, there were large differences—more than 20 dB in some cases—in thresholds between listeners for specific tokens. Because this was the fixed-masker condition, technically there should have been no uncertainty in the listener about which token would be presented and the only masking that should have occurred was energetic masking. However, it seems highly unlikely that differences in thresholds of that magnitude can be explained by differences in the peripheral auditory systems of the listeners. A similar finding was reported by Alexander and Lutfi (2004), who found differences between normal-hearing subjects as great as 30 dB in their no-uncertainty (fixed-frequency masker components) reference condition. These large intersubject differences in thresholds in no-uncertainty conditions pose a problem for attempting to separate energetic and informational masking. Either there was substantial uncertainty on the part of the listener even when there was no variability in the stimulus (cf. Green 1961), or simply the presence of target–masker similarity was sufficient to produce informational masking (usually similarity is thought of as increasing informational masking when the listening condition is uncertain).

The size of the set of maskers presented in a randomized design clearly affects the amount of informational masking. However, it is equally clear that other factors qualify that statement. Obviously, if the samples are homogeneous—not strongly different perceptually—then the effects of different draws of samples or size of sets should be minimal. The converse, then, should also be true. Randomizing a set of 10 broadband Gaussian noise bursts (of sufficient minimum duration) produces much less uncertainty than 10 samples of 2-tone random-frequency maskers. When the samples are very heterogeneous, simply mixing the samples can affect the overall amount of masking as well as the slope of under-lying psychometric functions. Furthermore, as found by Durlach et al. (2005), some subjects demonstrate large decrements in performance due to randomizing samples even after accounting for the mixing of heterogeneous stimuli.

# 6. Reducing Informational Masking by Perceptual Segregation of Sounds

There have been a number of demonstrations using a variety of experimental techniques that indicate that informational masking can be reduced by exploiting presentation schemes that perceptually segregate the signal from the masker. In fact, it is possible to conclude that many instances of informational masking inherently reflect a failure of segregation. The logic behind this argument is that if the signal can be perceived as an auditory object separate from the masker, then by definition it has been "detected." This argument is not as convincing for some suprathreshold tasks, as discussed in the sections that follow. Conversely, though, solving the detection task need not require that the signal be perceptually segregated from the masker. It could exert an influence on the overall quality of the sound in a way that informs the listener that the signal is present without segregating into a separate object. This topic has often been discussed, for example, in the context of detecting the mistuning of a partial in a harmonic complex (e.g., Moore et al. 1986; Hartmann et al. 1990; Darwin 1992). Nonetheless, it is very clear that segregating the signal from the masker in conditions under which there is a great deal of informational masking will often significantly improve detection performance. The following section reviews some of the evidence obtained using the multitone masking procedure, and a variation on that procedure, in support of the strong role of perceptual segregation in overcoming informational masking.

Neff (1995) examined how several stimulus manipulations intended to promote the perceptual segregation of sounds affected performance in the multitone masking experiment. Drawing on segregation cues known to be effective in many types of listening tasks (for a comprehensive review, the reader is referred to Bregman 1990), Neff demonstrated that perceptually segregating the target from the masker could provide a large release from informational masking. Figure 6.6

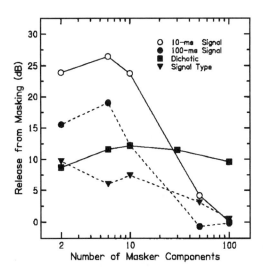

FIGURE 6.6. Release from masking in dB plotted as a function of the number of masker components due to stimulus manipulations intended to perceptually segregate the target from the masker. The data are group means. The different segregation cues are indicated in the key: shortened target duration (10- and 100-ms targets presented in a 200-ms masker), target dichotic with masker diotic, and qualitatively different target and masker types. (Reprinted with permission from Neff 1995.)

shows the amount of release from masking caused by some of these stimulus manipulations.

The abscissa is the number of frequency components in the multitone masker, and the ordinate is the amount of release from masking relative to a reference condition that was a version of the "standard" multitone masking paradigm. Two of the manipulations tested involved a target duration that was briefer than the masker duration; one manipulation involved dichotic presentation (target $\pi$ radians out of phase, masker in phase; $T_\pi M_0$) and the final manipulation was a qualitative difference between signal (narrowband noise) and multitone masker. Generally, the amount of release from masking decreased as the number of masker components increased beyond 8 or 10. In fact, except for dichotic presentation, no release from masking was apparent for 100 components. The interpretation is that the benefit of perceptual segregation is much greater for informational masking than for energetic masking. As noted above, as the number of masker components is varied over a range from very few to many, the proportion of energetic to informational masking is thought to increase so that little informational masking is present for the 100-component masker. The dichotic manipulation used by Neff (1995), $T_\pi M_0$, is an interesting case with respect to the discussion above. In $T_\pi$ presentation, the target is not heard as a compact image in the center of the head, as for $T_0$, but has greater apparent width and may be heard "at both ears." When embedded in diotic Gaussian noise, a common masking level difference (MLD) condition, the noise image is decorrelated, and again, the width of the image, and/or a timbre corresponding to the decorrelated frequency region, provides a cue to detection. For a tone in noise, the MLD is usually thought of as reducing energetic masking, that is, improving the T/M in the target's critical band. However, consider the threshold reduction due to $T_\pi M_0$ presentation relative to the definition of informational masking discussed above. If the physiological site at which energetic masking is defined is the auditory nerve, does that mean that the MLD reflects a reduction in informational masking? This example illustrates one reason that Durlach et al. (2003a) stressed that the definition of energetic masking should be tied to a given physiological locus, since it may be different at different points in the auditory system. In Neff's study, the $T_\pi$ presentation probably provided a fairly weak segregation cue when there were few masker components but because of binaural analysis continued to provide a release from masking as the number of masker components, and the concomitant energetic masking, increased.

Durlach et al. (2003b) also demonstrated that informational masking can be greatly reduced by cues thought to promote the perceptual segregation of sounds. They tested five separate manipulations that were intended to vary the strength of perceptual segregation. These manipulations included asynchronous onset, frequency sweep of signal in opposing direction to sweep of masker, dichotic presentation ($T_m M_0$), and two types of spectrotemporal patterns that varied the degree of relative coherence between target and masker. Averaged across all conditions and subjects, the benefit of the stimulus manipulations intended to segregate the target from the masker was about 17 dB, although large intersubject differences, and interactions with the specific manipulations, were observed, and the magnitude of the reduction in masking was a function of the exact stimulus parameters chosen. The important point of the Durlach et al. (2003b) study, however, was not simply the magnitude of the release from informational masking provided by these manipulations. Rather, the observed benefits were obtained, they argued, without substantially altering the variability in the masker that was intended to produce uncertainty in the listener. Instead, the differences in threshold were attributed to changes in target–masker similarity. The distinction between uncertainty and similarity in causing informational masking has been noted by others (e.g., Kidd et al. 2002a). Watson (2005), for example, has recently proposed that informational masking be divided into categories designated as resulting from uncertainty or from similarity. In any event, the findings of Durlach et al. emphasize the important role that target–masker similarity plays in informational masking.

Another technique that revealed how informational masking can be greatly reduced by perceptual segregation of target and masker was described by Kidd et al. (1994). They began with the single-burst multitone masking paradigm but modified it to incorporate a time-varying component. Their procedure involves presenting a sequence of multitone bursts in each observation interval. They examined two types of masker sequences. In one type of masker sequence, called "multiple-bursts same" (MBS), the random draw of frequencies constituting the masker in the first burst of the sequence was repeated in all of the subsequent bursts presented during that interval (with a new draw of frequencies occurring in the next interval). In the second type of masker burst sequence, referred to as "multiple-bursts different" (MBD), each burst within a given sequence was a different random draw of frequencies (also different draws across intervals and trials). Figure 6.7 illustrates these two types of masker sequences. The column labeled "S" illustrates conditions under which the target and masker are the "same," and the column labeled "D" represents target–masker "different" conditions.

First, consider the MBS stimulus illustrated in the upper left panel. Each multitone masker burst is 60 ms duration (8 bursts total). The onsets and offsets of each burst are lightly visible. In the first burst, 8 masker frequencies are drawn at random from the range 200–5000 Hz excluding the "protected region" of 820–1220 Hz. This first draw of frequencies is repeated in each burst in the interval. The target tone of 1000 Hz is also shown and is gated synchronously with the masker tones. Next, consider the MBD stimulus illustrated in the lower right panel. The first burst of the sequence was generated using exactly the same randomization rules as the first burst of MBS. However, instead of repeating the burst throughout the sequence, a new random draw occurred for the second burst, and the third burst, etc. However, the target tone of 1000 Hz is constant. Thus, the target is the only coherent frequency component in the stimulus, and it forms a perceptual stream that is easily segregated from the randomly varying masker bursts. In the Durlach et al. (2003b) study, the group mean amount of masking obtained using the MBS masker was nearly 25 dB greater than the

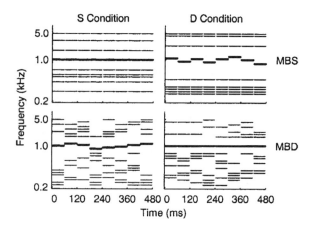

FIGURE 6.7. Schematic illustrations, in sound spectrogram form, of multiple-bursts same (MBS, top row) and multiple-bursts different (MBD, bottom row) maskers with targets (bolder lines near 1 kHz). In the left (S for Same) column, the target frequency over time varies/remains constant in a similar manner to the masker frequencies over time, whereas in the right (D for Different) column, the target frequency over time differs from the masker frequencies over time. (Reprinted with permission from Durlach et al. 2003b.)

corresponding amount of masking produced by MBD despite the expectation that the small amount of energetic masking produced by the maskers should be nearly the same (Kidd et al. 1994). The other two panels of the figure illustrate manipulations intended to vary target–masker similarity. This was accomplished by "jittering" the burst-to-burst frequency of the target within a narrow range of frequencies. The effect of this jitter is to diminish the sense of the target as a coherent stream (cf. Kidd et al. 1995; Richards and Tang 2006). However, the effect that this manipulation has on the amount of masking depends on the context provided by the masker. In the upper right panel, the jittered signal is presented in the MBS masker. Thus, it is the only *incoherent* "stream" in an otherwise coherent set of frequencies, and relative to the regular MBS case (upper left panel), the amount of masking observed was reduced by almost 20 dB. When frequency jitter was added to the target in the MBD masker, the amount of masking (relative to regular MBD, lower right) increased by about 10 dB, on average. Thus, the extent to which a particular target–masker combination produces informational masking is very dependent on contextual factors, such as target–masker similarity.

Kidd et al. (1994) speculated that the masking produced by the MBS masker illustrates the strength of a perceptual grouping cue. In everyday (nonlaboratory) listening, when frequency components are turned on and off together, it usually means that they arise from the same sound source. Thus, the natural tendency is to perceive them as a single auditory object, and determining whether one element of the object (i.e., the target frequency) is present may be very difficult. The MBD masker, on the other hand, promotes "analytic listening" (cf. discussion above concerning masking as a failure of analysis), because the target is the only coherent frequency component in an otherwise incoherent background. The strength and time course of this coherent frequency cue was examined by Kidd et al. (2003a). In their study, the number of bursts constituting the MBD masker, and the length of time between each burst (interburst interval, IBI), were varied systematically. The stimulus manipulations are illustrated in Figure 6.8. This figure shows the typical single-burst stimulus (upper left panel) extended to a sequence of eight contiguous bursts (upper right panel). The lower panel shows two 4-burst sequences having different IBIs. Kidd et al. were interested in two aspects of this experiment. First, they speculated that the pattern of masking resulting from changes in these two variables—number of bursts and IBI—would reveal how the perception of target stream coherence changed. And second, they evaluated the extent to which the MBD advantage relative to a single-burst (or MBS) stimulus could be attributed to "multiple looks" (cf. Viemeister and Wakefield 1991). The results, averaged across listeners, are shown in the left panel of the Figure 6.9.

The abscissa is the number of bursts in the sequence, while the different symbols indicate the length of time between bursts. The disconnected point (star) indicates the threshold for the single-burst stimulus. The results are quite orderly. As the number of bursts increased, the amount of masking decreased. As the delay between bursts increased, the amount of masking increased. For sequences consisting of eight bursts, increasing the delay between bursts from

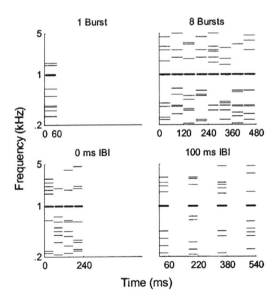

FIGURE 6.8. Schematic illustrations, in sound spectrogram form, of the MBD masker and target (bolder lines at 1 kHz) as the number of masker bursts is varied from 1 (single burst) to 8 (top row) and the interburst interval is varied for four masker bursts (bottom row). (Reprinted with permission from Kidd et al. 2003a.)

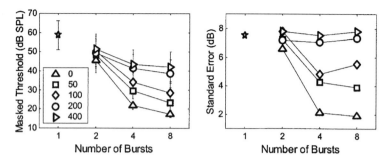

FIGURE 6.9. Group mean masked thresholds (left panel) for the MBD masker as the number of masker bursts is varied (abscissa) for different interburst intervals (symbol key in msec). The right panel shows the standard errors of the means corresponding to the data plotted in the left panel. The star not connected to the other symbols is for a single masker burst. (Reprinted with permission from Kidd et al. 2003a.)

0 ms to 400 ms increased thresholds by about 25 dB. Kidd et al. concluded that this increase in threshold with increasing delay between bursts was attributable to a breakdown in the perception of signal coherence. They concluded that those findings, and others described in the article, were inconsistent with the interpretation that the MBD advantage is primarily attributable to multiple looks. The right panel in Figure 6.9 plots the standard errors of the group means in the same manner as the thresholds in the left panel. There is a close correspondence between the general appearance of the functions in the two panels ($r = 0.9$) reinforcing the general observation that the greater the degree of informational masking, the greater the differences across subjects.

## 7. Discrimination and Nonspeech Identification Studies

To this point, the review of the informational masking literature has focused on the task of detection in studies employing the simultaneous multitone masking procedure and the MBS/MBD procedure described above. However, many other studies have been reported that used suprathreshold discrimination, identification, or speech-recognition tasks in which the energetic-informational masking contrast was examined. It should also be emphasized that the initial work on informational masking used discrimination or identification tasks with very little investigation of changes in detectability reported until the Neff and Green (1987) article (although there were a few; cf. Watson and Kelly 1981). Pollack (1976), for example, examined the interference caused by extraneous pulses in the identification of randomly generated pulse sequences. Most of the work on informational masking by Watson and colleagues was based on tasks that required judgments of differences in frequency or intensity of suprathreshold target tones.

Neff and Jesteadt (1996) measured pure-tone intensity discrimination for targets presented in broadband noise or in random-frequency multitone bursts.

They found that the presence of the multitone maskers greatly interfered with the ability to detect an intensity increment to the target tone. This interference was level-dependent and was greatest for the lower-level pedestals. In some cases, the listeners were unable to detect the intensity increment until it caused the target tone to exceed the level per component of the masker tones. Neff and Jesteadt were able to account for the trends in the data by considering the masked discrimination conditions as a special case of combined masking and successfully applied a model of additivity of masking (Lutfi 1983) to the results. Again, though, as in the studies above using the task of detection, much greater interference in performance was observed for the multitone maskers than would be attributable to energetic masking alone.

Another study that used a suprathreshold discrimination task to examine both energetic and informational masking was reported by Kidd et al. (2003b). The task of the listener was to choose which of two sounds, one presented in each of two observation intervals, consisted of tones that more nearly fell at an exact harmonic relation. The stimuli were 11-component harmonic complexes and jittered 11-tone inharmonic "foils," both of which had randomized fundamental frequencies. Kidd et al. measured the ability to distinguish between these two types of sounds in conditions in which random-frequency multitone maskers were added simultaneously. The multitone maskers, which consisted of 4, 8, or 12 frequencies, interfered with the ability of trained listeners to make the harmonic–inharmonic distinction. The simultaneous presentation of the multitone maskers undoubtedly created both energetic and informational masking. However, the results from a manipulation intended to promote the perceptual segregation of the target from the masker (target monotic masker diotic, $T_mM_0$) had a beneficial effect that depended on the number of masker components. For the 4-component masker, a dichotic advantage was found of about 8 dB. The advantage decreased as the number of masker tones increased with only about a 2-dB advantage found for the 12-component masker. The interpretation was that the ratio of energetic to informational masking increased with increasing number of masker tones, while the segregation cue reduced the informational—but not the energetic—component of masking. There were systematic changes in the slopes of the psychometric functions (shallower slopes for informational masking) as well that supported this interpretation.

Kidd et al. (1998) trained listeners to identify the members of a set of six narrowband nonspeech patterns. The patterns were created by arranging the frequencies of sequences of brief pure-tone bursts (constant, rising, falling, alternating, step-up, and step-down). Because the patterns were highly identifiable even when the variation in frequency was confined to a narrow range (e.g., 14% of nominal center frequency), it was possible to evaluate the effects of maskers that were directly overlaid on the targets, presumably causing large amounts of energetic masking, or were remote in frequency from the targets, presumably causing primarily informational masking. Kidd et al. argued that there were a number of benefits of using these stimuli and this type of approach. First, the task presumably was more complex than simple detection, requiring comparison of the stimulus or portions of the stimulus to reference patterns stored in memory. However, the complications of the linguistic content of speech were avoided. Furthermore, a number of secondary effects could be evaluated. Because it was the relationship among the frequencies of the elements in the sequence that defined the pattern, it was possible to measure performance across a wide range of center frequencies when both target and masker uncertainty were present. The main focus of the Kidd et al. (1998) study, though, was the effect of spatial separation of target and masker. The results of that study are shown in Figure 6.10.

This figure displays individual results obtained when the masker was intended to produce primarily energetic masking (broadband noise, upper row of panels) and when the masker was intended to produce primarily informational masking (random-frequency multitone complexes, lower row of panels). The abscissa is spatial separation, and the values on the ordinate are dB differences in midpoints computed from psychometric functions fit to percent correct identification. For the noise masker, the only large improvements with spatial separation occurred for the highest range of target frequencies, and follow the expected values for head shadow (in a mildly reverberant room) fairly closely. However, the improvements found for the random-frequency multitone masker were very different, showing a consistently large effect at low target frequencies and a wide range of improvements across subjects at high frequencies. The spatial release from masking was considerably less in the mid-frequency range,

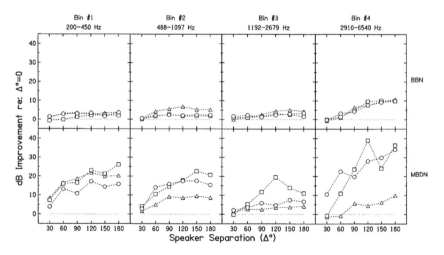

FIGURE 6.10. The improvement in nonspeech pattern identification due to spatial separation of target and masker. The abscissa is spatial separation in degrees azimuth, and the ordinate is the release from masking in decibels relative to target and masker at the same location. The data points are for three individual listeners. The upper row is for a broadband noise (BBN) masker, while the lower row is for a multitone (multiple-bursts different, narrowband, MBDN) masker. Columns indicate target frequency ranges. (Reprinted with permission from Kidd et al. 1998.)

where sound localization acuity is relatively poor. Kidd et al. concluded that spatially separating the target sequence from the masker caused, or significantly strengthened, the segregation of the target from the masker, allowing the listener to focus attention on the target, providing a substantial release from informational masking.

In an extension of the work discussed above using narrowband nonspeech patterns, Kidd et al. (2002a) examined whether changing the similarity between the set of target patterns and the masker could affect the amount of masking obtained. They employed two types of multitone maskers. In one case, masker frequencies were random draws on each burst throughout the sequence. This masker is essentially the MBD masker discussed above and used in detection experiments except that, because target frequency was randomized, the "protected region" followed the target. The second type of masker, which was intended to be more similar to the target patterns, was composed of sets of tones constrained to fall within narrow frequency regions as the burst sequence progressed. Thus, the masker was comprised of a set of narrowband "streams." There was also a high degree of both target and masker frequency uncertainty so that the listener had to monitor the entire frequency range to locate the target.

The results of this study indicated that the maskers consisting of sets of narrowband streams produced significantly more masking than did the MBD masker in which the masker tones were unrelated across bursts. The interpretation of that result was that the increased masking was a consequence of the greater similarity between the target and the masker. The similarity to the set of targets, in this case, was the arrangement of the masker into narrowband streams. Because of the high degree of frequency uncertainty, masker streams could plausibly sound like a target pattern, at least early in the burst sequence, increasing the likelihood of confusions with the target or loss of information due to misdirected attention.

Recently, Best et al. (2005) reported a study in which human listeners were trained to identify the calls of songbirds. The target songs, which were presented at a wide range of target-to-masker levels, were masked by sounds having properties that varied in their energetic/informational values. The three maskers they used were song-shaped noise, chorus-modulated noise, and birdsong chorus (multiple simultaneous bird songs). The amount of masking was measured for conditions in which target and masker were colocated vs. when they were spatially separated. They found large differences attributable to "better-ear advantages" (i.e., spatial separation of sources caused one ear to have a more favorable target-to-masker ratio than the other ear) in all conditions, but for the chorus masker, an additional 10 dB of release from masking was found that the authors attributed to a reduction in informational masking due to perceived differences in location. For the two noise conditions, the advantage due to spatial separation of sources appeared to be due only to "better ear advantages."

In summary, each of these studies indicates the effect of informational maskers on discrimination of some property of a target sound, or identification of the sound and/or its source, for targets presented at levels well above energetically masked thresholds. The key element of these studies seems to be that introducing stimulus manipulations that cause, or strengthen, the perceptual segregation of target and masker leads to reduced informational masking while having a much smaller effect on the amount of energetic masking. Because we typically listen to sounds well above detection threshold, these studies reveal factors likely to be important in realistic listening environments where the challenge for the listener is to select, attend to, and extract information from one audible sound source in the presence of other competing sound sources.

## 8. Informational Masking and Speech Recognition

It has long been appreciated that the task of comprehending the speech of one particular talker in the presence of other talkers is complex and performance may be affected by many diverse factors (e.g., Cherry 1953; Pollack and Pickett 1958; Schubert and Schultz 1962; also recent reviews by Yost 1997; Bronkhorst 2000; Ebata 2003). A review of that literature is far beyond the scope of this chapter (the reader is referred to the chapters on spatial hearing and speech recognition, Chapters 8 and 10, respectively, in this volume). Nonetheless, there are a number of recent studies of speech recognition that explicitly examined informational masking that deserve mention in the context of the topics covered here.

Freyman et al. (1999) devised a procedure in which the image of a masking talker was pulled away from that of a target talker, resulting in a significant improvement in performance, without decreasing the amount of energetic masking that was present. In the reference condition of their procedure, a target talker and masker talker were presented from the same location (0° azimuth) and the intelligibility of the target speech was measured. Then without disturbing the target and masker from the colocated location, an exact copy of the masker talker was presented from a second loudspeaker located off to one side of the listener. This copy of the masker talker slightly led the original masker talker in time. Because of the precedence effect, the result of this manipulation was to shift the apparent image of the masker away from the target and toward the temporally leading loudspeaker. The advantage of this perceptual effect, measured at a fixed point on the psychometric function, was about 8 dB. When the masker was speech-shaped noise, however, no advantage of the second masker was observed. The interpretation of this finding is that the ability to segregate and focus attention on the target talker was improved by exploiting a perceptual effect that caused the target and masker to appear to originate from different locations. This perceptual effect was not significant when the limitation on performance was primarily energetic masking.

An effective technique for designating which talker is the target talker (often a problem in multitalker masking experiments that can interact with the degree of uncertainty in the task) was developed and used extensively by Brungart and colleagues. They designed a closed-set forced-choice speech identification test, called the "coordinate response measure" (CRM; Bolia et al. 2000; see also Spieth et al. 1954), in which the listener must follow the speech of the talker uttering a specific "callsign" until two subsequent test words occurred. Masker talkers utter sentences having exactly the same structure with different callsigns and key words. Brungart (2001) showed that when a single masker talker was present, large differences in performance were found depending on how similar the two talkers were. The least similar case was when the talkers were different sexes while the most similar case was when the exact same talker uttered both target and masker sentences. This procedure has been used by a number of investigators to examine factors such as the effect of the number of masking talkers and their similarity to one another, binaural and spatial effects, and differences between speech and nonspeech maskers (e.g., Brungart et al. 2001; Gallun et al. 2005; Shinn-Cunningham et al. 2005).

One particularly interesting finding from this series of studies was reported by Brungart and Simpson (2002). Using the CRM test, they measured performance in a condition in which the target was presented to one ear and a single masker talker was presented to the contralateral ear. The contralateral masker did not affect performance, which was nearly perfect. Next, a target talker was presented to one ear and a single masker talker was presented to the same ear. Performance was degraded by the ipsilateral speech masker depending predictably on the T/M. Then a condition was tested in which the target talker was presented to one ear, one masker talker was presented to the contralateral ear, and a second, unrelated, masker talker was also presented to the same ear as the target. In this case, performance was much worse than under either of the previous conditions. A later study (Brungart and Simpson 2004) revealed that uncertainty about the semantic content of the ipsilateral speech masker was much more important in causing identification errors than uncertainty about the content of the contralateral speech masker. Brungart and Simpson (2002) concluded that this effect was related to limits on the ability of listeners to hold separate inputs to the two ears when there was an ipsilateral segregation task to be performed. A parallel finding by Kidd et al. (2003c) using MBS and MBD maskers indicated that this result was not unique to speech or speechlike stimuli.

Arbogast et al. (2002) used the CRM task and a processed version of the CRM corpus to attempt to separate energetic and informational masking in speech identification. The processing they imposed on the stimuli was a modified version of cochlear implant simulation speech (Shannon et al. 1995) in which the speech signal is reduced to a set of narrow frequency bands (see also Dorman et al. 1997; Loizou et al. 1999). As in the MBS/MBD-style experiments described above for detection and pattern identification, an advantage of reducing the stimulus to narrow bands is that it facilitates varying the amount of energetic and informational masking that is present in a controlled way. Arbogast et al. processed the CRM speech into sets of 15 very narrow frequency bands by extracting the envelopes of filtered narrow bands and using the envelopes to modulate carrier tones at the center frequencies of each band. Maskers consisting of sets of 15 corresponding narrow bands were generated that were other CRM sentences or were narrowband noises. The target speech was a randomly drawn subset of 8 of the 15 available bands on each trial. The maskers could be chosen to maximize energetic masking by superimposing noise bands on the selection of target bands (same-band noise masker), or maximize informational masking (minimizing energetic masking) by choosing masker bands of speech that were mutually exclusive with the target bands (different-band sentence). An energetic masking control for the latter consisted of nonoverlapping bands of noise (different-band noise). The spectrum of a processed speech target and speech masker is illustrated in Figure 6.11. Arbogast et al. presented these stimuli at various T/Ms in two spatial location conditions: target and masker colocated at 0° azimuth, and target at 0° azimuth with masker at 90° azimuth.

When target and masker were colocated, approximately 22 dB more masking was found for the speech masker than for the different-band noise masker. Subsequent studies (e.g., Kidd et al. 2005a), have generally supported this finding with one caveat - the fact that the masker bands vary in center frequency from trial to trial means that there is considerable frequency uncertainty in the maskers. So, the values for the different-band noise masker might be lower if uncertainty were removed. However, because the frequency uncertainty is the same for different-band noise and different-band sentence maskers, the 22 dB greater effect from speech cannot be attributed to uncertainty. The greater masking by the speech masker is likely due to fact that the masker, like the target, is intelligible speech. The two processed speech sources sound qualitatively very similar, and thus often may be confused. Indeed, analysis of the errors in this task supports the idea that target–masker confusions are common.

The advantage of spatially separating the target and masker depended on the type of masker that was present. For the same-band noise masker, which presumably produces primarily energetic masking, the spatial release from masking is likely due to a combination of a "better ear" advantage (the acoustics of the head improve the target-to-masker ratio in one ear when sources are

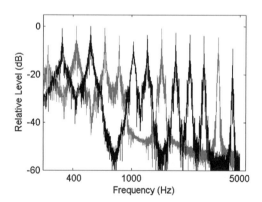

FIGURE 6.11. A schematic illustration of magnitude spectra for a single pair of a processed speech target (black) and masker (gray) used in the different-band sentence-masking condition. The spectra were computed over the entire length of the CRM sentence samples (see text).

separated) and binaural analysis, which for this group of subjects was about 7 dB. For the different-band sentence masker, a much greater release from masking of just over 18 dB was found. This difference—similar to some of the binaural/spatial work discussed in preceding sections (e.g., Kidd et al. 1998; Freyman et al. 1999, 2001)—is largely a result of perceptual factors. The two sound sources are perceived as emanating from different locations, facilitating the focus of attention toward the correct source.

In a study demonstrating the benefit of a priori information for speech-on-speech masking tasks, Freyman et al. (2004) reported that priming the target speech prior to presentation may provide a significant advantage in speech recognition when the target talker is masked by one other talker. The effect they found was diminished if the target and masker talkers were perceptually segregated or if the masker was continuous noise. Priming advantages were also observed when the priming talker was not the same as the target talker or the prime was provided in written form prior to the trial. The authors interpreted this performance advantage due to a priori information as providing an aid to the focus of attention on the target.

# 9. Age-Related Effects

There is a great deal of interest in determining how the ability to focus attention on one auditory source and exclude others changes with age. The development of this ability in children can have very important consequences in a variety of ways and may, for example, be crucial to success in educational settings. Werner and Bargones (1991) have shown that infants are highly distracted by background noise remote from the target frequency, causing significant elevations in threshold that cannot be attributed to sensory factors. If infants or children are susceptible to fairly low uncertainty distracters, what would the expectation be when the stimulus is highly uncertain, as in informational masking experiments? For the elderly, the ability to sort out acoustic environments may also be crucial to successful communication, but can be unusually challenging particularly when accompanied by hearing loss. Are there increases in susceptibility to informational masking that normally occur with advancing age? Is there an interaction between aging and hearing loss that explains the extreme difficulty older hearing-impaired listeners sometimes experience in complex and uncertain multisource listening environments?

While the literature on age-related effects in masking, at both ends of the lifespan, is vast and many studies are informative about processes involved in masking beyond energetic masking, here we confine our attention to those studies explicitly intended to separate energetic and informational factors. Allen and Wightman (1995) examined both target-frequency uncertainty and masker uncertainty in children and adults. The children ranged from 3 to 5 years of age. Their hypothesis was that children do not focus attention on a specific frequency region as well or selectively as adults. The consequences of unfocused or roving attention in the frequency domain, they reasoned, might be manifested in two contrasting ways: First, there should be reduced costs associated with single-channel vs. multichannel monitoring when target frequency is uncertain, and second, there should be greater susceptibility to masker-frequency uncertainty. Their findings generally supported this hypothesis: As a group, the children demonstrated higher thresholds in all conditions than those of the adults. However, most children failed to show a significant decrease in performance when the frequency of the *target* was randomized. In contrast, most adults did show such a performance decrement. Furthermore, using the multitone masking paradigm, they found that as a group, young children were significantly more susceptible to masker-frequency uncertainty. Allen and Wightman (1995) state, "Our data suggest that the majority of young children may be poorer listeners, in general, perhaps because the central mechanisms that subserve selective attention are immature" (p. 511).

Oh et al. (2001) obtained thresholds by number of components functions using the multitone masking procedure for a group of preschool children and a control group of adults. Their results are shown in Figure 6.12. The individual data are plotted as points, with the heavy solid line indicating group mean thresholds for children and the heavy dashed line the group mean thresholds for adults (the dotted line is for an earlier group of adults). The principal findings are that the difference between the two age groups varied significantly according to the number of masker components and, by inference, the proportion of energetic-to-informational masking. When energetic masking dominated (many masker components), the two groups exhibited similar amounts of masking. However,

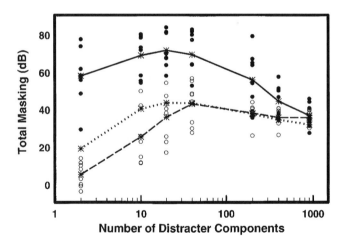

FIGURE 6.12. The amount of masking obtained in the multitone masking experiment is plotted as a function of the number of masker components for children (filled symbols) and adults (open symbols). The circles are thresholds for individual subjects, the asterisks connected by lines are mean data for children (solid line), and two adult groups (dashed and dotted lines). (Reprinted with permission from Oh et al. 2001.)

when there were few masker components, the children exhibited much more masking than the adults, with the largest difference of more than 50 dB found for the fewest number of components. These large differences are not likely to result from anatomical or physiological maturation, the authors conclude nor are they easily attributed to differences in auditory filter bandwidths. As with the studies above suggesting greater informational masking in children than adults, the likely explanation appears to be reduced selective listening capabilities.

In the preceding studies, the greater effects of masker uncertainty, and the lesser effects of signal-frequency uncertainty, were discussed in terms of the ability to selectively attend to frequency channels. However, the two ears may also be thought of as separate channels that, to some degree, may be selected and/or ignored. Wightman et al. (2003) reported an intriguing finding using a variation of the multitone masking experiment in which the masker components were presented to the ear contralateral to the target rather than to the same ear as the target. This extreme case of "spatial" or channel separation of target and masker produced virtually no masking in adults. In young children, however, a contralateral multitone masker was nearly as effective as the same masker presented ipsilaterally when considered with respect to a broadband noise masker reference. The implication is that the ability of adults to hold separate the inputs from the two ears in highly uncertain listening conditions is not fully developed in young children. It is interesting to consider these findings in the context of the work described in Section 8 above in which susceptibility to a contralateral informational masker was observed in normal-hearing adults, but only when the listener was faced with an ipsilateral segregation task. Thus, although the ability to ignore unwanted sounds in the ear contralateral to the primary focus of attention develops throughout childhood, even adults experience difficulty with that task in certain complex listening situations.

Hall et al. (2005) compared the ability of children 4–9 years of age and adults to "hear out" a target tone sequence embedded in MBS and MBD maskers. The MBS and MBD maskers consisted of two frequency components in each burst, one above and one below the target frequency. In addition to these two masked conditions, which were tested monaurally, a third monaural condition was tested in which the masker was MBS with a temporal fringe preceding the onset of the target. In a fourth condition, the masker was presented to both ears while the target was presented to only one ear ($T_m M_0$). Both groups—children and adults—exhibited significant amounts of masking when the masker was MBS and significant release from masking when the masker was MBD. In general, the children demonstrated more masking than the adults, with less of a release from informational masking observed in most conditions. In the dichotic condition, the adults demonstrated a moderate release from informational masking, while the children actually exhibited increased masking.

Wightman and Kistler (2005) examined the ability of children to identify speech using the CRM test when the target speech was masked by speech in the same ear and in some conditions when a second speech masker was presented contralaterally. In their study, 38 children 4–16 years of age were tested and their results compared to those of a group of 8 adults. Generally, as a group, the children required a higher T/M to achieve performance equivalent to that of the adults. Overall, the children demonstrated about 15 dB more informational masking than the adults. When a finer-grain analysis of performance by age was conducted, there was a monotonic improvement (reduction in target-to-masker ratio for constant performance level) as age increased, with a significant difference still apparent between the oldest children and the adults. Adding a second speech masker to the opposite ear increased masking by about 5 dB for both groups. The authors concluded that this study provided further evidence for the poorer selective listening abilities in children and hence their generally greater susceptibility to informational masking compared to adults. It is of interest to note that "selectivity" here is not selectivity along a simple stimulus dimension such as frequency or space but rather selection of one source of speech in the presence of competing sources.

In order to understand how susceptibility to informational masking changes across the lifespan, it is important to examine the performance of elderly listeners as well as children. While there are many studies that have attempted to determine the relationship between masking and age, there are relatively few that were specifically intended to examine the energetic/informational masking distinction. Of the few that have been published, the most relevant have been concerned with speech on speech masking, where there is the potential for informational masking to strongly influence the results. Li et al. (2004) suggested that while older listeners perform more poorly than younger listeners in complex environments, the differences can be described as a change in the T/M necessary for achieving a particular level of performance rather than a fundamental difference in the ability to use the cues that are present. Comparing speech maskers and noise maskers in a speech-recognition paradigm similar to that of Freyman et al. (1999), Li et al. also found greater release from masking for speech than for noise. The interesting result in terms of older listeners was that the relative improvement was the same for both older and younger listeners. What differed between the groups was the performance in the control condition (no perceived difference in location between target and masker), with younger listeners achieving better speech intelligibility than older listeners at a specified T/M. Based on these results, Li et al. concluded that any differences between the groups were based on age-related auditory declines and that "there is no evidence to suggest that age-related changes at the cognitive level are contributing to these difficulties" (p. 1088).

Tun et al. (2002) conducted a similar experiment but came to quite a different conclusion. Rather than measuring T/M, Tun et al. measured the ability of listeners to recall speech that had been presented in the presence of various distracters. For the older listeners, there was greater interference in target speech recall when the distracters were meaningful speech than when they were nonmeaningful speech. Younger listeners, however, did not exhibit a difference in the amount of interference from meaningful versus nonmeaningful distracters. In addition, the younger listeners were more likely to recall words from the distracting speech in an unexpected memory test. From these data, Tun et al. concluded that older listeners "may have a reduced ability in selective listening that would place them at a disadvantage in a listening environment that contained competing voices" (p. 465). The difference between the conclusions of these two studies is in large part based on the fact that in the Li et al. study, T/M was regarded as a confounding variable to be removed, whereas in the Tun et al. study, a reduction in performance at a given T/M constituted one of the central dependent variables. What is not well established is whether the increase in T/M necessary to equate the performance of younger and older listeners is due merely to sensory factors, in which case the susceptibility to informational masking may not be affected by increasing age. However, if there is a nonsensory component present, the converse may well be true.

## 10. Informational Masking and Hearing Loss

Is it reasonable to expect that listeners with peripheral hearing loss should demonstrate differences in their susceptibility to informational masking, or in making use of the cues that have been found to overcome it, as compared to their normal-hearing counterparts? Even if differences are found, to what extent can they be attributed purely to processes in the peripheral mechanism, such as reduced sensitivity or compression, or to broader filters? This section reviews the very limited and sometimes contradictory evidence available to date using procedures normally applied to the study of informational masking in listeners with normal hearing. An initial observation about this topic is that the effects of configuration of hearing loss may well be quite different than for energetic masking. This is because informational masking occurs due to masker energy remote in frequency from the target. Thus, hearing loss may affect the target and masker differently in ways that are difficult to predict.

Micheyl et al. (2000) and Alexander and Lutfi (2004) have both reported findings from listeners with sensorineural hearing loss using the multitone masking procedure. Although there are several differences in their studies (e.g., Micheyl et al. tested subjects with various configurations of loss including asymmetrical, Alexander and Lutfi used a modification of the more common procedure to vary uncertainty, etc.), the general conclusion from both studies was that sensorineural hearing loss did not cause greater susceptibility to informational masking. And in some cases, listeners with hearing loss demonstrated less-than-normal informational masking. As Alexander and Lutfi point out, this finding actually is predicted by the CoRE model (Lutfi 1993) because of the reduced dynamic ranges of the listeners with hearing loss. Recall from above that the CoRE model predicts that informational masking will increase as the variability of the outputs of the peripheral filters, as computed by statistical summation across trials, increases. A large contributor to the variability, normally, is the difference between output levels on trials in which masker components fall inside a filter and trials in which masker components do not fall inside a filter. In the latter case, the output of the filter is very low (e.g., the value for quiet threshold). If thresholds are increased due to hearing loss, and the range of possible output levels is reduced, the predicted amount of informational masking is also reduced. Alexander and Lutfi (2004) found support for this prediction in cases in which the thresholds for normal-hearing listeners and hearing-impaired listeners were obtained at equal masker sound-pressure levels. However, the differences between groups were reduced when threshold comparisons were made for maskers presented at equal sensation levels.

Kidd et al. (2002b) tested a group of 46 listeners with sensorineural hearing loss on the MBS and MBD multitone maskers described earlier. They also obtained auditory filter shape measurements from a subset of the subject group. They modified the procedure described earlier to attempt to limit the influence of peripheral factors in the measurements. Their findings are summarized in Figure 6.13. This figure shows target-to-masker level in dB at masked threshold as a function of hearing loss. The open and filled symbols show T/Ms at threshold for MBS and MBD maskers, respectively. Each point is for a different subject. The lines were fit to the results using a least-squares criterion. The most striking result is the increase in T/M as hearing loss increased for the MBD masker. The increase in T/M based on the fit is about 50 dB for an increase in threshold of 60 dB. Thus, performance in the MBD masker was strongly affected by hearing loss. The interpretation of performance in the MBD masker is that listeners achieve low thresholds by detecting the spectrotemporal coherence of the target in an otherwise random background. One possible interpretation for this large

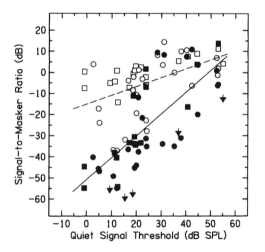

FIGURE 6.13. Signal-to-masker ratio in dB at masked threshold as a function of quiet signal threshold for a group of 46 listeners with sensorineural hearing loss. Each point represents a value for an individual subject. The solid symbols are for the MBD masker, while the open symbols are for the MBS masker. The lines are least-squares fits to the data. (Reprinted with permission of Springer Science+Business Media from Figure 4, p. 114, in Kidd et al. 2002b.)

effect is that listeners with sensorineural hearing loss are poorer at analytic listening in complex backgrounds. The auditory filter measurements could not account for the results. A less obvious effect was found for the MBS masker with T/Ms changing much less over the same range.

Arbogast et al. (2005) conducted a study of speech identification in a group of listeners with sensorineural hearing loss and an age-matched group with normal hearing. The stimuli and methods were very similar to those used in Arbogast et al. (2002) discussed above. The targets and speech maskers were processed multiband CRM sentences, and same-band and different-band noise maskers were used. Both colocated and spatially separated conditions were tested in a sound field. A summary of the group mean results (target-to-masker ratios in dB at midpoints of psychometric functions taken from their Tables II and III) is shown in Figure 6.14. Normal hearing (open symbols) and hearing-impaired (filled symbols) listener results are plotted side by side for each of the three types of maskers. For the colocated (0°) condition, the largest difference in performance between the two groups is for the different-band noise masker, where the T/Ms for the normal-hearing group are about 13 dB lower than for the hearing-impaired group. Arbogast et al. interpreted this result as indicating greater energetic masking in the hearing-impaired listeners due to wider auditory filters. The additional increase in T/M (with respect to different band noise) when the different-band speech was presented was about 10 dB greater for the normal-hearing listeners than for the hearing-impaired listeners. Thus, more informational masking (difference in T/M for the two types of maskers) was found for the normal hearing group. However, the ability to use spatial cues to overcome the masking produced by the different-band speech was reduced by about 6 dB in the hearing-impaired group compared to the normal hearing

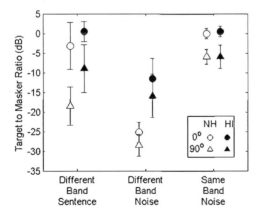

FIGURE 6.14. Group mean target-to-masker ratios at threshold for three types of maskers (abscissa) for listeners with normal hearing (open symbols) and with sensorineural hearing loss (filled symbols). Circles indicate values for signal and masker presented from the same location, while triangles indicate values when the signal and masker were spatially separated by a 90° azimuth.

group. Comparison of the group-mean results from normal-hearing and hearing-impaired listener groups may be summarized as follows: First, in the highly energetic (same band) masker, no differences were found in any condition; second, more masking was found for the hearing-impaired group in all different-band masker conditions; third, the ratio of energetic-to-informational masking (as estimated by these stimuli and procedures) was greater for the hearing-impaired group; and fourth, the hearing-impaired listeners were less able to take advantage of the spatial separation of target and masker to reduce masking in the different-band speech masker.

It is tempting to conclude that no general statements may be made about informational masking and hearing loss based on the few studies in the literature specifically intended to examine the issue. However, even based on the few studies described above, some consistencies are apparent. The well-known consequence of sensorineural hearing loss of broader-than-normal auditory filters means that spectrally sparse stimuli, independent of whether they are tones or even bands of speech, will interact more with one another in the periphery, causing greater-than-normal energetic masking. There are examples given in sections above (cf. Oh and Lutfi 1998; Kidd et al. 2003b), where increasing the amount of energetic masking in a stimulus results in a concomitant decrease in informational masking. Furthermore, two studies (Kidd et al. 2002b; Arbogast et al. 2005) found that listeners with sensorineural hearing loss were less able than normal-hearing listeners to use perceptual cues to reduce the informational masking that was present.

## 11. Summary and Concluding Remarks

This chapter has provided a review of selected topics and articles related to informational masking. The scope of the review was necessarily limited given the large number and diversity of the relevant studies. These studies included experiments employing detection, discrimination, and nonspeech and speech identification tasks. The common theme shared by this wide array of studies is the conclusion that the masking observed was not attributable to overlapping patterns of excitation in the auditory periphery. Although not reviewed in any detail here, this work is closely related to the original informational masking studies of Pollack (e.g., Pollack 1976) and of Watson and colleagues (Watson and Kelly 1981; Watson 1987, 2005).

### 11.1 What Is Informational Masking?

In the introductory section of this chapter, two primary problems were identified with the term "informational masking." First, defining informational masking by using energetic masking as a reference requires both a clear definition of energetic masking and accurate ways of estimating energetic masking (e.g., by modeling or empirical work). Some of the difficulties with the latter condition, at least, were reviewed above. The suggestion by Durlach et al. (2003a) that energetic masking be defined by an Ideal Observer analysis applied to a specific physiological site seems reasonable and ultimately may be achievable with sufficiently accurate physiological models. Second, masking that occurs beyond energetic masking (which has been identified with many labels, such as central masking, remote masking, excess masking, perceptual masking, cognitive interference) may be due to many mechanisms or limitations on processing. Concluding that the phenomenon is informational masking, then, does not necessarily distinguish among them. Does this mean that informational masking is not a useful term or concept?

Attention, grouping and segregation, memory, general processing capacity—all are factors that are related to producing informational masking or in causing release from informational masking. Unfortunately, these terms are also not always well defined or have generally agreed-upon meanings, and often factors interact in difficult-to-determine ways in particular experiments. Uncertainty about which stimulus or stimulus features will be presented on a given observation, or similarity among stimuli, is often associated with causing informational masking. However, those descriptors are insufficient, too. Is the adverse effect of uncertainty on performance always attributable to misdirected attention, or could it as easily involve a failure to segregate sources or a breakdown in the process of storage in memory? We would argue that informational masking as a term, and the energetic-informational masking distinction, are useful in describing various types of masking phenomena. What appears to be needed next, then, is more examination and discussion of the various mechanisms that cause informational masking. In this section we consider some of the possibilities and attempt to provide examples.

If we assume that in accordance with the suggestion of Durlach et al., informational masking occurs even when there is sufficient information about the target in the neural representation at a given physiological site, then what are the possible reasons the human observer fails to solve the task? Durlach et al. (2003a) speculate that the representation of a target at one physiological level might be sufficient, i.e., not energetically masked, yet at a higher level, that would not be true. Are there any examples in the literature that would support such an interpretation? One recent study has reported results that would be consistent with such an interpretation. Kidd et al. (2005a) used a speech identification procedure in which the speech target consisted of a set of very narrow frequency bands and the speech masker consisted of a mutually exclusive set of very narrow frequency bands (see Figure 6.11). When broadband noise was presented simultaneously to the ear opposite the target and masker, there was no effect on performance. However, when the contralateral noise was a set of narrow frequency bands that exactly corresponded to the speech masker bands, the speech masker was significantly diminished in effectiveness. Both the target and the masker had to be present for this contralateral noise effect to occur. The subjective impression, which is consistent with the results, is that the speech masker and contralateral noise bands were simply "added" at a site where binaural input occurs. Our recent observations indicate that a similar diminishing of the intelligibility of the target speech may be produced by corresponding contralateral bands of noise. Although there may be other explanations for this effect besides binaural summation, if confirmed, this finding would provide evidence that energetic masking can increase as the physiological site progresses to higher levels in the CNS.

Another possible cause of informational masking is the combination (or "grouping") of the outputs of the neural elements representing the target with those elements representing irrelevant stimuli at a given physiological site. This could happen for a variety of reasons, such as target and masker being "mixed" along a particular stimulus dimension and presented synchronously. One possible illustration of informational masking due to a failure of segregation was reported by Kidd et al. (2003b) (also discussed in Section 7). In the relevant experiment, a harmonic complex was masked by a random-frequency four-component masker. When both the target and masker were presented simultaneously to a single ear, a single auditory object was perceived. However, if the masker was also presented to the contralateral ear, two images were heard: The target was heard in the test ear, while the masker was heard near the midline, and the ability of the listener to discriminate the harmonic property of the target improved substantially. The interpretation of this finding that is germane here is that there was sufficient information in the monaural condition to solve the task, because no additional information about the target was provided by the contralateral masker, and attention was presumably fully focused on the monaural stimulus. However, the task could not be reliably solved because of the obligatory (in this case) grouping of target and masker elements. When the masker image was segregated from the target by dichotic presentation, there was a substantial improvement in performance because the representation of the target could be separated from that of the masker.

Informational masking, consistent with the definition proposed by Durlach et al. (2003a), could also be caused by an incorrect selection of the available neural elements—either to enhance or to suppress—at a given physiological site. For example, if both the target and masker were fully represented at a given site, but the responses of the elements representing the target were attenuated or eliminated, then errors in the subsequent processing of the stimulus would occur. In this case, errors occur not because the target and masker are combined into an inseparable object, but because the response to the target is suppressed and does not propagate to higher neural levels. It has been shown in many studies, dating at least from the probe-signal experiment of Greenberg and Larkin (1968), that observer expectation can affect detectability (or other tasks) at a point along a simple stimulus dimension, such as frequency. Higher-level tasks, such as speech recognition in multitalker environments, also are highly influenced by expectation and are vulnerable to misdirected attention (e.g., Brungart et al. 2001; Kidd et al. 2005b). Thus, misdirected attention can cause errors in performance even though the neural representation is sufficient to solve the task.

For sequences of related sounds, or "streams," it can also occur that the connection between successive elements is lost because of another (masking) stimulus. Connected speech is one such example, in which the ongoing comprehension of information requires maintaining the integrity of the perceptual stream of sounds emanating from a specific sound source. As an example of possible errors in this process, Kidd et al. (2005b) found evidence for the "loss of stream segregation" in a three-talker environment for spatially separated talkers. They interpreted "mixing errors" in which the key words reported in a sentence identi-fication task were a blend of target and masker key words, as indicating that the target speech stream was not held separate from that of the masker. This putative breakdown in stream segregation could have resulted from a variety of factors, including an inability to follow the vocal characteristics of the target talker over time, or perhaps from directing the focus of attention toward the wrong spatial location. Evidence supporting the important role of focus of attention in stream segregation has recently been reported by Carlyon et al. (2001).

And finally (although there are doubtless many other possibilities not considered here), limitations on the short-term storage and retrieval of sounds in memory, or interruptions in the processing of stored sounds, can produce informational masking, consistent with the definition we are applying here. Pollack's early work in this area was concerned with the interruption of the processing of "strings of acoustical sequences" by other similar sounds (Pollack (2001), personal commu-nication). His work was influenced by that of Massaro (1975a, b), in which the processing of a stored "preperceptual image" of a speech sound could be inter-rupted by another sound presented immediately after it (referred to as "backward recognition masking"). The demonstrations of "masking" produced in recognition of these speech sounds, or statistical sequences, were consistent with (and indeed inspired the term) informational masking. Another more recent example falling in this category was provided by Conway et al. (2001), who found that some subjects monitoring a primary source of speech and told to ignore irrelevant speech in the opposite ear demonstrated a lapse in performance immediately after their own name was presented in the "unattended" ear. The susceptibility to this form of contralateral interference was generally greater in subjects with relatively low estimated working memory capacity, suggesting that such listeners are less able to selectively attend to target sounds in the presence of distracting irrelevant sounds.

*Acknowledgments.* The authors thank several colleagues who generously contributed to this chapter by allowing reprinting of their figures and/or by comments and suggestions about the text: Laurel Carney, Robert Lutfi, Eunmei Oh, Donna Neff, and Beverly Wright. We also express our appreciation to our current and former collaborators, especially Tanya L. Arbogast, Virginia Best, H. Steven Colburn, Antje Inlefeld, Nicole Marrone, and Barbara Shinn-Cunningham. The authors gratefully acknowledge the support of NIH/NIDCD and AFOSR in the preparation of this chapter and a portion of the work described here.

# References

Alexander JM, Lutfi RA (2004) Informational masking in hearing-impaired and normal-hearing listeners: Sensation level and decision weights. J Acoust Soc Am 116:2234–2247.

Allen P, Wightman F (1995) Effects of signal and masker uncertainty on children's detection. J Speech Hear Res 38:503–511.

ANSI (1994) S1.1-1994 American National Standard Acoustical Terminology. New York: Acoustical Society of America.

Arbogast TL, Mason CR, Kidd G, Jr. (2002) The effect of spatial separation on informational and energetic masking of speech. J Acoust Soc Am 112:2086–2098.

Arbogast TL, Mason CR, Kidd G, Jr. (2005) The effect of spatial separation on informational masking of speech in normal-hearing and hearing-impaired listeners. J Acoust Soc Am 117:2169–2180.

Best V, Ozmeral E, Gallun FJ, Sen K, Shinn-Cunningham BG (2005) Spatial unmasking of birdsong in human listeners: Energetic and informational factors. J Acoust Soc Am 118:3766–3773.

Bilger RC (1959) Additivity of different types of masking. J Acoust Soc Am 31:1107–1109.

Bolia RS, Nelson WT, Ericson MA, Simpson BD (2000) A speech corpus for multitalker communications research. J Acoust Soc Am 107:1065–1066.

Bos CE, de Boer E (1966) Masking and discrimination. J Acoust Soc Am 39:708–715.

Bregman AS (1990) Auditory Scene Analysis. Cambridge, MA: MIT Press.

Bronkhorst AW (2000) The Cocktail Party Phenomenon: A review of research on speech intelligibility in multiple-talker conditions. Acta Acustica 86:117–128.

Brungart DS (2001) Informational and energetic masking effects in the perception of two simultaneous talkers. J Acoust Soc Am 109:1101–1109.

Brungart DS, Simpson BD (2002) Within-ear and across-ear interference in a cocktail-party listening task. J Acoust Soc Am 112:2985–2995.

Brungart DS, Simpson BD (2004) Within-ear and across-ear interference in a dichotic cocktail party listening task: Effects of masker uncertainty. J Acoust Soc Am 115:301–310.

Brungart DS, Simpson BD, Ericson MA, Scott KR (2001) Informational and energetic masking effects in the perception of multiple simultaneous talkers. J Acoust Soc Am 110:2527–2538.

Buus S, Schorer E, Florentine M, Zwicker E (1986) Decision rules in detection of simple and complex tones. J Acoust Soc Am 80:1646–1657.

Carhart R, Tillman TW, Greetis E (1969) Perceptual masking in multiple sound backgrounds. J Acoust Soc Am 45:694–703.

Carlyon RP, Cusack R, Foxton JM, Robertson IH (2001) Effects of attention and unilateral neglect on auditory stream segregation. J Exp Psychol [Hum Percept Perform] 27:115–127.

Carney LH, Heinz MG, Evilsizer ME, Gilkey RH, Colburn HS (2002) Auditory phase opponency: A temporal model for masked detection at low frequencies. Acust Acta Acust 88:334–347.

Cherry EC (1953) Some experiments on the recognition of speech, with one and with two ears. J Acoust Soc Am 25:975–979.

Colburn HS, Carney LH, Heinz MG (2003) Quantifying the information in auditory-nerve responses for level discrimination. J Assoc Res Otolaryngol 4:294–311.

Conway ARA, Cowan N, Bunting MF (2001) The cocktail party phenomenon revisited: The importance of working memory capacity. Psychonom Bull Rev 8:331–335.

Creelman CD (1960) Detection of signals of uncertain frequency. J Acoust Soc Am 32:805–810.

Darwin CJ (1992) Listening to two things at once. In: Meh S (ed) The Auditory Processing of Speech: From Sounds to Words. Berlin: Mouton de Gruyter, pp. 133–147.

Dau T, Puschel D, Kohlrausch A (1996) A quantitative model of the "effective" signal processing in the auditory system. I. Model structure. J Acoust Soc Am 99:3615–3622.

Delgutte B (1990) Physiological mechanisms of psychophysical masking: Observations from auditory-nerve fibers. J Acoust Soc Am 87:791–809.

Delgutte B (1996) Physiological models for basic auditory percepts. In: Hawkins HL, McMullen TA, Popper AN, Fay RR (eds) Auditory Computation. New York: Springer-Verlag, pp. 157–220.

Dorman MF, Loizou PC, Rainey D (1997) Speech intelligibility as a function of the number of channels of stimulation for signal processors using sine-wave and noise-band outputs. J Acoust Soc Am 102:2403–2411.

Durlach NI, Braida LD, Ito Y (1986) Towards a model for discrimination of broadband signals. J Acoust Soc Am 80:63–72.

Durlach NI, Mason CR, Kidd G, Jr, Arbogast TL, Colburn HS, Shinn-Cunningham BG Jr (2003a) Note on informational masking (L). J Acoust Soc Am 113:2984–2987.

Durlach NI, Mason CR, Shinn-Cunningham BG, Arbogast TL, Colburn HS, Kidd G Jr (2003b) Informational masking: counteracting the effects of stimulus uncertainty by decreasing target–masker similarity. J Acoust Soc Am 114:368–379.

Durlach NI, Mason CR, Gallun FJ, Shinn-Cunningham B, Colburn HS, Kidd G Jr (2005) Informational masking for simultaneous nonspeech stimuli: Psychometric functions for fixed and randomly mixed maskers. J Acoust Soc Am 118:2482–2497.

Ebata M (2003) Spatial unmasking and attention related to the cocktail party problem. Acoust Sci Tech 24:208–219.

Erell A (1988) Rate coding model for discrimination of simple tones in the presence of noise. J Acoust Soc Am 84:204–214.

Fletcher H (1929) Speech and Hearing. New York: Van Nostrand.

Fletcher H (1940) Auditory patterns. Rev Mod Phys 12:47–65.

Freyman RL, Helfer KS, McCall DD, Clifton RK (1999) The role of perceived spatial separation in the unmasking of speech. J Acoust Soc Am 106: 3578–3588.

Freyman RL, Balakrishnan U, Helfer KS (2001) Spatial release from informational masking in speech recognition. J Acoust Soc Am 109:2112–2122.

Freyman RL, Balakrishnan U, Helfer KS (2004) Effect of number of masking talkers and auditory priming on informational masking in speech recognition. J Acoust Soc Am 115:2246–2256.

Gallun FJ, Mason CR, Kidd G, Jr. (2005) Binaural release from informational masking in a speech identification task. J Acoust Soc Am 118: 1614–1625.

Gilkey RH (1987) Spectral and temporal comparisons in auditory masking. In: Yost WA, Watson CS (eds) Auditory Processing of Complex Sounds. Hillsdale, NJ: Lawrence Erlbaum 26–36.

Glasberg BR, Moore BCJ (1990) Derivation of auditory filter shapes from notched-noise data. Hear Res 47:103–138.

Green DM (1961) Detection of auditory sinusoids of uncertain frequency. J Acoust Soc Am 33.

Green DM (1967) Additivity of masking. J Acoust Soc Am 41:1517–1525.

Green DM (1983) Profile Analysis: A different view of auditory intensity discrimination. Am Psychol 38:133–142.

Green DM (1988) Profile Analysis: Auditory Intensity Discrimination. New York: Oxford University Press.

186    G. Kidd, Jr., et al.

Green DM, Swets JA (1974) Signal Detection Theory and Psychophysics. Mebourne, FL: Krieger Publishing Company.

Greenberg GZ, Larkin WD (1968) Frequency-response characteristic of auditory observers detecting signals of a single frequency in noise: The probe-signal method. J Acoust Soc Am 44:1513–1523.

Hall JW, III, Grose JH (1988) Comodulation masking release: Evidence for multiple cues. J Acoust Soc Am 84:1669–1675.

Hall JW, III, Buss E, Grose JH (2005) Informational masking release in children and adults. J Acoust Soc Am 118:1605–1613.

Hartmann WM (1997) Signals, Sound, and Sensation. New York: Springer-Verlag.

Hartmann WM, McAdams S, Gerzso A, Boulez P (1986) Discrimination of spectral density. J Acoust Soc Am 79:1915–1925.

Hartmann WM, McAdams S, Smith BK (1990) Hearing a mistuned harmonic in an otherwise periodic complex tone. J Acoust Soc Am 88:1712–1724.

Heinz MG (2000) Quantifying the effects of the cochlear amplifier on temporal and average-rate information in the auditory nerve. PhD thesis, Massachusetts Institute of Technology.

Heinz MG, Colburn HS, Carney LH (2002) Quantifying the implications of nonlinear cochlear tuning for auditory-filter estimates. J Acoust Soc Am 111:996–1011.

Kidd G Jr, Mason CR, Green DM (1986) Auditory profile analysis of irregular sound spectra. J Acoust Soc Am 79:1045–1053.

Kidd G Jr, Mason CR, Brantley MA, Owen GA (1989) Roving-level tone-in-noise detection. J Acoust Soc Am 86:1310–1317.

Kidd G Jr, Uchanski RM, Mason CR, Deliwala PS (1993) Discriminability of narrow-band sounds in the absence of level cues. J Acoust Soc Am 93:1028–1037.

Kidd G Jr, Mason CR, Deliwala PS, Woods WS, Colburn HS (1994) Reducing informational masking by sound segregation. J Acoust Soc Am 95:3475–3480.

Kidd G Jr, Mason CR, Dai H (1995) Discriminating coherence in spectro-temporal patterns. J Acoust Soc Am 97:3782–3790.

Kidd G Jr, Mason CR, Rohtla TL, Deliwala PS (1998) Release from masking due to spatial separation of sources in the identification of nonspeech auditory patterns. J Acoust Soc Am 104:422–431.

Kidd G Jr, Mason CR, Arbogast TL (2002a) Similarity, uncertainty, and masking in the identification of nonspeech auditory patterns. J Acoust Soc Am 111:1367–1376.

Kidd G Jr, Arbogast TL, Mason CR, Walsh M (2002b) Informational masking in listeners with sensorineural hearing loss. J Assoc Res Otolaryngol 3:107–119.

Kidd G Jr, Mason CR, Richards VM (2003a) Multiple bursts, multiple looks, and stream coherence in the release from informational masking. J Acoust Soc Am 114:2835–2845.

Kidd G Jr, Mason CR, Brughera A, Chiu C-YP (2003b) Discriminating harmonicity. J Acoust Soc Am 114:967–977.

Kidd G Jr, Mason CR, Arbogast TL, Brungart DS, Simpson BD (2003c) Informational masking caused by contralateral stimulation. J Acoust Soc Am 113:1594–1603.

Kidd G Jr, Mason CR, Gallun FJ (2005a) Combining energetic and informational masking for speech identification. J Acoust Soc Am 118:982–992.

Kidd G Jr, Arbogast TL, Mason CR, Gallun FJ (2005b) The advantage of knowing where to listen. J Acoust Soc Am 117:2537.

Leek MR, Watson CS (1984) Learning to detect auditory pattern components. J Acoust Soc Am 76:1037–1044.

Leek MR, Brown ME, Dorman MF (1991) Informational masking and auditory attention. Percept Psychophys 50:205–214.

Li L, Daneman M, Qi JG, Schneider BA (2004) Does the information content of an irrelevant source differentially affect spoken word recognition in younger and older adults? J Exp Psychol [Hum Percept Perform] 30:1077–1091.

Licklider JCR (1951) Basic correlates of the auditory stimulus. In: Stevens SS (ed) Handbook of Experimental Psychology. New York: John Wiley & Sons, pp. 985–1039.

Loizou PC, Dorman M, Tu Z (1999) On the number of channels needed to understand speech. J Acoust Soc Am 106:2097–2103.

Lutfi RA (1983) Additivity of simultaneous masking. J Acoust Soc Am 73:262–267.

Lutfi RA (1993) A model of auditory pattern-analysis based on component-relative-entropy. J Acoust Soc Am 94:748–758.

Mason CR, Kidd G, Jr, Hanna TE, Green DM (1984) Profile analysis and level variation. Hear Res 13:269–275.

Massaro DW (1975a) Backward recognition masking. J Acoust Soc Am 58:1059–1065.

Massaro DW (1975b) Experimental Psychology and Information Processing. Chicago: Rand McNally.

Micheyl C, Arthaud P, Reinhart C, Collet L (2000) Informational masking in normal-hearing and hearing-impaired listeners. Acta Otolaryngol 120:242–246.

Moore BCJ, Glasberg BR, Peters RW (1986) Thresholds for hearing mistuned partials as separate tones in harmonic complexes. J Acoust Soc Am 80:479–483.

Neff DL (1995) Signal properties that reduce masking by simultaneous, random-frequency maskers. J Acoust Soc Am 98:1909–1920.

Neff DL, Callaghan BP (1988) Effective properties of multicomponent simultaneous maskers under conditions of uncertainty. J Acoust Soc Am 83:1833–1838.

Neff DL, Dethlefs TM (1995) Individual differences in simultaneous masking with random-frequency, multicomponent maskers. J Acoust Soc Am 98:125–134.

Neff DL, Green DM (1987) Masking produced by spectral uncertainty with multicomponent maskers. Percept & Psychophys 41:409–415.

Neff DL, Jesteadt W (1996) Intensity discrimination in the presence of random-frequency, multicomponent maskers and broadband noise. J Acoust Soc Am 100:2289–2298.

Neff DL, Dethlefs TM, Jesteadt W (1993) Informational masking for multicomponent maskers with spectral gaps. J Acoust Soc Am 94:3112–3126.

Oh EL, Lutfi RA (1998) Nonmonotonicity of informational masking. J Acoust Soc Am 104:3489–3499.

Oh EL, Lutfi RA (2000) Effect of masker harmonicity on informational masking. J Acoust Soc Am 108:706–709.

Oh EL, Wightman F, Lutfi RA (2001) Children's detection of pure-tone signals with random multitone maskers. J Acoust Soc Am 109:2888–2895.

Oxenham AJ, Fligor BJ, Mason CR, Kidd G, Jr. (2003) Informational masking and musical training. J Acoust Soc Am 114:1543–1549.

Patterson RD, Nimmo-Smith I, Weber DL, Milroy R (1982) The deterioration of hearing with age: Frequency selectivity, the critical ratio, the audiogram, and speech threshold. J Acoust Soc Am 72:1788–1803.

Pfafflin SM, Mathews MV (1966) Detection of auditory signals in reproducible noise. J Acoust Soc Am 39:340–345.

Pollack I (1975) Auditory informational masking. J Acoust Soc Am 57:S5.

Pollack I (1976) Identification of random auditory waveforms. III. Effect of interference. J Acoust Soc Am 60:680–686.

Pollack I, Pickett JM (1958) Stereophonic listening and speech intelligibility against voice babble. J Acoust Soc Am 30:131–133.

Richards VM (1992) The detectability of a tone added to narrow bands of equal-energy noise. J Acoust Soc Am 91:3424–3435.

Richards VM, Neff DL (2004) Cuing effects for informational masking. J Acoust Soc Am 115:289–300.

Richards VM, Tang Z (2006) Estimates of effective frequency selectivity based on the detection of a tone added to complex maskers. J Acoust Soc Am 119:1574–1584.

Richards VM, Tang Z, Kidd G, Jr. (2002) Informational masking with small set sizes. J Acoust Soc Am 111:1359–1366.

Richards VM, Huang R, Kidd G, Jr (2004) Masker-first advantage for cues in informational masking. J Acoust Soc Am 116:2278–2288.

Schafer TH, Gales RS, Shewmaker CA, Thompson PO (1950) The frequency selectivity of the ear as determined by masking experiments. J Acoust Soc Am 22:490–496

Schubert ED, Schultz MC (1962) Some aspects of binaural signal selection. J Acoust Soc Am 34:844–849.

Shannon RV, Zeng F-G, Wygonski J, Kamath V, Ekelid M (1995) Speech recognition with primarily temporal cues. Science 270:303–304.

Shinn-Cunningham BG, Ihlefeld A, Satyavarta, Larson E (2005) Bottom-up and top-down influences on spatial unmasking. Acust Acta Acust 91:967–979.

Spiegel MF, Green DM (1982) Signal and masker uncertainty with noise maskers of varying duration, bandwidth and center frequency. J Acoust Soc Am 71: 1204–1210.

Spiegel MF, Picardi MC, Green DM (1981) Signal and masker uncertainty in intensity discrimination. J Acoust Soc Am 70:1015–1019.

Spieth W, Curtis JF, Webster JC (1954) Responding to one of two simultaneous messages. J Acoust Soc Am 26:391–396.

Tang Z, Richards VM (2003) Examination of a linear model in an informational masking study. J Acoust Soc Am 114:361–367.

Tanner WP (1958) What is masking? J Acoust Soc Am 30:919–921.

Tun PA, O'Kane G, Wingfield A (2002) Distraction by competing speech in young and older adult listeners. Psychol Aging 17:453–467.

Veniar FA (1958a) Signal detection as a function of frequency ensemble. II. J Acoust Soc Am 30:1075–1078.

Veniar FA (1958b) Signal detection as a function of frequency ensemble. I. J Acoust Soc Am 30:1020–1024.

Viemeister NF, Plack C (1993) Time analysis. In: Yost W, Popper A, Fay R (eds) Human Psychophysics. New York: Springer-Verlag 116–154.

Viemeister NF, Wakefield GH (1991) Temporal integration and multiple looks. J Acoust Soc Am 90:858–865.

Watson CS (2005) Some comments on informational masking. Acust Acta Acust 91: 502–512.

Watson CS (1987) Uncertainty, informational masking, and the capacity of immediate auditory memory. In: Yost WA, Watson CS (eds) Auditory Processing of Complex Sounds. Hillsdale, NJ: Lawrence Erlbaum 267–277.

Watson C, Kelly W (1981) The role of stimulus uncertainty in the discrimination of auditory patterns. In: Getty DJ, Howard JH (eds) Auditory and Visual Pattern Recognition. Hillsdale, NJ: Lawrence Erlbaum, pp. 37–59.

Watson CS, Wroton HW, Kelly WJ, Benbassat CA (1975) Factors in the discrimination of tonal patterns. I. Component frequency, temporal position, and silent intervals. J Acoust Soc Am 57:1175–1185.

Watson CS, Kelly WJ, Wroton HW (1976) Factors in the discrimination of tonal patterns. II. Selective attention and learning under various levels of stimulus uncertainty. J Acoust Soc Am 60:1176–1186.

Wegel RL, Lane CE (1924) The auditory masking of one sound by another and its probable relation to the dynamics of the inner ear. Phys Rev 23:266–285.

Werner LA, Bargones JY (1991) Sources of auditory masking in infants: Distraction effects. Percept and Psychophys 50:405–412.

Wightman FL, Kistler DJ (2005) Informational masking of speech in children: Effects of ipsilateral and contralateral distracters. J Acoust Soc Am 118:3164–3176.

Wightman FL, Callahan MR, Lutfi RA, Kistler DL, Oh E. (2003) Children's detection of pure-tone signals: Informational masking with contralateral maskers. J Acoust, Soc Am 113:3297–3303

Wright BA, Saberi K (1999) Strategies used to detect auditory signals in small sets of random maskers. J Acoust Soc Am 105:1765–1775.

Yost WA (1997) The cocktail party problem: Forty years later. In: Gilkey R, Anderson TR (eds) Binaural and Spatial Hearing in Real and Virtual Environments. Mahwah, NJ: Lawrence Erlbaum 329–348.

Zwicker E (1970) Masking and psychological excitation as consequences of the ear's frequency analysis. In: Plomp R, Smoorenburg GF (eds) Frequency Analysis and Periodicity Detection in Hearing. Leyden: Sijthoff 376–396.

# 7
# Effects of Harmonicity and Regularity on the Perception of Sound Sources

Robert P. Carlyon and Hedwig E. Gockel

## 1. Introduction

This chapter provides an overview of the strong influence of harmonic structure, spectral regularity, and temporal regularity on the perception of sound sources. Following a brief review of pitch perception, two aspects of the segregation of sound sources are covered: first, the segregation of simultaneously presented sounds and their perception as multiple sources (concurrent sound segregation); second, the perception of rapid sequences of sounds that may be perceived either as coming from a single sound source or as coming from more than one source (sequential sound segregation).

One way of assessing the perception of sound sources is to ask the listener how many sources have been perceived and with what specific characteristics. However, more objective approaches have been used that measure listeners' performance in a task that is assumed to be affected by the number of perceived sources. The chapter covers both approaches, and discusses how and why the demands of the specific task used can influence the pattern of results obtained.

## 2. Pitch of a Single Source

Many of the sound sources that we encounter in everyday life are periodic, and have a clear pitch. Such sounds include the voiced portions of speech and of animal calls, musical notes, and of man-made devices ranging from the buzz of an electric toothbrush to the loud whine of a jumbo jet. The focus of this chapter will be on how the auditory system exploits this periodicity to assign appropriate portions of the signal to the correct sound sources. This is not merely an academic exercise, but has practical implications for the design of real-world devices, including digital hearing aids and cochlear implants (Moore and Carlyon 2005). A first step is to understand how the pitch of a single sound is derived, a topic that is the subject of many decades of research, entire book chapters, and even whole books (Plomp 1976; Moore 2003; Moore and Carlyon 2005;

Plack et al. 2005). Here, we simply summarize a few key facts that will aid the reader in interpreting our later discussion of source segregation.

1. The waveform of a periodic sound, such as a vowel, repeats at a rate produced by the source (e.g., vocal fold vibration). Its frequency spectrum consists of components ("harmonics") that are integer multiples of a common fundamental (F0). Pitch corresponds to F0 for most harmonic sounds, even when the component whose frequency equals F0 is missing or masked (Licklider 1956). Some exceptions occur; for example, when the components correspond to nonconsecutive harmonics of a low F0 (e.g., 1830, 2030, 2230 Hz), the pitch is not equal to the true F0 (e.g., 10 Hz), but in this instance, about 203 Hz (Schouten et al. 1962). Further discussion of these effects can be found in Moore (2003).

2. As a first approximation, the ear can be modeled as a bank of overlapping bandpass filters. The bandwidths of these auditory filters, expressed in Hertz, increase with increasing center frequency (CF), whereas the spacing between consecutive harmonics is constant. Therefore, for a given F0, more harmonics will pass through a high-frequency filter than through a lower-frequency filter. The low-numbered harmonics of a complex sound are *resolved*, such that the outputs of individual filters are dominated by one harmonic. Auditory filters centered on higher, *unresolved* harmonics have an output influenced by several harmonics and that repeat at a rate equal to F0.

3. The resolved harmonics dominate the pitch of a complex sound (Plomp 1964; Moore et al. 1985).

4. Numerous models of pitch perception have been proposed. Pattern-recognition models (e.g., Goldstein 1973) assume that the brain tries to fit a best-matching harmonic series to the components present in a complex. These components could be encoded via place-of-excitation cues and/or in the temporal pattern ("phase-locking") of auditory nerve ("AN") responses to each component.

5. The most popular class of models assume that a temporal analysis is performed on the output of each auditory filter and that these individual analyses are combined to form a summary representation from which the pitch estimate is derived. In one very influential model, the temporal analysis takes the form of autocorrelation (Licklider 1951; Meddis and Hewitt 1991; Meddis and O'Mard 1997), but other forms have also been proposed (Moore 1982; de Cheveigné 1993; Patterson et al. 1995).

6. Although there is evidence that phase-locking is important for pitch perception, the way in which this information is analyzed is yet to be fully determined. Some stimuli, such as electric pulse trains presented to cochlear-implant listeners, or bandpass-filtered groups of unresolved harmonics presented to normal-hearing listeners, would be expected to produce highly accurate phase-locking, but result in quite weak pitch percepts (Shannon 1983; Carlyon and Deeks 2002; Moore and Carlyon 2005). One

hypothesis is that phase-locking is optimally used only when it "matches" the center frequency (CF ) of the fibers that convey it. (Moore 1982; Oxenham et al. 2004; Bernstein and Oxenham 2005; Moore and Carlyon 2005). Another idea is that the brain exploits the change in relative timing of neural impulses that occurs near the peak of the traveling wave (Kim et al. 1980; Shamma 1985; Oxenham et al. 2004; Moore and Carlyon 2005; Bernstein and Oxenham 2005). It is likely that both of these cues would be degraded with cochlear-implant stimulation and for bands of unresolved harmonics, consistent with the relatively weak pitch of these stimuli.

## 3. Concurrent Sounds

The two most important cues that the auditory system uses to segregate concurrent sounds are differences in onset time and deviations from harmonicity (Darwin and Carlyon 1995). The use of onsets in source identification is discussed by Sheft (Chapter 9 of this volume). Here, we focus on inharmonicity, whose definition here includes situations in which a signal contains components of two or more F0s. Evidence for the importance of this cue comes from a wide range of paradigms, using stimuli that range from the highly simple, which permit a great deal of experimental control, to more realistic sounds that may contain multiple cues.

## 3.1 Source Segregation of a Single Mistuned Component

Perhaps the simplest demonstration of the powerful effect of harmonicity comes from experiments in which one component of an otherwise harmonic complex is mistuned from the rest. Moore et al. (1986) presented listeners with a 410-ms complex sound that might, or might not, have one of its partials mistuned. The listeners were required to report whether they heard a single sound with a single pitch, or, alternatively, a mixture consisting of a complex sound and a component that sounded like a pure tone. The results depended strongly on which harmonic was mistuned. For the low-numbered resolved harmonics (up to about the sixth), listeners reliably identified mistuned stimuli corresponding to two sources, even for mistunings as small as 1%–3%. Thresholds were slightly higher for the lowest two harmonics than for the third to fifth harmonics. However, when one of the higher-numbered unresolved harmonics (above the sixth) was mistuned, listeners frequently could not perform the task reliably. The results of subsequent studies generally confirm this observation, but also show that absolute frequency plays a role, with subjects' ability to match the frequency of a mistuned harmonic deteriorating with increasing frequency (Hartmann et al. 1990; Roberts and Brunstrom 2001). This latter finding has been attributed to a deterioration in AN phase-locking with increasing frequency, and to the importance of phase-locking in sound segregation (Hartmann et al. 1990; Moore and Ohgushi 1993).

## 3.2 Source Segregation of Complex Tones Occupying Different Frequency Regions

In real life, it is rare for a harmonic complex tone with one partial missing to be accompanied by a pure tone that arises from another source and with a frequency close to that of the missing partial. Rather, the ear is more likely to be presented with a mixture of different complex sounds arising from separate sources. A step closer to this situation can be studied in experiments in which subjects are presented with pairs of complex sounds, presented simultaneously and filtered into separate frequency regions, and have to report whether they differ in F0. When subjects successfully perform this task, they typically hear a sound with two F0s as consisting of two separate sources. In this situation, too, the resolvability of the harmonics in each group has a profound effect on performance. Carlyon and Shackleton (1994) presented subjects with such pairs of complexes, and found that performance was much better when both complexes were resolved than when one was unresolved.

The greater role for resolved than for unresolved complexes in F0-based segregation extends to speech tasks. Darwin (1992) presented subjects with a four-formant vowel, which was cleverly designed so that removal of the second formant changed the percept from /ru/ to /li/. He found that a similar change in phonetic percept could be achieved by presenting the second formant on a different F0 to the others, but only when the F0 of the second formant was such that its components would be resolved by the peripheral auditory system. More recently, Deeks and Carlyon (2004) modified the popular "vocode" simulation of cochlear implant hearing so that the envelope in each of six frequency bands modulated an unresolved harmonic complex filtered into that same region (Figure 7.1). In one condition, they separately stimulated the odd-numbered bands with a masker sentence, and the even-numbered bands with a

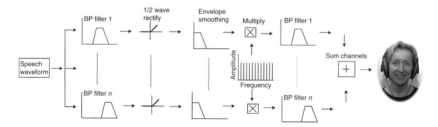

FIGURE 7.1. Schematic of the modification of the "vocode" simulation of cochlear implant hearing used by Deeks and Carlyon (2004). The stimulus is passed through a bank of bandpass analysis filters. The envelope at the output of each analysis filter is extracted and modulates a harmonic complex, which is then passed through the same analysis filter. The output in each channel resembled a pulse train. All channels were summed and presented to subjects over headphones. To avoid resolved harmonics, the lowest analysis filter used by Deeks and Carlyon was centered on 1089 Hz, and the harmonic complexes had low F0s whose perceived pitch was doubled by summing them in alternating phase.

target sentence (or vice versa). They found no consistent advantage when the F0 of the unresolved harmonic complexes differed between the masker and target channels, compared to when the same F0 was used for all channels. As we shall see, this result differs from the substantial advantage seen for identification of speech sounds that contain resolved harmonics.

## 3.3 Source Segregation of Spectrally Overlapping Complex Tones

A situation even more typical of everyday listening occurs when two complex sounds are presented simultaneously and have frequency spectra that overlap (Figure 7.2). A simplified version of this situation was studied by Carlyon (1996), who presented subjects with two harmonic complexes filtered in the same frequency region. In one experiment, the F0 of one of the complexes was 210 Hz, and that of the other differed from 210 Hz by either ±1% or by a larger amount of up to ±16%. When the complexes were filtered between 20 and 1420 Hz, so that the harmonics within each complex were resolved, subjects reliably reported that the mixture containing the larger ΔF0 sounded less fused, with this tendency increasing monotonically with increasing ΔF0. However, when the complexes were filtered between 3900 and 5400 Hz, so that the partials within each complex were unresolved, no such tendency was observed, and subjects heard a unitary percept with a "crackle" quality. In another experiment, subjects detected a ΔF0 between two complexes, presented simultaneously and filtered in the range 20–1420 Hz and 3900–5400 Hz, respectively. Performance was only slightly affected when a continuous harmonic masker was added to the lower (resolved) frequency region, but fell to chance when a masker was added to the higher region, where the harmonics were unresolved.

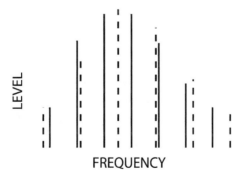

FIGURE 7.2. Schematic of two bandpass-filtered harmonic complexes, in which the harmonics in each complex presented alone are resolved but adjacent harmonics of the two complexes can be unresolved from each other. This figure illustrates the case in which the two complexes are passed through the same bandpass filter, but the same principle would apply to the case of two formants having slightly different center frequencies and different F0s.

Both of the experiments described above are consistent with the conclusion, drawn in our earlier subsections, that resolvability crucially modulates the effect of F0 differences on source segregation. It is, however, important to note that in the lower-frequency region, although the harmonics in each group were resolved *from each other*, they were not necessarily resolved from the harmonics of the other complex (Figure 7.2). The fact that subjects could nevertheless segregate sounds with different F0s is consistent with a finding reported by Beerends and Houtsma (1989), who presented listeners with simultaneous diotic pairs of two-tone complexes. Their listeners could identify the pitches of the two pairs, provided that some harmonics were resolved, even when adjacent partials from the two pairs were very close. For example, when complexes of 800 + 1000 and 1068 + 1335 Hz were mixed, subjects could reliably select the two correct pitches of 200 and 267 Hz from a set of five possible pitches (200, 225, 250, 267, and 300 Hz), despite the proximity of the 1000- and 1068-Hz components. One way of explaining this ability is to assume that when two components are very close, their frequencies are averaged in a way that is influenced by their relative amplitudes (cf. Feth 1974). The resulting average frequency might then be incorporated in the estimation of the pitch of both complexes. In Beerends and Houtsma's study, all components had equal amplitudes, and the average of 1000 and 1068 Hz is 1034 Hz. Combined with the 800-Hz partial, 1034 Hz forms a reasonable match to the fourth and fifth harmonics of 204 Hz (e.g., 816 and 1020 Hz), and combined with 1335 Hz, it reasonably matches the fourth and fifth harmonics of 263 Hz (1052 and 1315 Hz). These matches, although not perfect, are better than could be formed by the other pitches that Beerends and Houtsma's listeners were allowed to choose from, and so could explain the ability of their listeners to perform the task. An interesting remaining question is whether the amplitude modulation produced by the two adjacent components somehow "encourages" the auditory system to assign the average frequency of the two components to more than one source.

## 3.4 Source Segregation of Spectrally Overlapping Speech Sounds

The situation in which components from competing sources have very similar frequencies (Figure 7.2) is also relevant to the identification of pairs of vowels that are presented simultaneously. Performance in this task improves as the difference between the F0s of the two vowels is increased (Scheffers 1983), and, for long (e.g., 200 ms) vowels, a substantial improvement occurs even when $\Delta$F0 is no greater than half a semitone (Assmann and Summerfield 1990; Culling and Darwin 1994). Culling and Darwin (1994) showed that the improvement seen over this range of $\Delta$F0s was due to the beating that occurs between these pairs of adjacent components. The result of this beating is that, relative to the amplitudes of the other components, the amplitude of each adjacent pair will sometimes reflect that of the component belonging to one source, and sometimes that of the other. When several such pairs are present, the composite spectrum, once smoothed by the auditory filter bank, will sometimes reflect that of one vowel and sometimes that of the other. Listeners might then identify one or the other vowel by basing their response on different times during the sound. When this beating cue is less likely to be available, such as for shorter (50-ms) vowel pairs (Culling and Darwin 1994) or for sentences (Brokx and Nooteboom 1982), performance improves more gradually as $\Delta F0$ between competing sounds is increased up to three or four semitones. In Section 4 we describe evidence that the advantage gained from F0 differences in sentence-length sounds is likely to be dominated by the improved segregation of *simultaneous* portions of speech from the two sources, rather than from listeners identifying the F0 of one voice and tracking it over time (Darwin and Hukin 1999; see also Chapter 8).

## 3.5 Effect of F0 Differences on Interference Between Sound Sources in the Central Auditory System

There is some evidence that when two concurrent sounds have different F0s, it is easier selectively to process one sound without interference from the other. In Chapter 9, Sheft describes the phenomenon of modulation detection interference (MDI), in which the detection of AM or FM applied to a "target" sound is disrupted when modulation is applied to a simultaneous "interferer," even when the interferer and target occupy well-separated frequency regions. Lyzenga and Carlyon (1999) showed that when the target and interferer are synthetic formant-like sounds, the amount of MDI can be significantly reduced when they have different F0s. This in turn provides further evidence that MDI occurs at a fairly late stage of processing, and suggests that the mechanisms responsible for MDI receive input from those processes that extract F0.

More recently, Gockel et al. (2004, 2005) showed that two simultaneous sounds in different frequency regions can interfere in the pitch domain. They required subjects to discriminate an F0 difference between two sequentially presented "target" complex tones in the presence or absence of an "interferer" (Figure 7.3a). The interferer consisted of another complex tone, filtered into a separate remote frequency region, and in one condition, was turned on and off with each target. The smallest detectable difference in the targets' F0 ("F0DL") was elevated by the presence of the interferer. This "pitch discrimination inter-ference (PDI)" was tuned for F0, so that the greatest deterioration occurred when the interferer's F0 equaled the mean of the two targets to be discrimi-nated (Figure 7.3b). Under such conditions, subjects reported hearing a single source with one pitch. When the harmonics of the interferer were resolved, they would have dominated the pitch of this source (Plomp 1967; Moore et al. 1985), and indeed, the amount of PDI was greater when the interferer was resolved than when it was unresolved. As the interferer F0 was increased by about 20%, subjects reported hearing the interferer and target as separate sounds, but even when the interferer's F0 was 30% higher than that of the targets, it still significantly impaired performance. Further evidence that the pitches of complex sounds can interfere centrally, even when heard separately, comes from

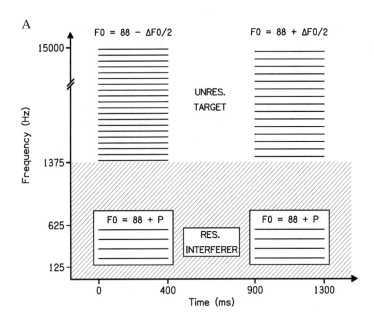

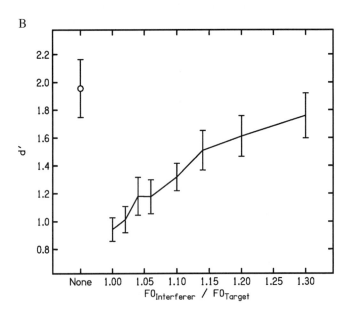

FIGURE 7.3. (**A**) Schematic spectrogram of the stimuli used by Gockel et al. (2005) to study pitch discrimination interference. The diagonally hatched area represents a lowpass noise that was presented to mask combination tones produced by the unresolved complex. (**B**) Mean performance across subjects as a function of the ratio of the interferer's F0 to the geometric mean of the targets' F0s.

the finding that the F0DL for an unresolved target was slightly but significantly elevated by a continuous interferer, which would have been easily segregated from it.

## 3.6 Harmonicity: A Special Case of Spectral Regularity?

Roberts and his colleagues have argued that source segregation based on deviations from a common F0 reflect a special case of a more general sensitivity to deviations from spectral regularity (Roberts and Bregman 1991; Roberts and Brunstrom 1998; Brunstrom and Roberts 2000; Roberts and Brunstrom 2001, 2003; Roberts 2005). One study (Roberts and Bregman 1991) showed that when a harmonic complex consists of only odd-numbered harmonics, with the exception of a single even-numbered harmonic, this even harmonic stands out perceptually from the rest. A subsequent set of experiments used stimuli whereby the component frequencies of a perfectly harmonic complex were shifted by a fixed value, or "stretched" by applying a cumulative increment to the frequency of each harmonic (Roberts and Brunstrom 1998). When required to match a pure tone to the frequency of a component whose frequency was "mistuned" from these spectrally regular, but inharmonic, stimuli, performance was comparable to that obtained for mistuning from a harmonic complex.

Roberts (2005) interpreted these data by suggesting that the auditory system is tolerant to "local" deviations from harmonicity, such as might occur between neighboring partials. Accordingly, a stimulus in which the spacing between adjacent components changes smoothly across frequency will be perceptually fused because there are no local deviations from this regularity. They give an example of how this might be implemented in the popular autocorrelogram model of pitch perception (Meddis and Hewitt 1991; Meddis and O'Mard 1997). The left-hand panel of Figure 7.4 shows the autocorrelograms ("ACFs") in several frequency channels to a 200-Hz harmonic complex whose fourth component is mistuned by -4%, with a summary autocorrelogram ("SACF") shown underneath. The ACF in each channel, except that driven by the mistuned fourth harmonic (768 Hz), has a peak at 5 ms, corresponding to the reciprocal of the F0, and the mistuned harmonic produces a local deviation from this central "spine." This deviation is indicated by a filled dot in the figure. The middle and right panels show similar plots for a "shifted" and a "stretched" complex respectively. In these two cases, the locations of the peaks in the individual autocorrelograms change smoothly across channels, with the exception of a local irregularity produced by the mistuned component. It is this irregularity that, it is proposed, causes the enhanced perceptual segregation of the mistuned component. Note that according to this scheme, the processes of pitch extraction and of sound segregation are identical until the final stage, where the ACFs in individual channels are either combined to produce a pitch estimate or compared to decide whether a component has been mistuned.

The demonstration that listeners are sensitive to deviations from spectral regularity, even for inharmonic sounds, is an important contribution to our

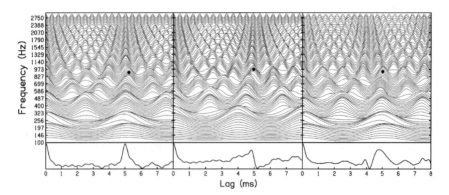

FIGURE 7.4. Roberts's implementation of an autocorrelogram model (Meddis and O'Mard 1997) to account for listeners' sensitivity to the mistuning of the fourth component of a harmonic, a spectrally shifted complex, and a spectrally stretched complex (left, middle, and right panels respectively). The local discontinuity produced by the mistuning is shown by a filled circle in each panel. The SACF is shown at the bottom of each panel. (Reproduced from Roberts 2005).

understanding of concurrent sound segregation. However, all of Roberts's experiments investigated the mistuning of a *single* component from a spectrally regular complex. In this section we have described the auditory system's sensitivity to deviations from a common F0 under a wide range of conditions, including spectrally distinct and overlapping complex tones and mixtures of speech sounds. An interesting remaining question is whether performance in these tasks is specific to harmonic complex tones or, alternatively, could also be explained in terms of a sensitivity to deviations from a more general form of spectral regularity.

## 4. Sequences of Sounds

Rapid sequences of sounds may be perceived either as coming from a single sound source, i.e., perceived as a single coherent stream (fusion), or as coming from more than one source, i.e., the sound sequence splits perceptually into concurrent subsets of coherent streams (fission) (Miller and Heise 1950; Bregman and Campbell 1971; van Noorden 1975; Bregman 1990). As stimulus parameters are changed, the perceptual organization of the sound sequence may change. Tonal sequences have been used frequently to investigate the nature of the underlying streaming processes. Fission tends to occur with large frequency separations between successive tones and high rates of presentation, while small frequency differences and low presentation rates lead to fusion. Large physical frequency differences between successive tones will of course be reflected both by differences in which auditory nerve fibers respond to each tone and in the phase-locking response of those fibers, leading to clear pitch differences (see Section 2). For pure tones or complex tones containing harmonics that are resolved in the auditory periphery, frequency differences also result in clear differences between the excitation patterns evoked in the cochlea by these tones. For some time, the predominant opinion was that streaming depends mainly on peripheral channeling (Hartmann and Johnson 1991; Beauvois and Meddis 1996; McCabe and Denham 1997). This means that in order for fission to occur, there has to be a certain amount of difference between the excitation patterns of those tones that are perceived in different subsets of streams. For example, Hartmann and Johnson (1991) studied streaming using an interleaved melody identification task. The notes of the two melodies were presented in alternation, and subjects had to identify both melodies. This task is easier if the notes from the two melodies are perceived in separate streams. Various conditions were tested, including notes from the two melodies differing in ear of presentation, interaural time delay, spectral composition, and temporal envelope. Performance was best in those conditions where most peripheral channeling would be expected, i.e., when the notes were presented to opposite ears or differed in spectrum. Hartmann and Johnson concluded that "peripheral channeling is of paramount importance" in streaming. Similarly, van Noorden (1975) reported that listeners always reported hearing two streams when they were presented with a sequence that alternated between a pure tone and a complex tone with an F0 corresponding to the frequency of the pure tone. Thus, a correspondence between pitch values was not sufficient to lead to the percept of a single sound source. He concluded that contiguity "at the level of the cochlear hair cells" was a necessary (but not sufficient) condition for fusion to occur.

While peripheral channeling, or differences between the excitation patterns of sequential sounds, clearly plays a role, more recent evidence suggests that differences in pitch without differences in peripheral channeling can lead to the perceptual segregation of a stream of acoustic events into several subsets. Indeed, it has been suggested that, depending on the task demands, any sufficiently *salient* perceptual difference can lead to fission (for an overview, see Moore and Gockel 2002). In what follows, first, evidence for the role of pitch in streaming when accompanied by differences in excitation pattern will be summarized briefly. Then, more recent evidence on the role of pitch based on unresolved harmonics, i.e., pitch derived from temporal information while differences in excitation patterns are minimized, will be reviewed and limitations outlined.

## 4.1 Pitch Differences with Concomitant Differences in Peripheral Channeling

Many studies (e.g., Miller and Heise 1950; van Noorden 1975; Rose and Moore 1997) have used sequences of rapidly alternating pure-tone stimuli to determine the amount of frequency difference needed for the listener to perceive two streams, i.e., to perceive the sounds as coming from two sources. Figure 7.5A illustrates a typical stimulus sequence and possible percepts. The frequency of tone A is fixed. The frequency of tone B starts below that of tone A and is

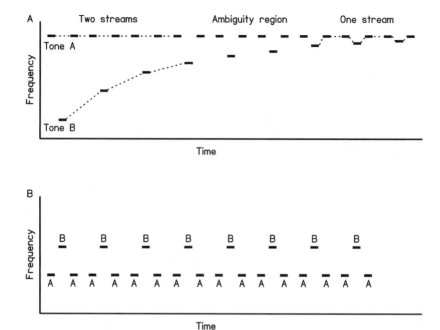

FIGURE 7.5. Stimulus configurations used in experiments on streaming of sequential sounds. (**A**) Schematic illustration of how the percept of a tone sequence depends on the frequency separation of the tones. With a large frequency separation between the tones **A** and **B**, the two tones will be heard as separate streams. When the frequency separation between the two tones is small, a single coherent stream with a galloping rhythm can be heard. (**B**) Schematic of the stimuli used by Vliegen and Oxenham (1999).

increased for each presentation toward that of A. At first, listeners hear two streams, one consisting of low B tones and the other consisting of high A tones, with the repetition rate of the high tones being twice that of the low tones. Toward the end of the sequence, when the frequencies of the two tones are similar, listeners hear one stream with a galloping rhythm. For intermediate frequency separations (the ambiguity region) either one or two streams might be heard. In this region, the tendency to hear two streams increases with exposure time (Bregman 1978; Cusack et al. 2004) but stabilizes after about 10 s (Anstis and Saida 1985); a brief silent period can then cause perception to revert to a single-stream, fused form (Bregman 1978). Attention to the sequence seems to affect the build-up of the tendency for fission to occur (Carlyon et al. 2001; see also Hafter et al., Chapter 5), although alternatively, switching attention to the sequence (away from something else) might have a similar resetting effect as a sudden change in physical characteristics of the sequence (Moore and Gockel 2002; Cusack et al. 2004).

The fission boundary is defined as the frequency separation at which a listener who is instructed to hear two streams as long as possible cannot do so, i.e., below the fission boundary, only a single stream can be heard. The temporal coherence boundary is defined as the frequency separation beyond which a listener who is instructed to hear one stream as long as possible consistently perceives two streams (van Noorden 1975). Fission that occurs even though the listener tries to hear one stream has also been called "obligatory" or "primitive" stream segregation (Bregman 1990; Vliegen et al. 1999). The fission boundary varies only slightly with repetition period of the tones, while the temporal coherence boundary increases markedly with increasing repetition period (van Noorden 1975). In a study in which subjects tried to hear a single-stream percept, Bregman et al. (2000) concluded that the time interval between successive tones of the same frequency was the most important temporal factor.

Using tone sequences similar to the one illustrated in Figure 7.5 (except that tone B started above tone A), Rose and Moore (1997) measured the fission boundary for various frequencies of tone A. The aim was to test whether the frequency separation at the fission boundary corresponds to a constant difference in ERB number, where ERB stands for *equivalent rectangular bandwidth* of the auditory filter for normally hearing listeners (Glasberg and Moore 1990). The results showed that at the fission boundary, the frequency separation between the A and B tones was approximately constant, as predicted by the model of Beauvois and Meddis (1996), and was about 0.5 (expressed as the difference in ERB number between tones A and B). At low frequencies, this value of 0.5 is a factor of about 25–30 times that of the frequency separation needed to *discriminate* between the two tones (Moore 2003).

Singh (1987) and Bregman et al. (1990) both used *complex* tones in order to investigate the role of F0 in streaming. For example, Singh (1987) used complex tones containing four successive harmonics. She independently varied F0 and the number of the lowest harmonic present in the complex. Singh (1987) and Bregman et al. (1990) found that differences in F0 and differences in spectral shape (an overall shift in the spectral locus of energy) independently can influence streaming. However, both of them used complex tones with resolved harmonics, and so changes in F0 also led to more local changes in the excitation pattern. This complicates the interpretation of their results, and it has been questioned whether it was the difference in F0 per se or the difference in excitation patterns that caused the observed effects on streaming (Vliegen and Oxenham 1999).

## 4.2 Effects of Fundamental Frequency of Complex Tones with Unresolved Harmonics

Vliegen and Oxenham (1999) investigated the effectiveness of F0 differences in the perceptual organization of sequential sounds in the absence of excitation pattern cues. In their first experiment, the basic stimulus was a sequence of 11 tone triplets (ABA ABA ABA ...) similar to that shown in Figure 7.5B. The F0 of tone A was always fixed at 100 Hz. Within a given sequence, the F0 difference between the A and B tones was kept constant. Across trials, the F0 of tone B was varied in such a way that the F0 difference between tones A and B corresponded to either 1, 2, 3, 4, 5, 7, or 11 semitones. Listeners were instructed to try to hear the B tones separately from the A tones, i.e., to perceive the A and B tones as originating from two different sources. After the sequence was presented, listeners indicated whether they perceived one or two streams. In three conditions, different types of tones were used: (1) complex tones containing only unresolved harmonics (above the tenth), leading to the perception of pitch in the absence of excitation pattern cues based on purely temporal information, (2) complex tones containing resolved harmonics, and (3) pure tones. As expected, the data generally showed an increasing percentage of segregation with increasing F0 difference. Interestingly, for all spectral conditions, the F0 difference needed for segregation in 50% of the trials was about four semitones. Thus, for this subjective judgment, source segregation based on temporal regularity only (the complexes containing only unresolved components) was not weaker than when additional spectral cues were available.

In a second experiment, Vliegen and Oxenham (1999) used an objective task, in which performance was expected to improve with increasing perceptual segregation (following Dowling 1973). Listeners were presented with an unfamiliar (random) short atonal melody that consisted of five tones. In the first interval, these tones were presented alone. In the second interval, either the same or a different melody was presented. This time, following each tone that was part of the melody, a random distracter tone was presented, keeping the presentation rate of the tones that formed part of the melody constant. All tones were either complex tones containing only unresolved harmonics, or pure tones. There were two conditions. In the first, the distracter tones were drawn from the same F0 (or frequency) range as the tones that formed the melody. In the second, the distracter tones were drawn from an F0 range that was 11 semitones above that for the melody notes. Listeners had to indicate whether the melody in the second interval was the same as or different from that in the first. Performance was better when the distracter tones were from a different F0 or frequency range than when they were from the same F0 range as the melody tones. Importantly, the advantage that was observed when the melody tones and distracter tones were drawn from two different ranges was at least as large for the complex tones containing only unresolved components as for the pure tones. So, in agreement with their first experiment, temporal (F0) cues were as effective as spectral differences in a task in which segregation was advantageous.

Vliegen et al. (1999) used a temporal discrimination task to determine whether this also holds if the nature of the task is such that it is advantageous to perceive the sequence of tones as originating from a single source rather than two independent sources. In each trial, listeners were presented with two ABA–ABA sequences, one after the other. In one sequence, the B tone was temporally centered between the A tones, while in the other, the B tone was delayed. Listeners had to indicate in which of the two sequences the B tones were shifted from the midpoint, and the smallest detectable temporal shift was measured. In such a task it is advantageous to perceive all tones as part of a single stream. If fusion occurs, listeners hear a regular rhythm when the B tones are centered and an irregular "jerky" rhythm when the B tones are delayed. Detection of a temporal shift between the A and B tones is poorer when the tones are perceived in different streams as the jerky rhythm is no longer heard and the B tones seem to float around in time with respect to the A tones (van Noorden 1975). With this task, Vliegen et al. (1999) found that differences between the F0s of complex tones containing only unresolved components slightly increased the smallest detectable temporal shift. However, this effect was clearly smaller than that observed for pure tones or for complex tones which had the same F0 but differed in spectral content. They concluded that spectral information is dominant in inducing obligatory segregation, but that differences in temporal regularity in the absence of spectral cues can play a role too. They also concluded that the role of periodicity information per se seems to be modulated by task demands. In their first study, segregation of the tones was advantageous. So all information available might be used to achieve this and improve performance. In their second study, perceiving the tones as originating from a single source was advantageous. The authors argued that periodicity differences without concomitant spectral cues might be easier to ignore than spectral differences. One way in which this could happen is if two streams were heard when different populations of neurons responded selectively to the A and B tones, and that this selectivity could occur in different domains, including spectral frequency and F0 (Carlyon 2004). Because spectral analysis occurs at the earliest stage of processing, streaming by this feature is likely to be compulsory. Other features, such as F0, are likely to be encoded at later stages of processing, and depending on task requirements, the auditory system might be able to select which stage of representation to process.

Grimault et al. (2000) performed an experiment similar to the first experiment of Vliegen and Oxenham (1999) described above. They measured subjective streaming scores for ABA sequences of harmonic complex tones. The F0 of tone A was fixed across trials (in all sequences) at either 88 or 250 Hz, while that of the B tones varied across trials between 88 and 352 Hz in half-octave steps. By varying the frequency region into which the complex tones were filtered, the authors manipulated the resolvability of the components in each complex. Like Vliegen and Oxenham (1999), they found that listeners reported hearing two streams when the difference in F0s was large enough, even if the tones differed only in F0 and contained no resolved harmonics. However, in contrast to Vliegen and Oxenham, they found less segregation for tones containing only unresolved harmonics than for those containing resolved components. As has been pointed out by Grimault et al., a possible reason for these different findings is the difference in the instruction to the listeners. While the listeners of Vliegen and Oxenham (1999) were instructed to try to hear the B tones separately from the A tones, probably measuring something related to the fission threshold, the listeners of Grimault et al. received the more neutral instruction to indicate whether they heard a single stream or two streams at the end of the sequence. Another possible reason, pointed out by the authors, was the larger range of F0 differences used by Grimault et al. that somehow "may have promoted the emergence of influences of resolvability." However, a third possibility is that the difference in the sequence length (4 s in Grimault et al. compared to 10 s in Vliegen and Oxenham) is responsible. The build-up of the tendency for fission to occur might be quicker for tones containing resolved components than for tones containing only unresolved components due to the less salient pitch produced by the latter (see Section 2). This might lead to differences in segregation judgments for the two types of tones after a short sequence is presented, which might be absent after presentation of a longer sequence, where the tendency for fission to occur could fully build up for both types of tones.

## 4.3 More-Complex Conditions

In more-complex situations, differences in periodicity have limited efficacy in sequential sound segregation in the absence, and occasionally, even in the presence, of spectral cues.

### 4.3.1 Complex Conditions in the Absence of Spectral Cues

Evidence for a limited role for F0 differences conveyed by unresolved harmonics comes from a study by Micheyl and Carlyon (1998). They measured the F0DL for two sequentially presented 100-ms harmonic target complexes under various conditions. Either the target complexes could be presented alone, or each of them could be temporally flanked (preceded and followed) by harmonic complex tones (the fringes) each of 200-ms duration. While the F0 of the target varied between the two observation intervals, the F0 of the fringes was always held constant (within a given condition). The targets had a nominal F0 of either 88 or 250 Hz. The F0 of the fringes was also either 88 or 250 Hz and was either identical to or different from the nominal F0 of the target. The target and the fringes were (independently) filtered into three different frequency regions, which, as in the study of Grimault et al., affected the resolvability of the components of each complex. In some conditions, but not all, the presence of the fringes significantly increased F0DLs for the target tones. In order for the impairment to occur, the target and fringes had to be filtered into the same frequency region, and additionally they usually, needed to have the same nominal F0. There was one exception: When both the fringes and targets contained only unresolved harmonics, the presence of the fringes markedly increased F0DLs, even when their F0 differed markedly from that of the targets. The authors discussed these findings in terms of auditory streaming. When the tones were filtered into different frequency regions, the target and fringes were perceived as coming from two different sources, and F0DLs were unaffected by the fringes. When the target and the fringes were filtered into the same frequency region, a large difference in F0 between target and fringes was sufficient for segregation when the complex tones contained resolved harmonics. However, when the tones contained only unresolved harmonics, the same large difference in F0 was not sufficient to allow segregation. This was attributed to the fact that the pitch evoked by unresolved harmonics is weaker than that evoked by resolved harmonics (Plomp 1967;

Houtsma and Smurzynski 1990). An alternative explanation is that the absence of spectral cues in the unresolved complexes was responsible. However, in a control experiment, Micheyl and Carlyon (1998) showed that the reduced interference occurring for resolved harmonics, observed when the F0 of target and fringes differed, was not due mainly to spectral differences. They found that when different subsets of harmonics of the same F0 were used for the target and fringe (with both target and fringe containing resolved harmonics), a large impairment was still observed, despite the large spectral differences between the two. Hence, in this situation, it appears that F0 differences produce more sequential segregation with resolved than with unresolved harmonics, and that this is due to the stronger pitch difference in the former case, rather than to the presence of excitation-pattern cues.

In a more recent study, Gockel et al. (1999) showed that the impairment caused by the presence of the fringes can be reduced by a difference in the perceived location of the target and fringes, produced by interaural time and intensity differences. The interesting point for the current topic is that the release from impairment for unresolved targets and fringes was larger when they differed in F0 than when they had the same F0. Thus, even though F0 differences without spectral differences were not sufficient to produce a significant improvement (Micheyl and Carlyon 1998), combining them with another cue for perceptual segregation, i.e., a difference in lateralization (see Chapter 8 in this volume), led to some additional effect. Furthermore, Gockel et al. (1999) found that with extended practice, listeners were able to take some advantage of the pitch differences between target and fringes even when they contained only unresolved harmonics; F0DLs were less impaired when target and fringes differed in F0 than when they had the same F0. Overall, these two studies seem to indicate that for the task investigated, differences in periodicity for unresolved harmonics lead to little segregation, but that their effect might be enhanced in combination with other cues for segregation or by prolonged practice.

### 4.3.2 Complex Conditions in the Presence of Spectral Cues

Evidence for the limited role of F0 differences even in the likely presence of spectral cues comes from a study by Darwin and Hukin (1999); see also Chapter 8 of this volume which is described in more detail in Darwin, Chapter 8. They instructed listeners to track a particular sound source over time. In one experiment, listeners were presented simultaneously with two sentences. In different conditions, the interaural time differences (ITDs) between the two sentences and their F0 differences were varied. Listeners were instructed to listen to one particular sentence. At a certain point, the two sentences contained two different target words, aligned in time and duration. The ITDs and the F0s of the target words were varied independently from those of the sentences. Listeners had to indicate which of the two target words they heard in the attended sentence. Listeners most often reported the target word that had the same ITD as the target sentence, even when target word and sentence differed in F0, rather than reporting the word that had the same F0. This study showed that when listeners were specifically instructed to try to track a particular sound source over time, differences in F0 were of little importance compared to location cues. This contrasts markedly with the case for concurrent sound segregation, where it appears that harmonicity is much more important than ITD (Hill and Darwin 1996; Darwin 1997). It also contrasts with the results from auditory streaming experiments, where we have seen that F0 differences play an important role. Hence, even within the domain of sequential sound organization, the role of F0 differences depends on the level of processing and on the specific form of the organization required.

## 5. Summary

When the frequency components of a sound arise from the same periodic source, they typically consist of harmonics of a common F0. Deviations from this pattern are used by the auditory system in source segregation, both when those components are presented simultaneously and when they alternate in time. In both cases, segregation is modulated by the resolvability of those components.

For concurrent source segregation, deviation from a common F0 is useful in a range of tasks including the detection of mistuning of a single harmonic, detection of $\Delta$F0s between complexes that are either spectrally overlapping or discrete, and identification of mixtures of speech sounds such as vowels and competing sentences. Source segregation under these conditions can be understood in terms of the signal being broken down into a number of frequency channels, and with each channel being assigned to one or other source. Selecting which components go with which source is probably accomplished by exploiting common temporal patterns in the outputs of individual auditory filters. When a given filter is driven by two closely spaced components, there is evidence that the frequencies of those two components can be averaged, and for them to be treated as a single component that can be assigned to more than one source. When the output of an auditory filter consists of a mixture of two inharmonically related periodicities, for example produced by a mixture of two unresolved harmonic complexes, there is no evidence that subjects can break down the resulting complex temporal pattern into its underlying F0s.

For sequential source segregation, as measured in auditory streaming paradigms, the usefulness of F0 differences depends primarily on the pitch strength of each individual sound. Because stimuli with resolved harmonics have stronger pitches than those having only unresolved harmonics, source segregation is usually superior in the former case. However, this conclusion depends somewhat on the task, and streaming based on unresolved harmonics can be as strong as that with resolved harmonics, provided that the task is designed to encourage segregation. When instructions are more neutral, or when the task encourages integration into a single stream, segregation is strongest with resolved harmonics. In a task where the listener has to assign a word to one of two prior phrases, presented concurrently with each other, F0 plays a surprisingly weak role, even when resolved harmonics are likely to be present (Darwin 1997; Darwin and Hukin 1999).

Finally, it is worth pointing out that although we have discussed simultaneous and sequential source segregation separately, there are many situations in which they interact. Dannenbring and Bregman (1978) alternated a harmonic complex with a pure tone having a frequency equal to one of its components, here termed the "target." Their listeners could sometimes hear the pure tone repeat at a rate faster than that of the complex, consistent with it "streaming out" the target from the complex. This streaming effect was greatest when the target was turned on or off asynchronously with the rest of the complex. Using a more objective measure, experiments from our laboratory have shown that the detection of mistuning applied to a low-numbered target harmonic can be substantially disrupted when a narrowband stimulus, centered on the target, is presented immediately before and after the complex (Carlyon 1994; Gockel and Carlyon 1998). This finding was attributed to the narrowband stimulus streaming the target out of the complex, so that it is heard as a separate source regardless of whether or not it is mistuned. Similarly, for speech sounds, Darwin et al. (1989) showed that a sequence of tones at a frequency corresponding to added energy in a vowel sound perceptually removed this energy from the vowel and thereby changed the position of the phoneme boundary. The interaction of simultaneous and sequential grouping processes is broadly consistent with a computational model proposed by Godsmark and Brown (1999), which also accounts for interactions with the spatial aspects of sound.

This chapter has focused largely on the successes of the past few decades of research into the role of harmonicity in source segregation: The stimulus features that the auditory system can and cannot exploit are well understood, and some useful computational models have been developed. In addition, there is emerging knowledge of both the dependence of source segregation on higher-level processes such as attention (Alain et al. 2001; Carlyon et al. 2001; Sussman et al. 2002) and of some neural correlates of source segregation (Alain et al. 2001; Fishman et al. 2001; Micheyl et al. 2005). Developments in these latter areas are still in their infancy, and are likely to form the focus of much future research into source segregation by the auditory system.

## References

Alain C, Arnott SR, Picton TW (2001) Bottom-up and top-down influences on auditory scene analysis: Evidence from event-related brain potentials. J Exp Psychol [Hum Percep Perform] 27:1072–1089.

Anstis S, Saida S (1985) Adaptation to auditory streaming of frequency-modulated tones. J Exp Psychol [Hum Percep Perform] 11:257–271.

Assmann P, Summerfield Q (1990) Modeling the perception of concurrent vowels: Vowels with different fundamental frequencies. J Acoust Soc Am 88:680–697.

Beauvois MW, Meddis R (1996) Computer simulation of auditory stream segregation in alternating-tone sequences. J Acoust Soc Am 99:2270–2280.

Beerends JG, Houtsma AJM (1989) Pitch identification of simultaneous diotic and dichotic two-tone complexes. J Acoust Soc Am 85:813–819.

Bernstein JGW, Oxenham AJ (2005) An autocorrelation model with place dependence to account for the effect of harmonic number on fundamental frequency discrimination. J Acoust Soc Am 117:3816–3831.

Bregman AS (1978) Auditory streaming is cumulative. J Exp Psychol [Hum Percep Perform] 4:380–387.

Bregman AS (1990) Auditory Scene Analysis: The Perceptual Organization of Sound. Cambridge, MA: Bradford Books, MIT Press.

Bregman AS, Campbell J (1971) Primary auditory stream segregation and perception of order in rapid sequences of tones. J Exp Psychol 89:244–249.

Bregman AS, Liao C, Levitan R (1990) Auditory grouping based on fundamental frequency and formant peak frequency. Can J Psychol 44:400–413.

Bregman AS, Ahad PA, Crum PAC, O'Reilly J (2000) Effects of time intervals and tone durations on auditory stream segregation. Percept Psychophys 62:626–636.

Brokx JPL, Nooteboom SG (1982) Intonation and the perceptual separation of simultaneous voices. J Phonet 10:23–36.

Brunstrom JM, Roberts B (2000) Separate mechanisms govern the selection of spectral components for perceptual fusion and for the computation of global pitch. J Acoust Soc Am 107:1566–1577.

Carlyon RP (1994) Detecting mistuning in the presence of synchronous and asynchronous interfering sounds. J Acoust Soc Am 95:2622–2630.

Carlyon RP (1996) Encoding the fundamental frequency of a complex tone in the presence of a spectrally overlapping masker. J Acoust Soc Am 99:517–524.

Carlyon RP (2004) How the brain separates sounds. Trends Cogn Sci 8:435–478.

Carlyon RP, Deeks JM (2002) Limitations on rate discrimination. J Acoust Soc Am 112:1009–1025.

Carlyon RP, Shackleton TM (1994) Comparing the fundamental frequencies of resolved and unresolved harmonics: Evidence for two pitch mechanisms? J Acoust Soc Am 95:3541–3554.

Carlyon RP, Cusack R, Foxton JM, Robertson IH (2001) Effects of attention and unilateral neglect on auditory stream segregation. Exp Psychol [Human Percept Perform] 27: 115–127.

Culling JF, Darwin CJ (1994) Perceptual and computational separation of simultaneous vowels: cues arising from low-frequency beating. J Acoust Soc Am 95:1559–1569.

Cusack R, Deeks J, Aikman G, Carlyon RP (2004) Effects of location, frequency region, and time course of selective attention on auditory scene analysis. J Exp Psychol [Hum Percept Perform] 30:643–656.

Dannenbring GL, Bregman AS (1978) Streaming vs. fusion of sinusoidal components of complex tones. Percept Psychophys 24:369–376.

Darwin CJ (1992) Listening to two things at once. In: Schouten B (ed) Audition, Speech, and Language. Berlin: Mouton-De Gruyter, pp. 133–148.

Darwin CJ (1997) Auditory grouping. Trends Cogn Sci 1:327–333.

Darwin CJ, Carlyon RP (1995) Auditory grouping. In: Moore BCJ (ed) Hearing. San Diego: Academic Press, pp. 387–424.

Darwin CJ, Hukin RW (1999) Auditory objects of attention: The role of interaural time-differences. J Exp Psychol [Hum Percept Perform] 25:617–629.

Darwin CJ, Pattison H, Gardner RB (1989) Vowel quality changes produced by surrounding tone sequences. Percept Psychophys 45:333–342.

de Cheveigné A (1993) Separation of concurrent harmonic sounds: Fundamental frequency estimation and a time-domain cancellation model of auditory processing. J Acoust Soc Am 93:3271–3290.

Deeks JM, Carlyon RP (2004) Simulations of cochlear implant hearing using filtered harmonic complexes: Implications for concurrent sound segregation. J Acoust Soc Am 115:1736–1746.

Dowling WJ (1973) The perception of interleaved melodies. Cogn Psychol 5:322–337.

Feth L (1974) Frequency discrimination of complex periodic tones. Percept Psychophys 15:375–379.

Fishman YI, Reser DH, Arezzo JC, Steinschneider M (2001) Neural correlates of auditory stream segregation in primary auditory cortex of the awake monkey. Hear Res 151: 167–187.

Glasberg BR, Moore BCJ (1990) Derivation of auditory filter shapes from notched-noise data. Hear Res 47:103–138.

Gockel H, Carlyon RP (1998) Effects of ear of entry and perceived location of synchronous and asynchronous components on mistuning detection. J Acoust Soc Am 104:534–3545.

Gockel H, Carlyon RP, Micheyl C (1999) Context dependence of fundamental-frequency discrimination: Lateralized temporal fringes. J Acoust Soc Am 106:3553–3563.

Gockel H, Carlyon RP, Plack CJ (2004) Across-frequency interference effects in fundamental frequency discrimination: Questioning evidence for two pitch mechanisms. J Acoust Soc Am 116:1092–1104.

Gockel H, Carlyon RP, Moore BCJ (2005) Pitch discrimination interference: The role of pitch pulse asynchrony. J Acoust Soc Am 117:3860–3866.

Godsmark D, Brown GJ (1999) A blackboard architecture for computational auditory scene analysis. Speech Commun 27:351–366.

Goldstein J L (1973) An optimum processor theory for the central formation of the pitch of complex tones. J Acoust Soc Am 54:1496–1516.

Grimault N, Micheyl C, Carlyon RP, Arthaud P, Collet L (2000) Influence of peripheral resolvability on the perceptual segregation of harmonic complex tones differing in fundamental frequency. J Acoust Soc Am 108:263–271.

Hartmann WM, Johnson D (1991) Stream segregation and peripheral channeling. Music Percept 9:155–184.

Hartmann WM, McAdams S, Smith BK (1990) Hearing a mistuned harmonic in an otherwise periodic complex tone. J Acoust Soc Am 88:1712–1724.

Hill NJ, Darwin CJ (1996) Lateralization of a perturbed harmonic: Effects of onset asynchrony and mistuning. J Acoust Soc Am 100:2352–2364.

Houtsma AJM, Smurzynski J (1990) J.F. Schouten revisited: Pitch of complex tones having many high-order harmonics. J Acoust Soc Am 87:304–310.

Kim DO, Molnar CE, Matthews JW (1980) Cochlear mechanics: Nonlinear behavior in two-tone responses as reflected in cochlear-nerve-fiber responses and in ear-canal sound pressure. J Acoust Soc Am 67:1704–1721.

Licklider JCR (1951) A duplex theory of pitch perception. Experientia 7:128–133.

Licklider JCR (1956) Auditory frequency analysis. In: Cherry C (ed) Information Theory. New York: Academic Press, pp. 253–268.

Lyzenga J, Carlyon RP (1999) Center frequency modulation detection for harmonic complexes resembling vowel formants and its interference by off-frequency maskers. J Acoust Soc Am 105:2792–2806.

McCabe SL, Denham MJ (1997) A model of auditory streaming. J Acoust Soc Am 101:1611–1621.

Meddis R, Hewitt M (1991) Virtual pitch and phase sensitivity studied using a computer model of the auditory periphery I: Pitch identification. J Acoust Soc Am 89:2866–2882.

Meddis R, O'Mard L (1997) A unitary model of pitch perception. J Acoust Soc Am 102:1811–1820.

Micheyl C, Carlyon RP (1998) Effects of temporal fringes on fundamental-frequency discrimination. J Acoust Soc Am 104:3006–3018.

Micheyl C, Tian B, Carlyon RP, Rauschecker JP (2005) Perceptual organization of tone sequences in the auditory cortex of awake macaques. Neuron 48:139–148.

Miller GA, Heise GA (1950) The trill threshold. J Acoust Soc Am 22:637–638.

Moore BCJ (1982) An Introduction to the Psychology of Hearing, 2nd ed. London: Academic Press.

Moore BCJ (2003) An Introduction to the Psychology of Hearing, 5th ed. San Diego: Academic Press.

Moore BCJ, Carlyon RP (2005) Perception of pitch by people with cochlear hearing loss and by cochlear implant users. In: Plack CJ, Oxenham AJ (eds) Springer Handbook of Auditory Research: Pitch Perception. New York: Springer-Verlag, pp. 234–277.

Moore BCJ, Gockel H (2002) Factors influencing sequential stream segregation. Acust Acta Acust 88:320–333.

Moore BCJ, Ohgushi K (1993) Audibility of partials in inharmonic complex tones. J Acoust Soc Am 93:452–461.

Moore BCJ, Glasberg BR, Peters RW (1985) Relative dominance of individual partials in determining the pitch of complex tones. J Acoust Soc Am 77:1853–1860.

Moore BCJ, Glasberg BR, Peters RW (1986) Thresholds for hearing mistuned partials as separate tones in harmonic complexes. J Acoust Soc Am 80:479–483.

Oxenham AJ, Bernstein JGW, Penagos H (2004) Correct tonotopic representation is necessary for complex pitch perception. Proc Natl Acad Sci USA 101:1421–1425.

Patterson R D, Allerhand M, Giguère C (1995) Time-domain modelling of peripheral auditory processing: A modular architecture and a software platform. J Acoust Soc Am 98:1890–1894.

Plack CJ, Oxenham AJ, Fay RR, Popper AN (eds) (2005) Pitch: Neural Coding and Perception. New York: springer.

Plomp R (1964) The ear as a frequency analyzer. J Acoust Soc Am 36:1628–1636.

Plomp R (1967) Pitch of complex tones. J Acoust Soc Am 41:1526–1533.

Plomp R (1976) Aspects of Tone Sensation. London: Academic Press.

Roberts B (2005) Spectral pattern, grouping, and the pitches of complex tones and their components. Acust Acta Acust 91:945–957.

Roberts B, Bregman AS (1991) Effects of the pattern of spectral spacing on the perceptual fusion of harmonics. J Acoust Soc Am 90:3050–3060.

Roberts B, Brunstrom JM (1998) Perceptual segregation and pitch shifts of mistuned components in harmonic complexes and in regular inharmonic complexes. J Acoust Soc Am 104:2326–2338.

Roberts B, Brunstrom JM (2001) Perceptual fusion and fragmentation of complex tones made inharmonic by applying different degrees of frequency shift and spectral stretch. J Acoust Soc Am 110:2479–2490.

Roberts B, Brunstrom JM (2003) Spectral pattern, harmonic relations, and the perceptual grouping of low-numbered components. J Acoust Soc Am 114:2118–2134.

Rose MM, Moore BCJ (1997) Perceptual grouping of tone sequences by normally hearing and hearing-impaired listeners. J Acoust Soc Am 102:1768–1778.

Scheffers MTM (1983) Sifting vowels: auditory pitch analysis and sound segregation. PhD thesis, University of Groningen, Netherlands.

Schouten JF, Ritsma RJ, Cardozo BL (1962) Pitch of the residue. J Acoust Soc Am 34:1418–1424.

Shamma S (1985) Speech Processing in the Auditory System: II. Lateral inhibition and the central processing of speech evoked activity in the auditory nerve. J Acoust Soc Am 78:1622–1632.

Shannon RV (1983) Multichannel electrical stimulation of the auditory nerve in man. I. Basic psychophysics. Hear Res 11:157–189.

Singh PG (1987) Perceptual organization of complex-tone sequences: A tradeoff between pitch and timbre? J Acoust Soc Am 82:886–895.

Sussman E, Winkler I, Huotilainen M, Ritter W, Naatanen R (2002) Top-down effects can modify the initially stimulus-driven auditory organization. Cogn Brain Res 13:393–405.

van Noorden LPAS (1975) Temporal coherence in the perception of tone sequences. PhD thesis, Eindhoven University of Technology.

Vliegen J, Moore BCJ, Oxenham AJ (1999) The role of spectral and periodicity cues in auditory stream segregation, measured using a temporal discrimination task. J Acoust Soc Am 106:938–945.

Vliegen J, Oxenham AJ (1999) Sequential stream segregation in the absence of spectral cues. J Acoust Soc Am 105:339–346.

# 8
# Spatial Hearing and Perceiving Sources

CHRISTOPHER J. DARWIN

## 1. The Problem

Our subjective experience of a natural mixture of sound sources is that each source is heard in its proper spatial position. This perceptual experience is a major achievement of the human brain, but one that is only partially understood. Almost all of the experimental research on our ability to localize sound sources has addressed the problem of how we localize individual sounds presented in isolation. Substantially fewer studies have addressed the ecologically much more valid and practically much more significant question of how we localize multiple sound sources that are in different spatial positions. A related question concerns the extent to which we use spatial information from individual auditory frequency channels to help us to group together both across frequency and across time those channels that correspond to a particular sound source.

## 1.1 Initial Observations

Abundant work (Middlebrooks and Green 1991; Hafter and Trahiotis 1997) on the spatial localization of single, pure-tone sound sources has given us a substantial and physiologically based understanding of how the primary cues to spatial direction in the horizontal plane (interaural time and intensity differences) are extracted separately in each frequency channel and later combined, again in a frequency-specific way (Stern and Colburn 1978), to give a subjective spatial position for the sound. We also have some understanding of how individual complex broadband sounds such as speech are localized (Wightman and Kistler 1992). But such studies only go part of the way toward solving the problem of the localization of multiple sound sources (see Yost 1997).

With a single sound source, the listener could cavalierly use all the information that is present to estimate the position of that sound, combining across all frequencies the cue values (or positions) associated with each individual frequency channel (Trahiotis and Stern 1989; Shackleton et al. 1992). But when more than one sound source is present, the listener must be more selective (Buell and Hafter 1991; Woods and Colburn 1992). One cannot simply pool all the information; the result would be a useless and vague compromise. But how many sound sources are there? Which information should one use for which sound source? How does the brain solve the problem of deciding which frequency regions should contribute to the localization of which sound sources?

The brain could treat the spatial information itself as the best guide: For example, one could make histograms of the individual cues (or their combinations) across frequency and then treat each peak in the resulting distribution as a separate sound source. With this approach, sound sources are defined by their spatial attributes. The spatially selective algorithms used in some digital hearing aids implicitly follow this approach.

Alternatively (Woods and Colburn 1992; Hill and Darwin 1996), the brain could first decide on the basis of auditory scene analysis (Bregman 1990) which frequency components form the separate sound sources and then confine any combination of cue information across frequency to those channels identified as belonging to the same sound source. Auditory scene analysis can use a variety of different types of information ranging from low-level cues such as harmonicity and onset time (and indeed spatial cues) through to schematic knowledge about individual sound sources. With this approach, sound sources are first defined eclectically and then their spatial position determined.

Which would be the most intelligent design? The answer depends on the reliability of the different types of information. If the spatial information in *each* of the individual frequency channels of a complex sound were a reliable indicator of that sound source's direction, then there would be a great deal to be said for the first alternative: quick, simple, and independent of knowledge about what sound sources are like. But if it is *not* reliable, then the first alternative runs the risk of producing a distractingly fragmented perceptual experience that, conveniently if somewhat remarkably, does not occur: We do not hear the different frequency regions of a complex sound coming from different directions. Sounds maintain their spatial integrity, even in circumstances in which spatial information has been severely degraded by other auditory objects and by echoes and reverberation. So the second alternative, which involves the complex processing of auditory scene analysis in the apparently simple task of localizing sounds, becomes a contender.

In this chapter we review some of the main aspects of the intriguing relationship between the perception of auditory objects and their spatial location. Section 2 concentrates on cues to azimuth. In it we first examine how reliable the various cues to localization are in normal listening environments in which there can be additional sound sources, along with echoes and reverberation. We then ask whether simple sounds are localized independently of others that are present at the same time. If space were a primary dimension for defining sound sources, we would expect such independent processing. The evidence shows that strong interactions can take place across widely different frequency regions, and some experiments suggest that auditory grouping cues such as simultaneity and harmonicity can influence the extent of these interactions.

In Section 3 we then ask the related question whether sounds can be grouped by spatial information. We see that spatial position provides good sequential grouping of alternating sound sources, but that spatial information has remarkable shortcomings for grouping when sounds with different spatial characteristics are present at the same time. Section 3.3 looks at how spatial differences influence the intelligibility of speech heard against a background of other sounds, and to what extent grouping by spatial location contributes to any improvement. Section 4 looks briefly at the much less extensive work on spatial separation in the median plane, where the relevant cues are very different from those in azimuth.

## 2.  Localization of Multiple Sounds in Azimuth

Although we are able to localize sound in all three spatial dimensions (normally arranged as the horizontal plane, or azimuth; the median plane; and distance), the most important and certainly the best-researched dimension is azimuth. Lord Rayleigh's duplex theory (1907) provides a basis for the azimuthal localization of ongoing pure tones. Natural low-frequency tones coming from one side produce unambiguous phase differences between the ears, but when more than a meter or so distant, they produce no useful intensity difference between the ears. Conversely, high-frequency tones produce large intensity differences (which vary with frequency) but ambiguous phase differences. Artificial differences in intensity between the ears can be used to shift the location of both low- and high-frequency tones, but we are insensitive to interaural phase differences of pure tones whose frequency is above about 1500 Hz. However, complex high-frequency tones do allow the use of interaural phase information: The relatively slow amplitude modulation that complex sounds can produce in high-frequency auditory filters produces unambiguous phase differences of the envelope, which listeners can use for localization (Henning 1974). A more general qualification to duplex theory, and one that is particularly relevant to the problem of localizing multiple dynamic sound sources, is that for many sounds, the *ongoing* phase and amplitude differences between the ears are less important than their value at the onset of the sound, a phenomenon referred to as *binaural adaptation* (Hafter et al. 1988). Its importance for the localization of multiple sounds is highlighted by the fact that it is reset by events such as short gaps and added tones and noise bursts (Hafter and Buell 1990; Freyman et al. 1997).

### 2.1  The Reliability of ITD

How reliable is the interaural time difference in a single auditory channel? Physical measurement of the binaural coherence of single sound sources in rooms indicates that ITD information is easily degraded in mildly reverberant listening conditions, so that information within a single frequency channel does not give reliable information about the azimuth of its sound source. Although such information *is* reliable under ideal anechoic conditions, it is easily adulterated by normal room acoustics so that from a broadband source, different low-frequency bands deliver substantially different values of ITD, while in the higher-frequency bands it is too difficult to estimate reliably where the true ITD lies (Hartmann et al. 2005; Shinn-Cunningham et al. 2005).

## 2.2 Interference in Lateralization Between Simultaneous Sounds

When different frequency regions have conflicting interaural information about lateral position, what should the brain's localization system do? Treat them as one sound source and pool the information, or treat them independently as different sound sources? The answer depends on the relationship between the sounds. Ideally, the system should treat them independently if there is clear evidence that they come from independent sound sources. The problem has been studied in experiments in which listeners have to judge the lateral position of one sound, either alone or in the presence of a second sound with a constant interaural time difference (ITD). McFadden and Pasanen (1976) showed that the threshold ITD needed to lateralize correctly a 4-kHz, 230-Hz-wide noise band roughly doubled when a diotic (i.e., the same sound in both ears) 500-Hz, 50-Hz wide, noise band was added simultaneously to it. However, the relationship was not symmetrical, a diotic upper band did not impair lateral position judgments of the lower band (cf. Bilsen and Raatgever 1973; Wightman and Kistler 1992). The interference of the low on the high band is substantially reduced when the low, interfering sound is continuous rather than gated to be simultaneous with the high, target sound (Trahiotis and Bernstein 1990). These results are compatible with the auditory system treating the simultaneously gated sounds as a single sound source but the continuous interferer as separate from the gated target (see also Sheft, Chapter 9).

Subsequent work has also shown considerable interference by low-frequency flanking sounds on low-frequency targets. Henning (1980) played his listeners a 250-ms complex sound consisting of a diotic center frequency of 600 Hz flanked by two sidebands at 300 Hz and 900 Hz to which he applied equal but opposite polarity ITDs. He measured how large these ITDs had to be for listeners to detect them compared to an entirely diotic sound. He found that thresholds were surprisingly large, around 500 μs. Listeners apparently combine the ITDs across the three frequencies so that all three components appear to come from the midline.

The sounds used in Henning's (1980) study were harmonically related. It is well known that harmonically related sounds are more likely to group together into a single sound source than are unrelated frequency components (for a review see Darwin 2005). Do harmonic sounds interfere more than unrelated ones? Experiments aimed at answering this question have shown mixed results: Some studies show that harmonically related components are more likely to interfere (Buell and Hafter 1991) and others show no effect (Stellmack and Dye 1993). A significant additional point (Stellmack and Dye 1993) is that there can be significant interference from a mistuned sound that is clearly heard as a separate sound source (cf. Hukin and Darwin 1995a).

More-direct evidence that the grouping of sounds by their harmonic relations is important in localizing complex sounds comes from experiments that have exploited an intriguing effect first noted by Jeffress (1972) and subsequently investigated by Stern et al. (1988). It is well known that a narrow band of noise centered on 500 Hz ($f_c$) and given an ITD of +1.5 ms ($t_i$)will be heard on the lagging (not the leading) side. Because of phase ambiguity, this stimulus is barely discernible from one that has the complementary ITD of –0.5 ms (i.e., $1/f_c - t_i$), and the auditory system prefers the shorter ITD. However, Jeffress discovered that if the bandwidth of the sound is gradually increased while the ITD is maintained at 1.5 ms, then the location of the noise moves across from the lagging to the leading side. Stern et al. replicated this effect and offered an interpretation in terms of the consistency of interaural time differences across frequency. As additional frequencies are added to the noise, the imposed ITD (+1.5 ms) stays constant, but the complementary ITD ($1/f_c - t_i$), being a function of the frequency concerned, varies. The only consistent ITD is then +1.5 ms, and this consistency eventually overcomes the auditory system's preference for short over long ITDs. This phenomenon is interesting, since it indicates that ITD information is being integrated across different frequencies in the calculation of lateral position (see also Shackleton et al. 1992). However, it makes sense to perform this integration only across those frequencies that make up a single auditory object; otherwise, sounds with different locations could be treated together to give a single average location, rather than separate locations for different objects. Hill and Darwin (1993) showed that harmonicity contributes to this grouping of sounds for across-frequency integration of ITD. They first replicated the Jeffress effect with harmonic sounds, starting with a single frequency component at 500 Hz and then adding additional harmonics of 100 Hz on either side of it. All components had an ITD of 1.5 ms. As with Jeffress's noise, when the flanking harmonics were added, the location changed away from the lagging side toward the leading side. What Hill and Darwin were then able to show was that mistuning the original 500-Hz harmonic by about 3% was sufficient to move it, as a separate sound source, back toward the lagging side. In other words, its location was being determined independently of the other frequency components by virtue of its mistuning.

These experiments (see also Best et al. 2007 for recent evidence of sequential grouping) offer some support to the view that interaural time differences are not used as a primary cue for auditory grouping. Instead, auditory objects are formed on the basis of scene analysis mechanisms, and these objects are localized. Against this view is the fact that interference in ITD discrimination clearly takes place when the interfering components are being heard as separate objects (Stellmack and Dye 1993). This latter result echoes the fact that a slightly mistuned frequency component of a complex sound can still make a full contribution to the pitch of the complex, even though it is clearly heard as a separate sound source by virtue its starting earlier than the rest of the complex (Darwin and Ciocca 1992). The concept of "sound source" is clearly not well understood, perhaps due to the hierarchical nature of sound sources and our ability to move our perception around that hierarchy (Cusack et al. 2004).

# 3. Azimuthal Cues in Object Formation

A corollary of the conclusion at the end of the previous section is that spatial cues should be rather weak at grouping together or segregating simultaneous auditory objects. The evidence that this section reviews supports this view. When only a *single* sound source is present at a time, so that the grouping problem is a sequential one, then spatial information can be very powerful. But spatial information appears to be much less effective at separating *simultaneous* sound sources, especially when there are no other grouping cues available. If other cues are available on which to base grouping, then spatial information can enhance the perceptual separation.

## *3.1 Sequential Grouping*

Spatial cues are most effective when only a single sound source is present at a particular time and the organizational problem is to allocate each successive sound to its appropriate source—sequential rather than simultaneous organization. The effectiveness of this spatial sequential grouping is substantially weakened when additional simultaneous sounds are present. This weakening is most likely related to the interference between the spatial information of simultaneous sounds discussed in Section 2.

One of the audio demonstrations on the CD produced by Bregman and Ahad (Bregman and Ahad 1995) shows that spatial separation—either by ILD or in the free field over loudspeakers—very effectively separates two alternating trains of amadinda (a West African xylophone) notes. A similar result can also be obtained by giving alternate notes different ITDs using interleaved melodies (Hartmann and Johnson 1991) or the Wessel illusion (Wessel 1979; Hukin and Darwin 2000). It seems likely that the effective property for this separation is the subjective spatial position of the individual sounds rather than the values of a particular spatial cue. Sach and Bailey (2004) used a phenomenon called rhythmic masking release (Bregman and Ahad 1995; Turgeon et al. 2002) whereby the rhythm of a series of monotonic notes that has been obscured by interleaving other similar notes is revealed by making the interleaved sequence distinctive through a difference in timbre or spatial position. Sach and Bailey showed that subjective spatial position, rather than simply a difference in the individual spatial cues, was responsible for the segregation that caused the reappearance of the original rhythm. The segregation produced by a difference in ITD between the two series of 500-Hz tones was abolished if an ILD was also introduced, but with opposite polarity, so that the subjective position of the two series of notes again became identical.

This clear segregation of two alternating sound sources by spatial location is seriously eroded if the sounds from the two sources are simultaneous or if other sounds are introduced that are simultaneous with the two alternating sources. Tchaikovsky knew about this effect (Butler 1979; Deutsch 1999). What would normally be the first violin part of the slow introduction to the last movement of his Sixth Symphony (*Pathétique*) is actually scored to give alternate notes to the first and second violins, who in Tchaikovsky's day sat to the left and right respectively of the conductor. The "second violin part" is similarly alternated, but with the opposite phase, so that each player's part alternates between the melodic and the accompanying line. The audience has no difficulty in hearing the melody rather than the actual note sequence that, say, the first violins are playing, though it may acquire a spatial vagueness that matches the wistful mood of the music.

Deutsch (1979) provides a formal demonstration of this weakening of spatial cues by simultaneous sounds. Identification of a melody is substantially impaired when its individual notes are allocated haphazardly to one or the other ear. However, if a constant lower-frequency drone note is played simultaneously with each melodic note, but in the opposite ear, performance dramatically improves.

## 3.2 Simultaneous Grouping

Although localization cues have been used quite extensively for the machine segregation of different talkers (Bodden 1996), one might expect the auditory system *not* to use spatial cues as a primary method for grouping simultaneous sound sources if the perceptual localization of one sound is significantly disrupted by the presence of another simultaneous sound. This expectation is borne out by experiments on the perception of pitch and on the perception of the vowels of speech.

### 3.2.1 Pitch

When two complex tones with sufficiently different fundamentals are played at the same time, it is a common observation that the two separate pitches are clear; yet how does the brain know which harmonics should be used for calculating each of the pitches (Goldstein 1973)? Naively, one might think that it would welcome some spatial information that would help in the sorting process. Yet most of the evidence indicates that spatial cues have surprisingly little effect (see also Carlyon and Gockel, Chapter 7).

If two consecutive harmonics from two different fundamentals are presented simultaneously, listeners are no better at identifying the two fundamentals when the appropriate pairs of harmonics are played to different ears than when each ear receives one harmonic from each fundamental (Beerends and Houtsma 1986). A similar lack of segregation by ear is seen in data gathered by Darwin and Ciocca (1992) measuring the pitch shift produced by a single mistuned harmonic. The pitch shift was very similar whether the mistuned harmonic went to the same ear as the rest of the complex or to the opposite ear. However, Gockel et al. (2005) have recently found that for very short tones (16 ms), there was a pronounced reduction in pitch shift when the mistuned component was presented only to the opposite ear.

### 3.2.2 Speech Sounds

It is well established that ITDs in the lower-frequency components are the dominant localization cue for a complex sound such as speech (Wightman and Kistler 1992). Hartmann et al. (2005) have recently pointed out a possible reason for this dominance:

> Because of the size of the human head, this frequency region corresponds to a minimum in coherence when the sound field is isotropic. If a listener is required to localize a source in the presence of an interfering reverberant field that is approximately isotropic, then any peak that occurs in this frequency region is likely to come from the direct sound from the source and not from the environment (p. 460).

Despite the dominance of ITDs in the region around 500 Hz, it turns out that ITDs are remarkably ineffective at segregating simultaneous sounds. This remarkable finding was reported by Culling and Summerfield (1995), who presented listeners with four narrowband noises at center frequencies appropriate for the first two formants of /i/ and /a/, or in different combination, the first two formants of /u/ and /er/. If these noise-band pairs were led to opposite ears (e.g., bands 1 and 4 give /i/, and 2 and 3 give /a/), listeners had no difficult in hearing /i/ in their left ear. However, if ITDs were manipulated rather than leading the bands to different ears, listeners were quite unable to do the task. We have, then, the apparently paradoxical result that the cue that is dominant for the localization of complex sounds is impotent to group simultaneous sounds. Subsequent work has partly confirmed and partly modified Culling and Summerfield's conclusions.

First, using just differences in ITD, some listeners can, with practice, learn to perform segregation by ITD (Drennan et al. 2003), a result that we have confirmed (Darwin 2002). Second, if the noise bands are not delivered over headphones but rather over spatially separated loudspeakers, so that listeners receive the full range of natural interaural cues, then grouping by these spatial cues becomes much easier (Drennan et al. 2003). The reason for this improvement may be that the natural combination of cues is seen by the brain as a more robust indicator of spatial position; when only a single cue is manipulated, the individual cue values are always in conflict with the neutral value of the other cues. In addition, the naturally present ILDs may give stronger lateralization for the high-frequency components than are given by their natural ITDs.

A difference in the F0 of simultaneous sounds can also help with their localization. We have already seen that localization cues can be ineffective for grouping simultaneous sounds. In particular, an interaural time difference gives virtually no improvement in the identification of two simultaneous steady vowels on the same F0 (Shackleton et al. 1994) or in the identification of the leftmost of two noise-excited vowel-like sounds (Culling and Summerfield 1995). However, if voiced vowels are given a difference in F0 (which itself helps in their identification), then an additional difference in ITD of 400 μs further improves identification (Shackleton et al. 1994), presumably by giving an additional spatial separation to the two sounds.

The interaction between ITD and other, both simultaneous and sequential, auditory grouping cues has been extensively investigated using a different paradigm, which involves the perceptual segregation of a single harmonic from a steady vowel sound. This paradigm uses speech categorization to measure the extent to which a harmonic has been perceptually removed from a vowel, by exploiting the fact that identifiable versions of the vowels /I/ and /ɛ/ can be made that differ only in their first formant frequency (F1). When a series of such sounds differing in F1 is synthesized, the phoneme boundary occurs at an F1 frequency of around 450 Hz. If, then, a harmonic at around 500 Hz is either reduced or physically completely removed from the vowel, there is a perceptual change in vowel quality that causes a change in the F1 frequency at the phoneme boundary of up to around 50 Hz. What is of interest is that a similar, or slightly smaller, shift can also occur if the harmonic is not physically, but *perceptually*, removed by giving it an earlier onset time than the rest of the vowel (Darwin 1984), by slightly mistuning it (Darwin and Gardner 1986), or by embedding the vowel in a series of 500-Hz tones similar to the to-be-removed 500-Hz harmonic of the vowel (Darwin et al. 1989). What effect do spatial cues have on this boundary shift?

If the 500-Hz harmonic is put into the opposite ear from the rest of the vowel, there is a shift in phoneme boundary equivalent to roughly a 6-dB physical reduction in its level. However, if the tone is given a substantially different ITD from the rest of the vowel (±666 μs), there is almost no shift in the phoneme boundary (Hukin and Darwin 1995b). This result is compatible with the double-vowel results of Culling and Summerfield (1995). But further experiments using the phoneme boundary shift paradigm qualified this conclusion. A difference in ITD *can* increase segregation if the listener has clear independent evidence that the 500-Hz harmonic can be a separate sound source. Such evidence does not have to be present on the same trial (such as the use of a difference in F0 in Shackleton and Meddis (1992)). It can also come from other trials in the same experimental block where the to-be-removed harmonic has a difference in onset time or is slightly mistuned (Darwin and Hukin 1998) or where the test vowel is embedded in a series of tones like the to-be-removed harmonic and all are presented with a different ITD from the rest of the vowel (Hukin and Darwin 1995b; Darwin and Hukin 1997).

These experiments have shown that exposure to the segregated object on a previous trial can increase segregation by ITD. An equivalent mechanism is probably at work in experiments that have used a very limited stimulus set and show a remarkable ability of the auditory system to separately localize complex sounds that differ in ITD.

If two simultaneous monosyllabic words ("bead" and "globe") are embedded in different carrier phrases and the two sentences given ITDs of ±90 µs respectively, they easily segregate into two spatially distinct auditory objects that can be readily attended to. For natural speech there are many cues (e.g., harmonicity, onset-time differences) that can help the auditory system to allocate individual frequency channels to the two different sound sources. What is surprising is that the impression of two separate objects survives when the two sentences are resynthesized (using a PSOLA-based algorithm) on exactly the same F0 and the two test words exactly synchronized (Darwin and Hukin 1999). For simpler sounds, such as steady vowels, the impression of two distinct sources with separate locations is destroyed by a common F0 (Darwin and Hukin 1999).

The conclusion from this work is that although ITD by itself provides only a weak basis for simultaneous segregation, two qualifications need to be made: first, with extended practice some listeners can learn to use it, and second, in the presence of other relevant information, such as primitive grouping cues, or knowledge about the segregated objects, a difference in ITD can augment perceptual and spatial segregation.

## 3.3 Spatial Separation and Speech Intelligibility

It is a longstanding observation that speech is more intelligible when it is spatially separated from competing sounds than when it is not (Cherry 1953). There is a variety of mechanisms responsible for this improvement (see the review Bronkhorst 2000), some of which involve auditory grouping, as has become clear from recent work using natural speech that has been constrained in various ways.

Work on automatic speech recognition in different noise backgrounds (Cooke et al. 2001; Cooke 2003; Wang 2005) has made important observations on the nature of both the speech signal itself and the problems for recognition posed by different noise backgrounds. The main distinction that is relevant to the role of spatial mechanisms in improving intelligibility is that between on the one hand a relatively consistent background sound such as either speech-shaped noise or a cafeteria-noise mixture of many voices, and on the other hand, a background source consisting of only one or two other voices. The perceptual and computational problems raised by these two types of background sound are quite different (Miller 1947; Bronkhorst 2000; Brungart 2001b; Assmann and Summerfield 2004).

For a speech signal against a steady noise background, performance is limited by our ability to detect the weaker components of the signal, and then to recognize the speech on the basis of the partial information that has been detected—spectrotemporally distributed "glimpses" (Miller and Licklider 1950; Cooke 2003; Assmann and Summerfield 2004; Cooke 2005; Wang 2005). There is little problem here in deciding what is signal and what is noise, because of the large qualitative difference between the two. Spatially related cues here contribute to the detectability of the speech features through mechanisms such as head shadow for the higher frequencies and binaural release from masking for the lower frequencies (see the review Bronkhorst 2000).

However, for a mixture of two voices at roughly equal levels, it turns out that there is little mutual masking, so that the main problem is not to *detect* features but to *allocate* each local spectrotemporal feature to the appropriate voice (Cooke et al. 2001; Cooke 2003; Brungart 2005; Wang 2005). Whether spatial properties of the two talkers are important now depends on whether there are other differences between the two talkers. The issues here are quite complex, and we need to consider both sequential and simultaneous grouping. The relative importance of these two types of grouping depends on the task and the detailed stimulus configuration.

Consider, for example, the task demands posed by listening to two simultaneous sentences taken from one of three different types of speech material: the Co-ordinate Response Measure (CRM, Bolia et al. 2000; Brungart 2001a), semantically anomalous nonsense sentences (Freyman et al. 1999), and conventional sentences such as the Harvard Sentence Lists or the Bamford–Kowal–Bench Standard Sentence Lists.

In the CRM task, the sentence frame is stylized (e.g., "Ready Baron go to green three now"), and the response set limited to a small number of call signs, colors and digits, so rather little information is needed to identify the individual items, but the sentence context provides no constraint on the response set. With two simultaneous sentences from the CRM set, the main problem facing the listener is to identify which of the two clearly heard colors and which number were spoken by the talker who produced the call sign. This task is thus a very good one for revealing the effectiveness of cues that can contribute to sequential grouping (Brungart 2001a; Darwin et al. 2003). It has been used to show that a difference in pitch between the two talkers or of vocal-tract length (Darwin and Hukin 2000; Darwin et al. 2003) or (even a reduction) of level (Egan et al. 1954; Brungart 2001b) can help the listener to track one voice in the presence of another otherwise similar voice. The issues here are similar to those in the stream segregation of simple alternating tones, where any sufficiently salient cue can lead to segregation (Moore and Gockel 2002).

The usefulness of this task for revealing cues for tracking a talker across time is won at the expense of naturalness. In normal speech, the vocabulary is less constrained (thank goodness!), so that the detection of speech features and their simultaneous grouping across frequency may be more important. Attending to one of two semantically anomalous sentences (Freyman et al. 1999) thus puts a greater burden on the identification of the individual words than does the CRM task, but still maintains the need to track the target talker across time. With the more natural Harvard and BKB sentences, semantic constraints shift the burden away from tracking the target talker. The semantic constraints may not be particularly powerful at predicting an upcoming word in a single sentence, such as "The birch canoe slid on the smooth planks," but may be very useful at deciding which of two words belongs to the attended sentence, for example against "Sickness kept him home the third week" to take two example Harvard sentences. One would therefore expect that sequential grouping cues might be less effective for tests involving pairs of the Harvard- and BKB-type of sentences than for pairs of CRM or semantically anomalous sentences (cf. Edmonds and Culling 2005).

In order to maximize the conditions for showing spatial effects in speech intelligibility that are not due to detectability, the task should be to identify the target speaker against a single competing speaker (in order to minimize problems of detectability), and there should be minimal differences between the speech tokens in level and in speaker characteristics (including F0); there should also be minimal semantic constraints within the two sentences (Darwin 2006). Many of these conditions are met in a study by Arbogast et al. (2002), which used the four male speakers from the CRM set as target and distracter. The speech of each talker was filtered into 15 frequency bands and the temporal envelope within each band used to modulate a sine wave at the center frequency of each band. The target always consisted of seven randomly chosen bands. The distracter consisted of six of the remaining seven bands and could be either another CRM sentence or, in another condition, noise that was also confined to those frequency bands. (One of the effects of this processing would have been to reduce the perceptual difference between the four male talkers by eliminating the pitch of their voices.) The target was always presented in front of the listener, but the distracter could be presented either straight ahead, or from 90° to the right. The spatial separation gave a much larger improvement in performance for the speech distracter (18 dB) than for the noise distracter (<10 dB) from a baseline 51% performance, which conveniently happened to give the target and distracter speech similar levels.

Although processing speech and distracter into different frequency bands in the Arbogast et al. study provides an objective way of minimizing mutual masking, there is in fact very little mutual masking without such processing in the normal CRM task with a single distracter talker (Brungart et al. 2006), a result that reflects the sparseness of speech (see above). However, masking of the target by the distracter rapidly increases as the number of distracters increases.

The Arbogast et al. study neatly separates out the effects of auditory grouping from those produced by masking, and shows that spatial cues can be important for grouping under appropriate conditions (see also Best et al. 2005). This conclusion can be refined somewhat thanks to an ingenious experiment by Freyman et al. (2001). They exploited the precedence effect to demonstrate that improved intelligibility by spatial separation was attributable to the perceived location of the target and the distracter, despite the absence of interaural cues that are necessary to give binaural release from masking. Their conclusion is echoed by Sach and Bailey (2004), described in Section 3.1.

# 4. Median Plane and Distance

## 4.1 Median Plane

How do we know where two different simultaneous sounds are in the median plane? When Romeo and Juliet are talking at the same time, which one is on the balcony? The elevation of a sound source is determined mainly by the frequencies of the spectral notches imposed on the high-frequency region (>5 kHz) by reflections off the external ear (or pinna) (Batteau 1967; Hebrank and Wright 1974). However, when more than one sound source is present, two problems arise. First, the notches may be filled in by the competing sound. Second, how does the auditory system know which notches belong with which sound source? When the different sounds come from different azimuthal positions, binaural information could potentially help to solve both of these problems, exposing "masked" spectral notches through head shadow and providing a common spatial position for grouping high-frequency regions with low frequencies from the same source. There is conflicting evidence as to whether in the absence of other cues, listeners can use just elevation cues to separate two simultaneous sound sources.

Best et al. (2004) played two broadband noises simultaneously from two different directions in a virtual auditory environment using individualized outer-ear filter functions. The bursts were either spatially coincident or separated horizontally, vertically or both. Listeners were unable to say reliably whether they perceived one or two source locations when the sounds differed only in elevation.

Using materials closer to the Romeo and Juliet problem, Worley and Darwin (2002) used the same task as that used by Darwin and Hukin (1999) described in Section 3.2.2. Listeners were asked to track one of two carrier sentences presented at different physical elevations in an anechoic chamber and report its target word. The words were synchronized, and the sentences were played on monotonic F0s that were either identical or differed by one, two, or four semitones. With an elevation separation of 31°, listeners performed almost perfectly even when the sentences were on the same F0. The spectrotemporal sparseness of speech may have helped listeners to detect the spectral notches, and the very limited response set may have helped listeners to allocate spectral notches to the appropriate target word. At smaller separations, listeners' performance improved as the F0 difference between the sentences increased. By switching the F0 of the target words but not their elevations, Worley and Darwin were able to estimate the extent to which this improvement was due to listeners using the difference in F0 directly to track a particular source, and the extent to which the F0 difference was improving the use of the difference in elevation. The effects that they found were all attributable to the direct use of F0; there was no evidence for the difference in F0 improving their ability to exploit a difference in elevation.

## 4.2 Distance

Sounds at different distances differ in a number of ways: More-distant sounds are less intense, have reduced high-frequency content, and have relatively more reverberant energy. In addition, even low-frequency sounds that are closer than about one meter have interaural level differences thanks to the inverse square law (Brungart and Rabinowitz 1999). There has been almost no work on the effectiveness of these cues in auditory grouping, although we do know that differences in level and spectral content (Egan et al. 1954) can help listeners to track a sound source over time.

One piece of work directly addresses this issue for changes in the distance of near-field sounds (Brungart and Simpson 2002). In the CRM task with same-sex talkers (see Section 3.3), listeners could exploit interaural level difference cues to help track a talker across time.

## 5. Summary

In this chapter we have looked at the effectiveness of spatially related cues for helping us to form and localize complex auditory objects. We have contrasted the use of spatial cues in sequential grouping with that in simultaneous grouping. Spatial cues can be very effective in sequential grouping when only one sound source is present at a time. In simultaneous grouping, where the problem is to group together those frequency channels that originate from the same auditory object, spatial cues are substantially weaker especially for interaural time differences despite their being the dominant cue for localizing complex wideband sounds. This weakness is probably related both to psychophysical interference effects between different frequency regions and to the vulnerability that interaural time differences show to reverberation.

An attractive model for how we localize simultaneous complex sounds proposes that the formation of auditory objects precedes decisions on their location. The model would allow pooling of location information across appropriate frequency channels in order to reduce the variability found in individual channels and so produce a percept with a stable location. However, the evidence in support of such a model is at present very limited, partly due to a paucity of relevant experimental data. We know remarkably little about how we localize simultaneous complex auditory objects such as speech.

Recent work on our ability to listen to one talker against a background of either another talker or noise has highlighted the difference between the masking of one sound by another and our ability to group together both across frequency and time the sounds of the attended source. Differences in the apparent spatial position of two talkers can be very beneficial in tracking one talker across time when other cues (such as voice quality, level, or semantic constraints) are insufficient.

## References

Arbogast TL, Mason CR, Kidd G Jr. (2002) The effect of spatial separation on informational and energetic masking of speech. J Acoust Soc Am 112:2086–2098.
Assmann PF, Summerfield Q (2004) The perception of speech under adverse conditions. In: Greenberg S, Ainsworth WA, Popper AN, Fay RR (eds) Speech Processing in the Auditory System. New York: Springer-Verlag, pp. 231–308.

Batteau DW (1967) The role of the pinna in human localization. Proc R Soc Lond B 168:158–180.

Beerends JG, Houtsma AJM (1986) Pitch identification of simultaneous dichotic two-tone complexes. J Acoust Soc Am 80:1048–1055.

Best V, van Schaik A, Carlile S (2004) Separation of concurrent broadband sound sources by human listeners. J Acoust Soc Am 115:324–336.

Best V, Ozmeral E, Gallun FJ, Sen K, Shinn-Cunningham BG (2005) Spatial unmasking of birdsong in human listeners: Energetic and informational factors. J Acoust Soc Am 118:3766–3773.

Best V, Gallun FJ, Carlile S, Shinn-Cunningham BG (2007) Binaural interference and auditory grouping. J Acoust Soc Am 121:1070–1076.

Bilsen FA, Raatgever J (1973) Spectral dominance in binaural lateralization. Acustica 28:131–132.

Bodden M (1996) Auditory demonstrations of a cocktail-party processor. Acustica 82: 356–357.

Bolia RS, Nelson WT, Ericson MA, Simpson BD (2000) A speech corpus for multitalker communications research. J Acoust Soc Am 107:1065–1066.

Bregman AS (1990) Auditory Scene Analysis: The Perceptual Organization of Sound. Cambridge, MA: Bradford Books, MIT Press.

Bregman AS, Ahad PA (1995) Compact Disc: Demonstrations of auditory scene analysis. Montreal: Department of Psychology, McGill University.

Bronkhorst AW (2000) The cocktail party phenomenon: a review of speech intelligibility in multiple-talker conditions. Acustica 86:117–128.

Brungart DS (2001a) Evaluation of speech intelligibility with the coordinate response measure. J Acoust Soc Am 109:2276–2279.

Brungart DS (2001b) Informational and energetic masking effects in the perception of two simultaneous talkers. J Acoust Soc Am 109:1101–1109.

Brungart DS (2005) A binary masking technique for isolating energetic masking in speech perception. J Acoust Soc Am 117:2484.

Brungart DS, Rabinowitz WM (1999) Auditory localization of nearby sources. Head-related transfer functions. J Acoust Soc Am 106:1465–1479.

Brungart DS, Simpson BD (2002) The effects of spatial separation in distance on the informational and energetic masking of a nearby speech signal. J Acoust Soc Am 112:664–676.

Brungart DS, Chang PS, Simpson BD, Wang DL (2006) Isolating the energetic component of speech-on-speech masking with ideal time-frequency regregation. J Acoust Soc Am 120:4007–4018.

Buell TN, Hafter ER (1991) Combination of binaural information across frequency bands. J Acoust Soc Am 90:1894–1900.

Butler D (1979) A further study of melodic channeling. Percept Psychophys 25:264–268.

Cherry EC (1953) Some experiments on the recognition of speech, with one and with two ears. J Acoust Soc Am 25:975–979.

Cooke M (2003) Glimpsing speech. J Phonet 31:579–584.

Cooke M (2005) Making sense of everyday speech: A glimpsing account. Divenyi PL (ed) Speech Separation by Humans and Machines. New York: Kluwer Academic, pp. 305–314.

Cooke MP, Green PD, Josifovski L, Vizinho A (2001) Robust automatic speech recognition with missing and unreliable acoustic data. Speech Commun 34:267–285.

Culling JF, Summerfield Q (1995) Perceptual separation of concurrent speech sounds: Absence of across-frequency grouping by common interaural delay. J Acoust Soc Am 98:785–797.

Cusack R, Deeks J, Aikman G, Carlyon RP (2004) Effects of location, frequency region, and time course of selective attention on auditory scene analysis. J Exp Psychol [Hum Percept Perform] 30:643–656.

Darwin CJ (1984) Perceiving vowels in the presence of another sound: constraints on formant perception. J Acoust Soc Am 76:1636–1647.

Darwin CJ (2002) Auditory streaming in language processing. In: Tranebjaerg L, Andersen T, Christensen-Dalsgaard J, Poulsen T (eds) Genetics and the Function of the Auditory System: 19th Danavox Symposium. Denmark: Holmens Trykkeri, pp. 375–392.

Darwin CJ (2005) Pitch and auditory grouping. In: Plack CJ, Oxenham AJ, Fay RR, Popper AN (eds) Pitch: Neural Coding and Perception. New York: Springer-Verlag, pp. 278–305.

Darwin CJ (2006) Contributions of binaural information to the separation of different sound sources. Int J Audiol 45: S20–S24.

Darwin CJ, Ciocca V (1992) Grouping in pitch perception: Effects of onset asynchrony and ear of presentation of a mistuned component. J Acoust Soc Am 91:3381–3390.

Darwin CJ, Gardner RB (1986) Mistuning a harmonic of a vowel: Grouping and phase effects on vowel quality. J Acoust Soc Am 79:838–845.

Darwin CJ, Hukin RW (1997) Perceptual segregation of a harmonic from a vowel by inter-aural time difference and frequency proximity. J Acoust Soc Am 102: 2316–2324.

Darwin CJ, Hukin RW (1998) Perceptual segregation of a harmonic from a vowel by inter-aural time difference in conjunction with mistuning and onset-asynchrony. J Acoust Soc Am 103:1080–1084.

Darwin CJ, Hukin RW (1999) Auditory objects of attention: The role of interaural time-differences. J Exp Psychol [Hum Percept Perform] 25:617–629.

Darwin CJ, Hukin RW (2000) Effectiveness of spatial cues, prosody and talker characteristics in selective attention. J Acoust Soc Am 107:970–977.

Darwin CJ, Pattison H, Gardner RB (1989) Vowel quality changes produced by surrounding tone sequences. Percept Psychophys 45:333–342.

Darwin CJ, Brungart DS, Simpson BD (2003) Effects of fundamental frequency and vocal-tract length changes on attention to one of two simultaneous talkers. J Acoust Soc Am 114:2913–2922.

Deutsch D (1979) Binaural integration of melodic patterns. Percept Psychophys 25: 399–405.

Deutsch D (1999) Grouping mechanisms in music. Deutsch D (ed) The Psychology of Music 2nd ed. San Diego: Academic Press, pp. 299–348.

Drennan WR, Gatehouse S, Lever C (2003) Perceptual segregation of competing speech sounds: the role of spatial location. J Acoust Soc Am 114:2178–2189.

Edmonds BA, Culling JF (2005) The spatial unmasking of speech: Evidence for within-channel processing of interaural time delay. J Acoust Soc Am 117:3069–3078.

Egan JP, Carterette EC, Thwing EJ (1954) Some factors affecting multi-channel listening. J Acoust Soc Am 26:774–782.

Freyman RL, Zurek PM, Balakrishnan U, Chiang YC (1997) Onset dominance in lateralization. J Acoust Soc Am 101:1649–1659.

Freyman RL, Helfer KS, McCall DD, Clifton RK (1999) The role of perceived spatial separation in the unmasking of speech. J Acoust Soc Am 106:3578–3588.

Freyman RL, Balakrishnan U, Helfer KS (2001) Spatial release from informational masking in speech recognition. J Acoust Soc Am 109:2112–2122.

Gockel H, Carlyon RP, Plack CJ (2005) Dominance region for pitch: effects of duration and dichotic presentation. J Acoust Soc Am 117:1326–1336.

Goldstein JL (1973) An optimum processor theory for the central formation of the pitch of complex tones. J Acoust Soc Am 54:1496–1516.

Hafter ER, Buell TN (1990) Restarting the adapted binaural system. J Acoust Soc Am 88:806–812.

Hafter ER, Trahiotis C (1997) Functions of the binaural system. In: Crocker MJ (ed) Encyclopedia of Acoustics. New York: John Wiley & Sons. pp. 1461–79

Hafter ER, Buell TN, Richards VM (1988) Onset-coding in lateralization: Its form, site and function. In: Edelman GM, Gall WE, Cowan WM (eds) Auditory Function: Neurobiological Bases of Hearing. New York: John Wiley & Sons, pp. 647–676.

Hartmann WM, Johnson D (1991) Stream segregation and peripheral channeling. Music Percept 9:155–183.

Hartmann WM, Rakerd B, Koller A (2005) Binaural coherence in rooms. Acust Acta Acust 91:451–462.

Hebrank J, Wright D (1974) Spectral cues used in the localization of sound sources on the median plane. J Acoust Soc Am 68:1829–1834.

Henning GB (1974) Detectability of interaural delay in high-frequency complex waveforms. J Acoust Soc Am 55:84–90.

Henning GB (1980) Some observations on the lateralization of complex waveforms. J Acoust Soc Am 68:446–454.

Hill NI, Darwin CJ (1993) Effects of onset asynchrony and of mistuning on the lateralization of a pure tone embedded in a harmonic complex. J Acoust Soc Am 93:2307–2308.

Hill NI, Darwin CJ (1996) Lateralization of a perturbed harmonic: Effects of onset asynchrony and mistuning. J Acoust Soc Am 100:2352–2364.

Hukin RW, Darwin CJ (1995a) Comparison of the effect of onset asynchrony on auditory grouping in pitch matching and vowel identification. Percept Psychophys 57:191–196.

Hukin RW, Darwin CJ (1995b) Effects of contralateral presentation and of interaural time differences in segregating a harmonic from a vowel. J Acoust Soc Am 98:1380–1387.

Hukin RW, Darwin CJ (2000) Spatial cues to grouping in the Wessel illusion. Br J Audiol 34:109.

Jeffress LA (1972) Binaural signal detection: Vector theory. Tobias JV (ed) Foundations of Modern Auditory Theory, Vol II. New York: Academic Press, pp. 349–368.

Lord Rayleigh (J.W. Strutt) (1907) On our perception of sound direction. Phil Mag 13:214–232.

McFadden D, Pasanen EG (1976) Lateralization at high frequencies based on interaural time differences. J Acoust Soc Am 59:634–639.

Middlebrooks JC, Green DM (1991) Sound localization by human listeners. Annu Rev Psychol 42:135–159.

Miller GA (1947) The masking of speech. Psychol Bull 44:105–129.

Miller GA, Licklider JCR (1950) The intelligibility of interrupted speech. J Acoust Soc Am 22:167–173.

Moore BCJ, Gockel H (2002) Factors influencing sequential stream segregation. Acust Acta Acust 88:320–332.

Sach AJ, Bailey PJ (2004) Some characteristics of auditory spatial attention revealed using rhythmic masking release. Percept Psychophys 66:1379–1387.

Shackleton TM, Meddis R (1992) The role of interaural time difference and fundamental frequency difference in the identification of concurrent vowel pairs. J Acoust Soc Am 91:3579–3581.

Shackleton TM, Meddis R, Hewitt MJ (1992) Across frequency integration in a model of lateralization. J Acoust Soc Am 91:2276–2279.

Shackleton TM, Meddis R, Hewitt MJ (1994) The role of binaural and fundamental frequency difference cues in the identification of concurrently presented vowels. Q J Exp Psychol 47A:545–563.

Shinn-Cunningham BG, Kopco N, Martin TJ (2005) Localizing nearby sound sources in a classroom: Binaural room impulse responses. J Acoust Soc Am 117:3100–3115.

Stellmack MA, Dye RH (1993) The combination of interaural information across frequencies: the effects of number and spacing of components, onset asynchrony, and harmonicity. J Acoust Soc Am 93:2933–2947.

Stern RS, Colburn HS (1978) Theory of binaural interaction based on auditory-nerve data. IV. A model for subjective lateral position. J Acoust Soc Am 64:127–140.

Stern RM, Zeiberg AS, Trahiotis C (1988) Lateralization of complex binaural stimuli: A weighted image model. J Acoust Soc Am 84:156–165.

Trahiotis C, Bernstein LR (1990) Detectability of interaural delays over select spectral regions: Effects of flanking noise. J Acoust Soc Am 87:810–813.

Trahiotis C, Stern RM (1989) Lateralization of bands of noise: Effects of bandwidth and differences of interaural time and phase. J Acoust Soc Am 86:1285–1293.

Turgeon M, Bregman AS, Ahad PA (2002) Rhythmic masking release: Contribution of cues for perceptual organization to the cross-spectral fusion of concurrent narrow-band noises. J Acoust Soc Am 111:1819–1831.

Wang DL (2005) An ideal binary mask as the computational goal of auditory scene analysis. In: Divenyi PL (ed), Speech Separation by Humans and Machines. New York: Kluwer Academic, pp. 181–197.

Wessel DL (1979) Timbre space as a musical control structure. Comp Music J 3:45–52.

Wightman FL, Kistler DJ (1992) The dominant role of low-frequency interaural time differences in sound localization. J Acoust Soc Am 91:1648–1661.

Woods WA, Colburn S (1992) Test of a model of auditory object formation using intensity and interaural time difference discriminations. J Acoust Soc Am 91:2894–2902.

Worley JW, Darwin CJ (2002) Auditory attention based on differences in median vertical plane position. In: Proceedings of the 2002 International Conference on Auditory Display, Kyoto, July 2–5, Japan. pp. 1–5.

Yost WA (1997) The cocktail party effect: 40 years later. In: Gilkey R, Anderson TR (eds) Localization and Spatial Hearing in Real and Virtual Environments. Hillsdale, NJ: Lawrence Erlbaum, pp. 329–348.

# 9
# Envelope Processing and Sound-Source Perception

Stanley Sheft

## 1. Introduction

The role of the auditory system is to process the acoustic information that originates from various environmental sources. Two general aspects pertain to this function. The first relates to sound-source determination, the perceptual organization of a complex sound field into images or entities that correspond to individual sources. The second aspect concerns the extraction of information from the received signals. For example, in a setting with multiple talkers, a listener must partition the complex sound field to group acoustic elements by source, at least to the extent of segregating a primary talker or message. The listener then must also be able to extract information from the segregated source. The distinction between these two aspects of source processing is that the first uses the acoustic information to derive a representation of the world or environment, a process termed auditory scene analysis by Bregman (1990), while the second attempts to extract meaning from the acoustic information, something specific to an individual source that often varies over time. This division of sound-source processing into two areas contrasts with the major requirements of vision perception. For the visual world, in which objects are relatively stable, scene analysis conveys much of the environmental information. For audition, where source output can change rapidly over time, scene analysis or source determination represents only a limited utilization of the sensory information.

By its nature, envelope or amplitude modulation (AM) can show involvement in both aspects of sound-source processing. For many sound sources, the pattern of output modulation is a fundamental and distinguishing characteristic of the source. Considerations of the acoustic cues used for sound-source determination have generally included temporal modulation (e.g., Hartmann 1988; Bregman 1990; Yost 1991; Darwin and Carlyon 1995). Spectral components that originate from a single source may exhibit coherent modulation, which could cue their common origin. When multiple sources are present in a single sound field, differences of modulation pattern among sources may aid segregation of the auditory images that represent the individual sources. These considerations apply to both AM and frequency modulation (FM). The current review focuses on envelope modulation.[1] Over the last 30 years, studies have attempted to evaluate the relationship between AM and source determination, using both direct and indirect psychophysical measures. In this context, direct measures refer to judgments of numerosity, salience, or extent of segregation, while indirect measures have inferred the relationship based on results obtained in detection and discrimination paradigms. This work is reviewed in Section 3.

By definition, temporal modulation is involved in the second aspect of sound-source processing, the extraction of information or meaning from the acoustic signal. Formally, information is quantified by the level of uncertain variation (i.e., entropy) in the received signal. This variation is a type of temporal modulation of the signal. Procedures used in early studies of auditory information processing required identification or labeling of stimuli (e.g., Garner and Hake 1951; Pollack 1952). In these studies, the level of information was quantified in bits, a logarithmic measure of uncertainty, with performance evaluated in terms of the amount of information transferred to the observer. Alternative to deriving information transfer rates, recent studies have used discrimination and masking paradigms characterized by stimulus uncertainty to assess auditory information processing (Kidd, Mason, Richards, Gallun, and Durlach, Chapter 6). Though sometimes calculating information transfer rates, the emphasis of these discrimination and masking studies has been on information processing in terms of the manner in which observers weight stimulus information and the relationships among stimulus dimensions or attributes in perceptual processing. Section 4 describes results from studies examining information processing of envelope modulation and also considers cross-spectral masking in terms of information processing. In recent years, the extent of information conveyed by envelope modulation has received most attention in studies of speech perception; this work is briefly reviewed in Section 5.

In a discussion of information processing of complex sounds, Lutfi (1990) commented on two requirements: the integration of information derived from the components of a target sound while segregating this information from that of irrelevant sources. These two requirements represent a restatement of the two general aspects of sound-source processing. Clearly, information extraction and source determination are not independent processes, with the division between the two admittedly somewhat arbitrary. Garner (1974) has argued that information and structure are identical terms. In that auditory scene analysis represents perception of structure, a similar argument can be made in the present case. However, a distinction does exist between consideration of envelope modulation as a factor affecting spectral grouping and segregation versus the modulation representing information content or message. It is in this regard that the potential involvement of envelope modulation in sound-source processing is evaluated.

---

[1]For review of aspects of FM processing relevant to sound-source perception, see Yost and Sheft (1993) and Darwin and Carlyon (1995).

## 2. Models of Envelope Processing

Envelope modulation is generally taken to mean slow variation, at least relative to carrier frequency, in stimulus amplitude. Before considering psychophysical models of envelope processing, a formal definition of stimulus envelope is needed. An arbitrary (real) band-limited signal can be expressed as

$$f(t) = r(t) \, \cos \, [\varphi(t)], \tag{9.1}$$

where $r(t)$ and $\varphi(t)$ modulate the envelope and phase, respectively, of a cosinusoidal function. The envelope function, or instantaneous amplitude, can be derived from the Hilbert transform $f_H(t)$ of $f(t)$. The Hilbert transform represents a filter that without affecting gain, shifts the phases of all positive frequency components by $-\frac{1}{2}\pi$ radians with a complimentary shift of the negative frequency components. The complex analytic signal $z(t)$ of $f(t)$ has $f_H(t)$ as its imaginary part with $f(t)$ the real part. The envelope function $r(t)$ is then defined as the magnitude of the analytic signal $z(t)$ (for the derivation of this relationship, see Hartmann 1998). The function $z(t)$ is a quadrature representation of the signal. As a quadrature (e.g., Fourier) signal representation, the phase function $\varphi(t)$ is $\tan^{-1}[f_H(t)/f(t)]$. Similar to $r(t)$ indicating instantaneous amplitude, a function can be found expressing instantaneous frequency, defined as the time rate of change of the phase of the analytic signal $z(t)$.

In auditory work, signals are often considered in terms of envelope and fine structure. Signal representation through amplitude and phase modulation of a sinusoid as in Eq. (9.1) has led to derivation of the two followed by separate manipulation and possible recombination (Hou and Pavlovic 1994; Smith et al. 2002; Xu and Pfingst 2003). Some caution is warranted. The envelope and phase functions are not independent, allowing for restoration of an altered envelope to an approximation of its original form by the phase function (Ghitza 2001; Zeng et al. 2004). The extent of restoration depends on analysis bandwidth, and in practice does not appear a factor if the audio range of roughly 80–8000 Hz is divided into eight or more analysis bands (Gilbert and Lorenzi 2006). A more general limitation to independent representation of envelope and fine structure is that due to the abrupt phase transitions of some multicomponent stimuli, fine structure defined by the phase function can itself exhibit fluctuation of its Hilbert envelope. Loughlin and Tacer (1996) noted that a complex signal is generally not specified by a unique combination of envelope- and phase-modulation functions, suggesting as an alternative to Eq. (9.1) a derivation of both instantaneous frequency and amplitude from the phase function. Envelope defined as the magnitude of the analytic signal assumes a real and nonnegative function. Atlas et al. (2004) showed the limitations of this form of the modulation envelope. If filtering the envelopes of stimulus subbands, these limitations result in undesirable distortion and a reduction in effective filter attenuation when envelope(s) and carrier(s) are recombined. As an alternative, Atlas and Janssen (2005) worked with the complex form of the envelope (i.e., including imaginary terms), extracting the envelope by coherent carrier detection.

Envelope defined by analysis shows no rate dependency. In contrast, auditory perception of AM often varies with the rate of modulation. The most widely studied psychophysical paradigm measures AM detection ability. In this procedure, AM depth is adjusted to find the threshold level at which listeners can just discriminate between a modulated and unmodulated sound. Based on AM detection thresholds, the psychophysical temporal modulation transfer function (TMTF) exhibits a lowpass characteristic with sensitivity declining with increasing rate. For a continuous wideband noise (WBN) carrier and signal duration of greater than 250 ms, the 3-dB down point of the TMTF is generally between 50 and 70 Hz with an attenuation slope of roughly 4 dB per octave (Viemeister 1979; Bacon and Viemeister 1985). In the spectral domain, amplitude modulation introduces sidebands about each carrier component with the frequency separation between each sideband and the carrier equal to the modulation rate. With sinusoidal carriers, sideband resolution can cue AM detection at higher AM rates. When high-frequency pure-tone carriers are used to minimize effects of sideband resolution, the cutoff frequency of the TMTF is over an octave higher than that obtained in WBN conditions (Kohlrausch et al. 2000). With either pure-tone or WBN carriers, a gated, instead of continuous, carrier presentation decreases sensitivity to low-rate AM, adding a highpass segment to the TMTF (Viemeister 1979; Yost and Sheft 1997). Results from studies concerned with sound-source perception are generally consistent with the lowpass characteristic of envelope processing, though in most cases with a lower cutoff rate than indicated by either the pure-tone or WBN TMTF. Some higher-rate results have been obtained, most notably in the case of stream segregation.

The envelope is not directly represented in the stimulus spectrum. Psychophysical models of AM processing are commonly based on envelope detection in which operation of a nonlinearity introduces the components of envelope modulation into the spectral domain. The most influential version presented by Viemeister (1979) consists of three stages: predetection filtering and half-wave rectification followed by lowpass filtering. If the carrier and modulation frequencies are well separated, the lowpass filter can remove the carrier components from the processed stimulus spectrum to retrieve the envelope. In practice, filter parameters are chosen to fit the WBN TMTF with Viemeister using a 2-kHz bandpass filter centered at 5 kHz and a 65-Hz first-order lowpass filter. The initial filter bandwidth is much broader than estimates of auditory filter bandwidth, while the lowpass filter can allow for significant passage of the carrier through the detector under some conditions. Implications of these filter settings for sound-source processing are discussed in Section 3.

Though the initial filter stage has some effect, temporal resolution is set primarily by the time constant of the lowpass filter. Results from several different paradigms indicate selectivity in the modulation domain that cannot be accounted for by processing based on a single time constant. Modulation-masking procedures measure thresholds for detecting probe modulation in the presence of masker modulation. Either a single carrier is used for the two modulators, or the probe and masker carriers spectrally overlap (e.g., independent samples of WBN). With either a sinusoid or narrowband noise (NBN) as the masker modulator, results show a bandpass masking pattern with greatest masking generally obtained when the probe and masker modulation rates are the same or overlap (Bacon and Grantham 1989; Houtgast 1989; Strickland and Viemeister 1996; Ewert and Dau 2000; Ewert et al. 2002). Threshold shifts due to masker modulation can exceed 15 dB, covering much of the dynamic range from unmasked threshold to a 100% depth of modulation. Due to the lowpass characteristic of the TMTF, dynamic range is reduced with increasing AM rate, this in part leading to smaller threshold shifts at higher probe AM rates. The tuning in the modulation domain is quite broad, with significant threshold elevation often present when the probe and masker modulation rates are separated by over two octaves.

Showing a smaller range of threshold elevation, a similar degree of tuning in the modulation domain is obtained in nonsimultaneous masking procedures in which the probe and masker modulations are not concurrent (Sheft 2000; Wojtczak and Viemeister 2005). Nonsimultaneous masking of AM detection was initially reported in studies of selective adaptation that measured modulation-threshold change following prolonged exposure to modulated stimuli (Kay 1982; Tansley and Suffield 1983). From these studies came the earliest suggestions of processing by channels tuned to modulation characteristics. However, nonsensory factors may have influenced results. Bruckert et al. (2006) showed that the effect on detection thresholds of prolonged exposure to an AM adaptor can be eliminated with extensive training. More generally, presentation of suprathreshold nonsignal AM may serve to focus attention away from threshold signals, regardless of whether presented as a long-duration adaptor or as a nonsimultaneous masker on every trial. Even when a forced-choice experimental procedure is used, an inappropriate perceptual anchor can increase the overall variance in signal processing, leading to threshold elevation. This speculation does not diminish the findings of modulation selectivity in nonsimultaneous procedures; rather it places the selectivity at the stage of decision processing.

A threshold shift related to modulation masking is observed in an AM detection task that uses NBN carriers. Narrowband noise is characterized by its intrinsic envelope fluctuation, which is reflected in the envelope power spectrum. For band-limited Gaussian noise, the ac component of the envelope power spectrum is approximately triangular in shape, with power decreasing linearly with envelope frequency (Lawson and Uhlenbeck 1950). The base of the triangle is equal to noise bandwidth; its height is $N_0\pi/4$ with $N_0$ the noise-band spectral density. If the overall level is fixed, the area under the triangle is constant across change in noise bandwidth. The envelope spectrum thus both narrows and increases in slope with decreasing noise bandwidth. Since the dc term of the envelope is a linear function of both bandwidth and spectral density, increasing spectral slope increases the modulation depth (the ratio of spectral-component amplitude to dc) of the lower-rate envelope components. Though changing with bandwidth, component modulation depth is independent of level.

Thus regardless of level, the prominence of low-rate envelope fluctuation is inversely related to noise bandwidth.

Similar to the interference obtained in modulation-masking procedures, the intrinsic envelope fluctuations of band-limited Gaussian-noise carriers elevate AM detection thresholds (Eddins 1993; Dau et al. 1997, 1999; Strickland and Viemeister 1997). Since the width of the envelope spectrum of the carrier is equal to carrier bandwidth, interference is not restricted to low AM rates coinciding with the prominent envelope fluctuations of NBN carriers. The change in TMTF shape with carrier bandwidth indicates selectivity in the modulation domain with threshold elevation following measures of carrier envelope power integrated over a limited range about the signal AM rate (Dau et al. 1999).

To account for selectivity to the rate of envelope fluctuation, Dau et al. (1997) proposed auditory processing by a modulation filterbank. Similar to the lowpass-filter approach, filterbank models incorporate envelope detection by a nonlinearity followed by lowpass filtering. Assumed to reflect the loss of phase locking with increasing frequency, the lowpass filter cutoff of the filterbank model is one to one and a half orders of magnitude higher than proposed by Viemeister (1979). Envelope detection is then followed by processing of channels selective to modulation rate. Reflecting results from modulation-masking studies, the channels of the model are broadly tuned with a constant filter Q [the ratio of filter center frequency (CF) to 3-dB bandwidth] and logarithmic spacing for filters above 10 Hz. This arrangement requires a dozen or fewer channels to span the range of rates over which human listeners can detect modulation. As mentioned above, the envelope spectrum of Gaussian noise broadens and flattens with increasing bandwidth. Channels that mediate detection of high-rate AM thus pass high levels of envelope fluctuation intrinsic to a WBN carrier. This reduction in the signal-to-noise (S/N) ratio with increasing AM rate leads to the filterbank model accurately predicting the lowpass characteristic of the WBN TMTF. Possible involvement in auditory grouping of processing by a modulation filterbank is considered in Section 3.

Modulation-masking studies and AM-detection experiments that vary carrier bandwidth both involve complex stimulus modulation.[2] As in the audio-frequency domain, where two components beat with a periodicity not represented in the stimulus spectrum, the envelope spectrum does not directly indicate beats and other complex interaction among modulation components. In a modulation-masking experiment, beating between the signal and masker can cue signal detection, while other second-order masker fluctuations can mask signal modulation (Strickland and Viemeister 1996; Ewert et al. 2002; Füllgrabe et al. 2005). AM detection with three-component modulators shows effects of component spacing and relative phase consistent with the presence of second-order envelope fluctuations in the modulation spectrum (Moore and Sek 2000;

---

[2] Apart from the previous discussion of envelope analysis, which includes real and imaginary components, the term *complex* is used throughout to refer to multicomponent or nonsinusoidal functions.

Lorenzi et al. 2001). A second nonlinearity following the half-wave rectification of envelope detection will achieve this type of representation of intermodulation or distortion among modulation components. Filtering or off-frequency processing of stimuli with complex envelopes can also introduce second-order modulation into the first-order modulation spectrum (Füllgrabe et al. 2005).

With multiple time constants and nonlinearities, modulation-filterbank modeling puts emphasis on the various stages of auditory processing of envelope modulation. Limitations of the single time constant of a lowpass-filter model can be overcome in the decision process that follows model stages. Strope and Alwan (2001) modeled AM detection and discrimination by following linear envelope detection with computation of running autocorrelation. By using a decision statistic based on peak differences in summary autocorrelograms at delays that vary with modulator period, in essence multiple time constants are introduced into their model. In the modeling of AM detection thresholds obtained in the presence of complex periodic modulation maskers, Viemeister et al. (2005) utilized a decision statistic that was windowed to match temporal properties of the signal.

The distinction between modulation selectivity established in the frequency versus temporal domain can in principle be reduced to one of method rather than result. A key difference in current applications is that modulation-filterbank models explicitly discard envelope-phase information above low modulation rates. For channels centered above 10 Hz, only the Hilbert envelope of the channel output is used to compute the decision statistic. The basis of this limitation is monaural discrimination of starting envelope phase with sinusoidal modulation. Sheft and Yost (2004), however, showed that this limitation is too restrictive; listeners were able discriminate envelope starting phase at rates an octave to an octave and a half higher than a cutoff in the teens. Apart from modeling concerns, a restrictive loss of envelope-phase information has several implications for sound-source processing. To the extent that cross-spectral envelope coherence affects auditory grouping, envelope phase is an important variable; a common rate and phase determines coherence. Studies evaluating the ability to detect cross-spectral envelope coherence are reviewed in Section 3.4. With complex modulation, component phase affects envelope pattern. Section 4 reviews studies of envelope-pattern discrimination with noise stimuli.

An issue related to rate-dependent retention of envelope-phase information concerns the uniformity of model decision statistic across modulation rate. A variety of decision statistics including envelope root-mean-square (rms) amplitude, the ratio of the peak-to-trough envelope values, and cross-correlation between the received signal and an internal template have been proposed to model AM detection. Most models, however, assume that a single statistic mediates temporally based detection at all modulation rates. Recent physiology offers evidence of two coding schemes for AM by the auditory system with slower fluctuations coded by temporal response pattern and faster ones by neural discharge rate (Schulze and Langner 1997; Giraud et al. 2000; Lu et al. 2001; Wang et al. 2003). The relevance to sound-source processing of distinct coding strategies for low versus higher modulation rates is that it can account for results that do not follow the form of the TMTF. Specifically, the dominance of very low AM rates in effects associated with auditory grouping and in conveying information as in speech may in part reflect the limited range of temporal coding of envelope fluctuation. Psychophysical results consistent with dual coding strategies of AM are considered in Section 4.

## 3. Envelope Processing in Perceptual Segregation and Auditory Grouping

From a formal standpoint, concern with information processing of AM treats the carrier as a means of conveying the information contained in the pattern of envelope fluctuation. An alternative approach especially relevant to auditory processing is to view envelope modulation as a carrier-gating function. In this regard, the pattern of carrier onsets and offsets conveys information represented by the presence or absence of the carrier. Along this line, involvement of auditory grouping and segregation in sound-source processing is through the perceptual organization of a complex stimulus spectrum. That is, it is the carriers, not their envelopes, that are accordingly structured (though see Section 5 regarding segregation of competing envelope-speech messages). The following section considers the role of envelope processing in the grouping and segregation of spectral components.

### 3.1 Onset and Offset Asynchrony

For complex stimuli, a gating disparity among stimulus components can percep-tually segregate the asynchronous component(s). The enhanced ability to "hear out" an asynchronous component from a complex is distinct from the ability to detect the presence of the asynchrony. Dependent on stimulus parameters, detection thresholds can be as small as a few milliseconds or less (Patterson and Green 1970; Zera and Green 1993), roughly corresponding to other measures of auditory temporal resolution including the time constant of the TMTF. In the context of sound-source processing, effects of asynchronous gating are not obtained until the temporal disparity is one to two orders of magnitude greater than detection thresholds. Two general approaches have been used to measure these effects. In the first, improvement in the ability to process some attribute of an asynchronous component is taken to indicate component segregation from the complex. This explanation assumes that masking of asynchronous-target attributes by the remaining complex is reduced by segregation. The assumption fails in limits; distinct sources can mask one another, and change in an individual component of a complex can be cued by variation in a unary percept such as timbre. In this case, interpretation of asynchrony resulting in segregation is based on result rather than process. The second approach measures the change in some percept associated with the complex (e.g., pitch, timbre, or vowel quality) as a function of the temporal disparity in gating stimulus components. Regarding this approach, the assumption is that the contribution of the asynchronous component to the percept diminishes with segregation from the complex.

Rasch (1978) measured masked thresholds for identifying the direction of pitch change of a complex tone (i.e., fundamental plus harmonics) in the presence of a masker that was also a complex tone. Delaying masker onset relative to the target lowered thresholds, with the improvement increasing with delay from 10 to 30 ms. In conditions with an onset asynchrony, masker level, reverberation, or temporal overlap between the target and masker had little effect on target-tone thresholds. With the target onset preceding the masker onset, the procedure essentially measured the backward masking of complex pitch discrimination. Rasch noted that with asynchronous gating, the target tended to stand out from the masker despite unawareness by listeners of the onset asynchrony. This observation highlights two aspects of segregation by gating asynchrony. The first is that asynchrony may enhance the perceptual clarity of a subset of components of a complex sound. The second aspect is that multiple sequential events may be perceived due to asynchrony. The relevance of the former was indicated in a complimentary study in which Rasch (1979) observed that listeners often judge ensemble performance as synchronous despite asynchronies of 30–50 ms. Relatively brief asynchronies may enhance the clarity of individual parts in polyphonic music or chord voicing (Rasch 1979; Goebl and Parncutt 2003), and may also be used as a deliberate expressive strategy by performers (Repp 1996). Though stimulus differences exist among studies, the range of gating disparity that does not alter judgment of synchronicity is well above thresholds for discriminating the temporal order of asynchronous events (e.g., Pastore et al. 1982), indicating influence of task requirements on measured results.

Effects of gating disparity on the processing of an asynchronous target have received most attention in studies of cross-spectral masking. Experiments that utilized modulated maskers are discussed in Section 3.5. Results from a number of studies show that the ability to detect change in a target component of a complex is at times either unaffected or impaired by gating asynchrony. These include studies of identification thresholds of masked signals (Gordon 2000), pitch processing (Carlyon 1994; Brunstrom and Roberts 2001), and profile analysis (Green and Dai 1992; Hill and Bailey 1997; Lentz et al. 2004). Though the independent variable in all cases was some characteristic of the target component, task performance was influenced by cross-spectral processing between the target and remaining (nontarget) stimulus components. Involvement of cross-spectral processing is most apparent in the profile-analysis studies. Due to randomization of overall stimulus level in these studies, detection of change in relative target level is based on discrimination of spectral shape or timbre. Green and Dai (1992) showed that either a 50-ms onset or offset asynchrony between target and nontarget components significantly elevated thresholds, with the effect increasing with gating disparity up to at least 250 ms. Subsequent work demonstrated that the deleterious effect of target asynchrony can be reduced by either reducing the frequency separation between stimulus components (Lentz et al. 2004) or adding captor tones concurrent with the leading temporal fringe of the target (Hill and Bailey 1997). The intent of the latter manipulation was to minimize target segregation from nontarget components by encouraging a perceptual organization in which the target fringe combines with the captor tones to result in two sequential events: target-fringe plus captors followed by target plus nontarget components.

Bregman and Pinker (1978) and Dannenbring and Bregman (1978) directly evaluated the role of gating asynchrony in affecting the tendency of a single spectral component of a series of acoustic events to either become part of a sequential stream or fuse with other concurrent components. In these two studies of stream segregation, a sequence of a pure tone followed by an asynchronously gated two- or three-tone complex was cyclically repeated (Figure 9.1). Two effects of the gating asynchrony were apparent. Once the gating disparity at either onset or offset was roughly 30 ms or greater, judgments of the richness of timbre of the complex tone diminished, consistent with segregation of the asynchronous component. The asynchronous component was also more likely to be perceived as part of a rhythmic sequence with the alternating pure tone. As in the profile-analysis work of Lentz et al. (2004), effects diminished with reduction in the frequency spacing between stimulus components.

Along with affecting timbre and sequence perception, a gating disparity can affect the contribution of an asynchronous component to the pitch or vowel quality of a complex sound. Darwin and Ciocca (1992) investigated the influence of an onset asynchrony on the contribution of a mistuned harmonic to the pitch

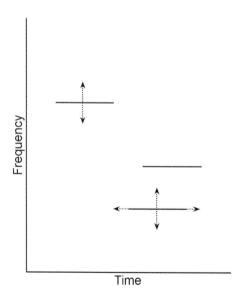

FIGURE 9.1. Schematic illustration of a single cycle of a stimulus configuration used to study stream segregation. Arrows represent change in component frequency or gating across conditions.

of a complex tone. With either ipsilateral or contralateral presentation, the contribution of the mistuned component to the pitch of the complex progressively diminished with increasing onset asynchrony from roughly 80 to 320 ms. In earlier work with speech stimuli, the effect of asynchrony was apparent with shorter onset disparities. A 32-ms onset or offset asynchrony of a single harmonic of a vowel sound was found to significantly alter the phoneme boundary, with the change in vowel quality continuing as asynchrony was increased to 240 ms (Darwin and Sutherland 1984). Though greater effect has been found with equivalent onset rather than offset asynchrony, the effect of an offset disparity on vowel quality indicates that adaptation alone cannot account for the results. As in the profile-analysis study of Hill and Bailey (1997), adding a captor concurrent with the leading temporal fringe of the asynchronous component reduced the effect of a gating disparity on both pitch and vowel perception (Darwin and Sutherland 1984; Ciocca and Darwin 1993), further indicating involvement of a process other than adaptation. Recently, Roberts and Holmes (2006) determined that the effect of a captor on vowel quality did not depend on either common onset or harmonic relationship between the captor and the leading fringe of the vowel component. This result led Roberts and Holmes to suggest that the effect of the captor may be based on broadband inhibition within the central auditory nervous system (CANS).

Using similar stimulus sets, Hukin and Darwin (1995) confirmed the slower time course for the effects of gating asynchrony on pitch versus vowel perception, leading the authors to note that auditory grouping and segregation are not absolutes, but instead can show relative strengths in part dependent on the type of perceptual classification required by a given experimental procedure. In an earlier study, Darwin (1981) found little effect of onset asynchronies on phoneme category if applied to formants rather than individual harmonics of a vowel. Darwin suggested that the results are consistent with the characteristics of natural speech, in which vowel formants are often asynchronous yet still combine into a common phonetic category. Across the range of studies that have evaluated effects of gating asynchrony, not only do results vary with task demands, but they also reflect the fact that grouping and segregation are themselves not processes, but rather the consequences of auditory and cognitive processing.

## 3.2 Stream Segregation

A gating asynchrony among stimulus components can lead to the perception of sequential auditory events. The perception of multiple auditory events as a coherent sequence, in essence the output of a single sound source, is referred to as streaming (Hafter, Sarampalis, and Loui, Chapter 5; Carlyon and Gockel, Chapter 7). A general characteristic of streaming is that pattern recognition is superior for elements that have arisen from a common source in contrast to multiple real or implied sources. The parametric study of Hartmann and Johnson (1991) set the context for considering stream segregation in relation to envelope processing. The experiment measured the accuracy of listeners at identifying interleaving melodies, with results used to evaluate a basis for stream segregation in peripheral channeling. The channeling model posits that stream coherence is established by processing of sequence elements through a common channel determined by either frequency content or ear of stimulation. Employing a dozen different stimulus manipulations, results indicated that channeling is a critical factor in determining stream segregation. Subsequent work has incorporated peripheral channeling into computational models of stream segregation (Beauvois and Meddis 1996; McCabe and Denham 1997). Following frequency-selective processing, envelope detection as described in Section 2 is applied to the output of each channel; envelope rather than fine structure is the basis for simulation of segregation. These models have proved successful in accounting for a variety of aspects of stream segregation, including van Noorden's (1975) distinction of boundary for sequence segregation and coherence, and also the buildup of stream segregation over time (e.g., Cusak et al. 2004).

Several recent studies have demonstrated stream segregation in the absence of peripheral-channeling cues (Grimault et al. 2002; Roberts et al. 2002). The basis of this work comes from previous illustrations of segregation despite spectral overlap of sequence elements (e.g., Dannenbring and Bregman 1976; Iverson 1995; Singh and Bregman 1997). The recent studies utilized sequence elements with identical long-term power spectra, distinguishing elements by envelope modulation. In the work of Grimault et al. (2002), sequence elements were 100-ms WBN pulses. In a procedure that estimated stream-segregation thresholds, the rate of sinusoidal amplitude modulation (SAM) of one sequence was set at 100 Hz, while the modulation rate of the other was varied between 100 and 800 Hz. Thresholds were in the region of an octave separation of modulation rates, with the strength of segregation increasing up to roughly a two-octave difference. To account for their findings, Grimault and coworkers suggested that a channeling basis of stream segregation may extend to the modulation domain to operate on the output of a modulation filterbank. In the work of Roberts et al. (2002), stimuli were 60-ms complex tones comprising unresolved harmonics of a 100-Hz fundamental. Cross-sequence differences in envelope modulation were obtained by manipulating the phase relationship among components of the complex tones. In conditions in which sequence elements shared a common long-term power spectrum, measures of stream segregation showed an effect of component phase. Roberts et al. noted that their results were consistent with a proposal of Moore and Gockel (2002) basing the extent of stream segregation on the degree of perceptual difference between sequence elements. An effect of element similarity on streaming had been previously suggested by McNally and Handel (1977), and is also supported by the work of Iverson (1995), who found that similarity and streaming judgments were highly correlated. While spectral differences are clearly a strong contributor to perceptual distinction, other factors, including envelope modulation, will also come into play. Iverson (1995) showed that for tonal sequences played by musical instruments, envelope dissimilarity encouraged tone segregation.

A basis of stream segregation on perceptual salience can be considered in terms of auditory information processing. Dependence of streaming on sequence-element similarity has the extent of streaming increasing with decreasing potential for information transmission. The trend in streaming tasks for a performance decrement due to segregation contrasts with results from informational-masking studies (Kidd, Mason, Richards, Gallun, and Durlach, Chapter 6). In informational-masking experiments, threshold elevation is often associated with similarity between target and masker. A number of studies have demonstrated a reduction in informational masking through stimulus manipulations intended to enhance dissimilarity or segregation among stimulus components or events (Kidd, Mason, Richards, Gallun, and Durlach, 2002, 2003; Durlach et al. 2003; Watson 2005). A key distinction between informational-masking and streaming tasks is that the former measures the ability to process change in a single event or target, while the latter evaluates patt- ern recognition for a sequence of events. Results from both types of studies indicate a limitation in auditory information-processing capacity. For informational-masking experiments, the limitation is seen in the inability to disregard extraneous information; in streaming studies, the limitation is observed as difficulty in attending to multiple concurrent streams. Kidd and his coworkers (2002) noted that for some stimulus configurations, either aspect may provide a reasonable account of experimental results.

In the stream-segregation experiments of both Grimault et al. (2002) and Roberts et al. (2002), effects are noted with relatively high rates of envelope modulation. Use of short pulse durations in streaming studies necessitates high AM rates. With auditory streaming defined as a perceived coherence among sequential events, it seems reasonable that slower modulation could also enhance sequence coherence. If a low-rate AM is present, the envelope of each sequence element would show only a partial period of the modulation with a full cycle evidenced across elements. Presumably, coherence of the continuous discourse of a single talker is enhanced by the low-rate envelope modulation of speech (Hawkins 1995).

## 3.3 Perceptual Prominence

Related to the consideration that salience of perceptual change influences stream segregation is the concept of perceptual prominence, a subjective measure of perceived distinction. The discussion of gating asynchrony noted an enhanced ability to "hear out" an asynchronous component. Measures of both stream segregation (Iverson 1995) and the effects of gating asynchrony (Rasch 1978) can be influenced by envelope rise time. Rise time, along with intensity level, is a primary determinant of the perceptual onset or attack time of musical notes (Vos and Rasch 1981; Gordon 1987).

The effect of envelope on perceptual prominence is not limited to stimulus onset. If the amplitude of a single component of a tonal complex is either briefly lowered or continuously modulated, that component tends to stand out from the complex (Kubovy and Daniel 1983; Kubovy 1987). Pitch segregation by a momentary amplitude disparity depends on the frequency spacing between the target and nontarget components, with the effect diminishing if adjacent components are separated by less than a critical bandwidth. An effect of frequency spacing suggests involvement of spectral factors. The demonstrations of Kubovy and Daniel (1983) are similar to studies of auditory enhancement in which results are at least in part a consequence of peripheral frequency-selective processing. Auditory enhancement refers to the perceptual prominence of a spectral component or region following prior stimulation in which the level in that spectral region was attenuated relative to adjacent signal components. Enhancement may be observed as a negative afterimage, or through change in masking effectiveness, loudness matching, or vowel identification (e.g., Wilson 1970; Viemeister and Bacon 1982; Summerfield et al. 1984). One explanation for the effect assumes differential adaptation to the initial stimulus, allowing for enhanced representation of the less well adapted components with subsequent stimulation. Some results, however, indicate signal amplification rather than just change in relative adaptation. These findings have led to speculation that auditory enhancement may arise from either adaptation of suppression (Viemeister and Bacon 1982) or enhancement in the coding of AM by units in the CANS (Summerfield et al. 1987).

Along with spectral effect in the audio-frequency domain, manipulations affecting perceptual prominence and auditory enhancement lead to change in the modulation spectrum. Sheft and Yost (2006) evaluated the perceptual prominence or salience associated with complex modulation of either tones or WBN. Complex modulators were defined by two terms: envelope slope or rise/fall time, which varied from 1 to 25 ms, and a variant of duty cycle, the ratio of peak to peak-plus-valley durations of a modulation cycle. Modulation rate was either 4 or 10 Hz. Manipulation of modulator slope and duty cycle led to large changes in the modulation spectrum, with the amplitude at the modulator fundamental varying by over 40 dB across the condition set. Salience judgments were analyzed with an ordinal individual-differences-scaling model with interpretation concerned with stimulus grouping in multidimensional space. For stimuli with high values of modulator rms amplitude, grouping in most cases was consistent with the amplitude of the fundamental of the modulator spectrum. The exception was with 10-Hz modulation of tonal carriers, where timbre variations may have influenced responses. For lower-level modulators, gross envelope wave shape set by the combination of modulation rate and duty cycle accounted for the organization of judgments of perceptual salience.

## 3.4 Grouping and Coherent Envelope Modulation

The counterpoint to segregation arising from gating asynchrony is grouping by envelope coherence. Along with a common onset and offset among spectral components, envelope coherence includes the ongoing fluctuations of the components, that is, their patterns of amplitude modulation. Though cross-spectral coherence of AM is an often assumed characteristic of sound sources, there has been relatively little study of this issue. Attias and Schreiner (1997) analyzed the temporal statistics of a large database of natural auditory scenes including speech, animal vocalizations, music, and environmental sounds. Each sample was passed through a filterbank with filter CFs logarithmically spaced within the range 100–11025 Hz. Typical filter bandwidth was 1/8 octave. Following the form of equation (1), each filter output was expressed as a band-limited signal with amplitude probability distributions computed for the individual envelope functions. Attias and Schreiner found that for each sound studied, distribution statistics were nearly identical across filter channels. While not demonstrating cross-spectral coherence of modulation, this result does indicate redundancy of modulation information across spectral location. A subsequent analysis of the modulation statistics of natural-sound ensembles by Singh and Theunissen (2003) confirmed the general distribution form reported by Attias and Schreiner, but without the finding of cross-spectral redundancy. Results from both studies indicate differences in modulation spectra according to source type. Though not directly affecting source segregation, these differences allow for involvement of modulation characteristics in source classification, an aspect of auditory information processing. A practical application of this characteristic is use of the modulation spectrum for separating speech signals from environmental noise (Miyoshi et al. 2004). Analyzing animal vocalizations and other natural sounds, Nelken et al. (1999) did find that low-rate AM was often prominent and coherent across spectral regions. A counterexample to this coherence comes from speech, where vowel sounds may exhibit a prominent frequency modulation. As modulated components move through the passbands of auditory filters, signal FM can result in AM of the filter outputs with the AM of adjacent filters showing different patterns and phases. Cross-filter envelope incoherence with speech is also observed with wider frequency separations, this due to the complex spectrotemporal nature of speech (Figure 9.2).

Involvement of cross-spectral envelope coherence in auditory grouping requires sensitivity to the coherence. Several groups of investigators have studied the ability to detect change in envelope correlation of concurrent modulated tones or NBNs at different CFs (Richards 1987, 1990; Wakefield 1987; Strickland et al. 1989; Yost and Sheft 1989; Moore and Emmerich 1990; Sheft and Yost 1990, 2005). In these studies, contrasting stimuli were generated either by varying envelope-modulator phase or through selection of independent amplitude and phase arguments for noise samples of equal bandwidth. With interest in assessing the ability to detect cross-spectral envelope synchrony, a variety of procedures have been used to restrict within-channel cues arising from the direct interaction of the concurrent envelopes. Common stimulus controls are the addition of either a wideband or narrowband masker, or use of level randomization. When using SAM-tone stimuli, the task is to detect a cross-spectral envelope-phase disparity with threshold measured as the delta phase applied to one of the two modulators (Figure 9.3). To avoid detection based on discrimination of starting phase, this value is often randomized. With noise stimuli, ability to discriminate

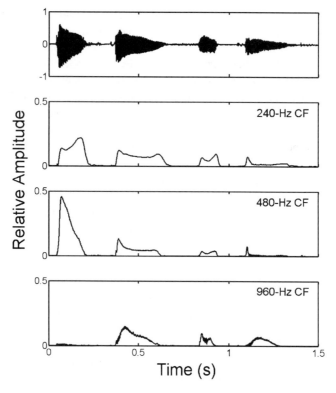

FIGURE 9.2. Temporal waveform (top panel) and analyses of the utterance "baseball hotdog" by a female talker. Panels 2–4 show the Hilbert envelope of the output of filters simulating peripheral auditory frequency selectivity with filter CF indicated in the top right corner of each panel. Across filters, envelopes are asynchronous.

between coherent and independent concurrent noise samples is generally assessed in terms of $P(c)$ or $d'$.

Results from initial studies indicated a lowpass function relating synchrony-detection threshold to SAM rate with low- to moderate-rate thresholds in the region 0.5–1.0 radian if modulation depth was 1.0 (Wakefield 1987; Strickland et al. 1989; Yost and Sheft 1989). More recently, Sheft and Yost (2005) found that threshold elevation began with increasing SAM rate above 10 Hz (Figure 9.4, squares). This low cutoff value in part related to the higher masker levels used to restrict within-channel cueing. When modulation depth decreased from 1.0 to 0.5, thresholds rose and function shape changed from lowpass to bandpass (Figure 9.5, triangles). For noise stimuli, modulation rate has been manipulated by varying noise bandwidth (see Section 2). Results obtained with a two-interval forced-choice (2IFC) procedure indicate that discrimination ability improves with noise bandwidth (Moore and Emmerich 1990), though the effect can be quite small (Figure 9.6, filled symbols; Sheft and Yost 1990). Since increasing bandwidth raises the upper cutoff of the modulation spectrum, this trend is

Signal          NonSignal

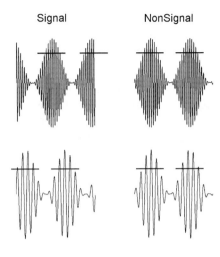

FIGURE 9.3. Schematic representation of a 2IFC synchrony-detection trial. The horizontal lines are intended to illustrate the effect of cross-spectral asynchrony on timing of events in the short-term amplitude spectrum.

opposite the lowpass result obtained with tonal stimuli. However, modifying the task by preceding each trial with a cue of the signal interval reverses the relatively small effect of bandwidth (Figure 9.6, open symbols).

By Fourier relationship, cross-spectral envelope asynchrony can be viewed as altering the timing of events in the short-term amplitude spectrum. The solid horizontal lines in Figure 9.3 are intended to illustrate the sequence in which each carrier of a synchrony-detection procedure achieves some criterion level. In this context, synchrony-detection procedures show similarity with measures of stream segregation that evaluate cross-spectral perception of temporal order.

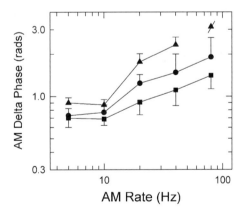

FIGURE 9.4. Squares indicate mean synchrony-detection thresholds from five listeners as a function of AM rate. Carrier frequencies were 2.0 and 3.6 kHz, and modulation depth was 1.0. In the remaining conditions, a 250-ms forward fringe was added to the upper carrier with the modulators either synchronously (circles) or asynchronously (triangles) gated. Error bars represent one standard error of the mean threshold. (From Sheft and Yost 2005.)

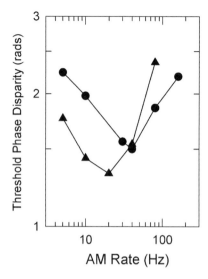

FIGURE 9.5. Triangles indicate mean synchrony-detection thresholds from three listeners as a function of AM rate. Carrier frequencies were 2.0 and 3.6 kHz, and modulation depth was 0.5. Modulation-masking of 20-Hz synchrony detection is indicated by the circles, with the masker added as a second component of the function modulating the pure-tone carriers. For these conditions, the independent variable is masker AM rate. (From Sheft and Yost 2005.)

As with streaming, when the carriers segregate, judgments of synchrony, or temporal order, become difficult. For the functions indicated with the circles and triangles in Figure 9.4, a forward temporal fringe was added to the upper tonal carrier so that it began 250 ms before the other, with both carriers gated off together. Either the modulation of the upper carrier was delayed by the

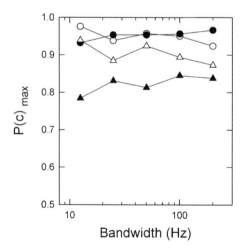

FIGURE 9.6. Mean synchrony-detection performance of three listeners as a function of the bandwidth of the noise stimuli. 2IFC results are shown with the filled symbols, while results from conditions in which a cue preceded each 2IFC trial are indicated with the open symbols. There were either two (CFs of 500 and 3125 Hz) or three (CFs of 500, 1250, and 3125 Hz) concurrent noise bands, with results indicated by the triangles and circles, respectively. (From Sheft and Yost 1990.)

duration of the forward fringe, so that despite the carrier-gating asynchrony, the modulators were synchronously gated (circles), or both modulators began at their respective carrier onset, leading to asynchronous gating of carriers and modulators (triangles). Especially for AM rates above 10 Hz, gating asynchrony made the task more difficult. The separation between the functions marked with the circles and triangles represents an uncommon situation that consistently distinguishes between a carrier and carrier-plus-modulator delay.

In relation to sound-source processing, the assumption is that synchrony detection relies on sensitivity to cross-spectral envelope correlation. Instead, synchrony detection could be based on a global percept associated with the modulation of multiple carriers. Terhardt (1974), Vogel (1974), and Pressnitzer and McAdams (1999) have shown that with modulation of two tonal carriers, the strength of the roughness percept associated with AM depends on the phase relationship between the two modulators. While change in the roughness percept may represent sensitivity to envelope correlation, it may also reflect a type of summation or wideband processing that recombines envelopes following peripheral filtering. If multiple sound sources are present, involvement of envelope coherence in source segregation requires that in some manner, the various fluctuation patterns that characterize the different sources are selectively processed. Summation or wideband processing would not establish the requisite selectivity. Results obtained from studies of modulation masking of synchrony detection distinguish between selective and wideband processing of envelope modulation.

Sheft and Yost (1990) measured the ability to detect envelope coherence among target noise bands in the presence of masking bands with all bands sharing a common bandwidth. Masking increased with noise bandwidth as the perception associated with the stimuli progressed from fluctuation to roughness. Especially with narrow stimulus bandwidths of 12.5–50 Hz, there was little effect of envelope coherence between masking bands on the detection of target-band synchrony. Both results indicate carrier selectivity in low-rate envelope processing. Sheft and Yost (2005) evaluated modulation masking of synchrony detection with tonal stimuli. Modulation-masker carriers were placed between the target carriers of 2.0 and 3.6 kHz, with the 400-ms targets temporally centered in the 1200-ms masker. Target SAM rate was 10, 20, or 40 Hz, and masker SAM rate ranged from 5 to 160 Hz. With a target modulation depth of 1.0, there was little if any modulation masking of synchrony detection, again demonstrating selectivity in cross-spectral envelope processing. When target AM depth was reduced to 0.5 and modulation maskers were introduced as a second component of the functions modulating the target carriers, significant masking was obtained (Figure 9.5, circles). The masking functions, however, did not show tuning in the modulation domain; regardless of target SAM rate, interference was greater with distal rather than proximal masker SAM rates. This absence of selectivity stands in contrast to modulation masking of AM detection, which does indicate tuning in the modulation domain (see Section 2). The difficulty in detecting cross-spectral envelope synchrony in the presence of a distal modulation rate suggests some limitation on the use of envelope coherence to segregate multiple sound sources.

Synchrony detection studies do not directly demonstrate involvement of envelope coherence in auditory grouping. To address this issue, experimental procedures similar to those used to evaluate gating asynchrony (see Section 3.1) have been applied to the study of envelope coherence. Replacing the gating asynchrony of the stimulus configuration illustrated in Figure 9.1 with modulation of sequence elements, Bregman and his coworkers (1985, 1990) studied the effect of AM on stream segregation and spectral grouping. Conditions used relatively rapid envelope periodicities of roughly 80 to 200 Hz, corresponding to the range of fundamental frequencies common in adult speech. Results showed that the perceptual fusion of two spectrally distant carriers was affected by coherence of modulation. When the waveforms modulating the two carriers differed in either rate or phase, spectral fusion was diminished. Effects at these relatively high AM rates contrast with the lowpass characteristic of synchrony-detection results. Darwin and Carlyon (1995) noted that in at least some of the Bregman conditions, the output of an auditory filter centered between the two carriers would show different patterns of FM dependent on modulator coherence, and that this difference could potentially cue listener response. Since effects were greatest with large cross-spectral differences in AM rate in the Bregman work, it also seems possible that timbre differences among sequence elements affected judgments of the extent of segregation.

Using a variation of the three-element repeating cycle of Figure 9.1 with low-rate AM, Bregman and Ahad (see Bregman 1990, pp. 287–289) and Rappold et al. (1993) found that coherence of modulation affected values of stimulus parameters needed for listeners to judge the sequences as segregated. In both studies, investigators inferred that segregation of the target stream was discouraged by coherent modulation enhancing auditory fusion of concurrent sequence elements. For profile analysis (see Section 3.1), the effect of cross-spectral modulation coherence is largest at low AM rates (Green and Nguyen 1988). If the modulation is applied to only the nonsignal components of the stimulus, profile-task performance is unaffected by AM coherence as long as components are in separate critical bands (Dai and Green 1991).

Direct evaluation of the role of envelope modulation in auditory grouping requires an identifier listeners can use to distinguish or label a target subset of a multicomponent stimulus. To this end, investigators have studied complex pitch and vowel identification with low-rate modulation of stimulus components. Sheft and Yost (1992a) used residue pitch as the target identifier. Stimuli consisted of two concurrent harmonic complexes that differed in fundamental frequency. In a 2IFC procedure, one harmonic subset was modulated during one observation interval of a trial, the other subset during the other interval. Without modulation, the stimuli presented during the two intervals were identical. The task was to select the interval in which the lower residue pitch was modulated. To avoid discrimination based on the modulation of individual harmonics, harmonics were randomly selected during each trial. Baseline performance was defined as the ability to perform the task with the harmonic subsets presented sequentially on each trial without modulation rather than concurrently with distinction by AM. In conditions with coherent modulation, performance progressively decreased below baseline levels as AM rate was increased from 5 to 23 Hz, though it was always above chance. Despite performance decrements, the ability to perform the task could be taken to indicate proper grouping of components due to coherence of modulation. An alternative view is that AM troughs simply provide unmasked "looks" at the contrasting harmonic subset, allowing its pitch to dominate. A further performance decrement obtained by Sheft and Yost with randomization of AM rate or phase is consistent with either explanation. Adding fringe tones to the stimuli also hindered performance. With fringe tones partially filling AM troughs, this final result supports involvement of modulation through dynamic modification of the intensity profile of the stimulus, rather than as a grouping factor per se. Absence of an effect of low-rate AM on grouping for pitch perception has also been reported by Darwin (1992) and Darwin et al. (1994). Results from these studies showed that the contribution of a mistuned harmonic to the pitch of a complex tone did not vary if modulation was applied to either the mistuned component or all stimulus components.

Results from speech studies also have failed to provide evidence of auditory grouping due to coherence of low-rate AM.[3] Summerfield and Culling (1992) measured the S/N ratio at which listeners could identify target vowels in the presence of a vowel masker. With 8-Hz target AM, thresholds were unchanged as masker AM rate was varied from 3.4 to 19 Hz. When both the target and masker were modulated at a lower rate of 2.5 Hz, thresholds were lower with antiphasic than with in-phase modulation. However, this effect could be accounted for by local variation in the S/N ratio due to antiphasic modulation. A large number of studies have demonstrated the ability of listeners to utilize brief windows of high S/N ratios when attending to speech signals in the presence of fluctuating maskers (e.g., Miller and Licklider 1950; Gustafsson and Arlinger 1994). The benefit obtained in the presence of fluctuating maskers can be severely disrupted if spectral information is compromised so that speech perception is reliant primarily on envelope cues (Kwon and Turner 2001; Nelson et al. 2003; Qin and Oxenham 2003; Stickney et al. 2004; Füllgrabe et al. 2006). The enhanced masking effect in conditions of spectral degradation in large part reflects the loss of redundancy in the natural speech signal.

Of relevance to consideration of envelope coherence is the finding that for speech-in-noise processing, the benefit of modulating maskers does not require cross-spectral synchrony of masker fluctuation (Howard-Jones and Rosen 1993; Buss et al. 2003). Complementary studies have shown good intelligibility despite introduction of cross-channel spectral asynchrony to the speech signal (Arai and Greenberg 1998; Fu and Galvin 2001; Buss et al. 2004). In these studies, the

---

[3]Working with synthetic speech that minimized stimulus envelope, Carrell and Opie (1992) found that addition of coherent high-rate AM at either 50 or 100 Hz did improve sentence intelligibility.

signal was filtered into contiguous frequency bands with asynchrony achieved by either time shifting or incoherently modulating the filter-channel outputs. Resilience of speech to imposed asynchrony is consistent with the asynchrony inherent in the natural signal, and also the dominance of low-rate AM in speech perception (see Section 5). Similar to the enhanced modulation masking of speech reliant on envelope cues, a greater effect of extraneous asynchrony on speech intelligibility is obtained when spectral, or audio-frequency, information is restricted (Greenberg et al. 1998; Fu and Galvin 2001). In some cases, the effect of asynchrony with degraded spectra indicates cross-spectral interference along with a limitation on information integration. Spectral asynchrony is evidenced in the cross-spectral modulation-phase spectrum; this is the relationship schematically illustrated in Figure 9.3. Varying the window duration in which speech segments were time-reversed, Greenberg and Arai (2001) found correspondence between speech intelligibility and representation of cross-spectral low-rate modulation phase.

Moore and Alcántara (1996) studied vowel identification using a stimulus configuration in which low-rate AM of a subset of components of a harmonic complex distinguished a target vowel from the background. As in the above studies, conditions that led to good performance levels were interpreted to indicate use of windows of high S/N ratios due to modulation. Moore and Alcántara further noted that a perceived continuity of interrupted masked components may serve to reduce the effects of asynchrony. Thus despite sensitivity to envelope synchrony, results from both pitch and speech studies do not indicate utilization of cross-spectral envelope coherence to enhance auditory grouping. Concerning involvement of modulation in sound-source processing, the frequent suggestion that effects relate to modulation introducing a dynamic S/N ratio raises a distinction between the Gestalt notion of common fate affecting auditory grouping and auditory processing of the consequences of envelope fluctuation.

## 3.5 Cross-Spectral Masking

Over the last twenty years, considerable effort has been devoted to the study of cross-spectral masking with envelope-modulated stimuli. The three major paradigms have been termed comodulation masking release (CMR), comodulation detection differences (CDD), and modulation detection or discrimination interference (MDI). All three involve presentation of concurrent modulated bands, with cross-spectral envelope coherence a primary experimental variable. Much of the past work has been concerned with the extent to which results reflect processing that may contribute to auditory grouping. Green (1993) and Hall et al. (1995) reviewed many of the major results. The present section will consider several issues relevant to sound-source processing; Section 4 discusses CMR and MDI in the context of information processing.

The basic CMR result is that the threshold of a signal centered in a narrowband masker can be lowered by adding coherently modulated bands spectrally remote from the signal frequency. CDD refers to the finding that the masked threshold of a narrowband signal can be higher if the signal and masker envelopes are coherent rather than independent. In the MDI paradigm, the ability to process the modulation of a probe tone can be severely disrupted by the addition of spectrally remote modulated maskers, with this observed as an elevation in AM detection and discrimination thresholds. Similarity among the three procedures is noted by the fact that with minor stimulus manipulation, the paradigm or result can shift from one to another (Yost and Sheft 1990; Moore and Jorasz 1992). Conditions are also possible that reflect a balance of cross-spectral interference and masking release (Moore and Jorasz 1992; Grose and Hall 1996; Kwon and Turner 2001).

Present issues of concern regarding cross-spectral masking are the role of envelope rate, complexity, and coherence, effects of gating asynchrony and stream segregation, and detection cues with emphasis on within-channel versus cross-channel processing. Concern with the effects of rate and complexity relates to the general lowpass nature of the involvement of envelope in sound-source processing and the need for selectivity when multiple sources are present. In the modulation domain, both CMR and MDI are lowpass effects; the extent of the masking release or interference diminishes with increasing modulation rate. With a common rate of sinusoidal modulation of the probe and masker, the largest change in MDI occurs as SAM rate is increased from 5 to 20 Hz (Yost et al. 1989; Bacon and Konrad 1993). In CMR tasks, the rate of masker fluctuation has been controlled by varying the rate of periodic modulators, the bandwidth of lowpass-noise modulators, or masker bandwidth itself (Buus 1985; Schooneveldt and Moore 1987, 1989; Hall et al. 1988; Eddins and Wright 1994; Bacon et al. 1997). Masking release due to cross-spectral envelope coherence generally decreases as envelope rate increases, though as a threshold change, the effect of envelope rate on CMR can depend on reference condition (see Carlyon et al. 1989). Once allowance is made for involvement of within-channel cues (see below), the lowpass effect of rate is more gradual for the stochastic fluctuation of NBN than with SAM. In contrast to both CMR and MDI, CDD shows little effect of envelope rate, observed as a constancy of thresholds as the common signal and masker bandwidth is increased from either 4 to 64 Hz (Fantini and Moore 1994) or 20 to 160 Hz (Moore and Borrill 2002).

Manipulation of cross-spectral envelope correlation has received the most attention in studies of MDI, primarily through variation of modulation rate or phase. When the probe and masker are modulated at different rates, interference declines with increasing separation of the probe and masker SAM rates (Yost et al. 1989, 1995; Moore et al. 1991; Bacon et al. 1995). This broad tuning in the modulation domain is roughly equivalent to the extent of selectivity obtained in studies of modulation and temporal masking of AM detection (see Section 2). Results from studies in which the probe and masker were sinusoidally modulated at the same rate have tended to show a relatively small effect of cross-spectral envelope-phase relationship, though in some cases with large individual differences noted (Yost and Sheft 1989, 1994; Moore et al. 1991; Bacon and Konrad 1993; Moore and Shailer 1994; Richards et al. 1997). When replacing the sinusoidal modulator of either or both the probe and masker with narrowband noise, the extent of MDI remains significant (Mendoza et al. 1995; Moore et al. 1995). While similarity of probe and masker envelope rate affects MDI when stochastic modulators are used, there is no effect depending on whether the two modulators are coherent or independent. A similar result of an effect of rate but not of coherence is obtained with periodic modulators (Shailer and Moore 1993).

In studies of CMR, cross-spectral envelope correlation between the on-signal-frequency and flanking maskers was first controlled through time delay of the flanking band. McFadden (1986) reported that with 100-Hz-wide noise bands, CMR was significantly reduced with delay of just a few milliseconds. Working with a greater range of stimulus parameters, Moore and Schooneveldt (1990) found an interaction of masker bandwidth and time delay, suggesting that the magnitude of CMR depends on the correlation among masker envelopes. Related is the report of Grose et al. (2005) that the magnitude of CMR drops when the masker/signal complex is preceded by presentation of a random temporal fringe. Results from both the fringe and flanker-delay conditions presumably reflect integration and smearing through the time constant(s) of auditory envelope processing. Using periodic masker modulation with variation of relative envelope phase and duty cycle, the data of Buss and Richards (1996) indicate that a large CMR can be obtained in conditions of cross-spectral envelope dissimilarity, that is, when correlation is low. Conversely, CMR can be obtained in conditions in which cross-spectral masker coherence is unaffected by the addition of the signal (Hall and Grose 1988), that is, when correlation is always high. A second stimulus manipulation of CMR studies that alters cross-spectral envelope correlation is the use of multiple concurrent patterns of envelope fluctuation. Though the need for spectral separation of pattern varied between studies, both Eddins and Wright (1994) and Grose and Hall (1996) found that multiple masking releases could be obtained, with each corresponding to one of the concurrent modulation patterns. Hall and Grose (1990) and Grose et al. (2001) evaluated how the addition of independently modulated (deviant) bands affects a single CMR. In baseline conditions, the presence of deviant bands reduced the masking release. Indicating interference due to the addition of extraneous envelope modulation, this result is analogous to MDI. Across conditions, however, CMR could be restored by either increasing the number of deviant bands, manipulating their spectral placement, or introducing a gating asynchrony between the deviant bands and bands comodulated with the on-signal-frequency masker. The investigators have interpreted the restored CMR to indicate perceptual segregation of the deviant bands from the remaining masker bands.

Use of intermediate values of cross-spectral envelope correlation has received little attention in studies of CDD. Wright (1990) investigated CDD using multiple signal and masker bands. Across conditions, the bands either had the same or different temporal envelopes. Combination of different temporal relationships for signals and maskers affects both overall and local values of cross-spectral envelope correlation. Variation in coherence among masker bands has also been studied by McFadden and Wright (1990), Fantini and Moore (1994), and Borrill and Moore (2002). Across studies, thresholds were lowest when the envelope of the signal band(s) differed from a common envelope of the masker bands, and were highest if all stimulus bands shared a single envelope pattern. Results varied by study when all masker envelopes were independent, with Borrill and Moore (2002) observing that when multiple maskers were present, thresholds were often determined by the single most effective masking band. A common concern regarding all studies that involve presentation of multiple modulation patterns in separate spectral regions is the intermingling of patterns following auditory processing. Possibility of within-channel interaction in all measures of cross-spectral masking is considered more fully below.

CMR and MDI show dramatic effects of gating asynchrony. In most conditions, gating asynchrony between the signal and maskers increases the magnitude of CMR, though the effect is lessened with either a large number of masking bands or use of a contiguous comodulated masker (McFadden and Wright 1992; Fantini et al. 1993; Hatch et al. 1995; Hall et al. 1996). With retention of comodulation despite asynchronous gating between on- and off-signal-frequency masking bands, a deleterious effect of gating asynchrony is obtained (Grose and Hall 1993; Dau et al. 2005). Asynchronous gating of the probe and masker carriers in the MDI paradigm reduces the extent of cross-spectral interference (Hall and Grose 1991; Moore and Jorasz 1992; Mendoza et al. 1995). In both AM detection and depth discrimination tasks, the reduction in MDI by gating asynchrony varies with asynchrony duration in a manner consistent with the effect of asynchrony on pitch perception (see Section 3.1). If the probe and masker carriers are turned on and off together with the asynchronous gating restricted to the two modulators, an effect of asynchrony on MDI has been reported only with sinusoidal and not NBN modulators (Oxenham and Dau 2001; Gockel et al. 2002). For CDD, asynchronous gating between the signal and masker bands has either no effect or one marked by large individual differences, especially in conditions of high uncertainty due to randomization of masker-band CF (McFadden and Wright 1990; Moore and Borrill 2002; Hall et al. 2006). The effects of stream segregation on CMR and MDI are consistent with those related to the introduction of a gating asynchrony; stream segregation encouraged by the addition of sequential stimuli reduces the magnitude of the cross-spectral effect (Grose and Hall 1993; Oxenham and Dau 2001; Dau et al. 2005).

Consideration of the diversity of result as a function of stimulus configuration suggests use by listeners of a variety of detection cues. As with synchrony detection, presence of within-channel cueing is a concern. Both McFadden (1975) and Wakefield and Viemeister (1985) have demonstrated for multicomponent modulated stimuli the possibility of temporal interaction within a single peripheral frequency channel despite wide frequency separation of components. In general, the possibility of within-channel interaction is enhanced in procedures that measure the ability to detect low-level signals in the presence of higher-level maskers. The effect of frequency separation between signal and maskers is often a major aspect of the argument either for or against involvement of within-channel interaction in cross-spectral masking (e.g., Cohen and Schubert 1987;

Schooneveldt and Moore 1987; Bacon and Konrad 1993; Borrill and Moore 2002). Hall et al. (1995), however, noted that cross-channel processing may show an effect of component proximity similar to that of within-channel inter-action, complicating a simple disposition of interpretation. For CMR and MDI, the presence of significant effect with dichotic stimulus presentation strongly supports at least partial involvement of cross-channel processing (e.g., Schoon-eveldt and Moore 1987; Yost et al. 1989; Fantini et al. 1993; Sheft and Yost 1997a).

As noted above, the effects of envelope rate and gating asynchrony in measures of CDD differ from those observed with either CMR or MDI procedures. To account for this distinction, Moore and Borrill (2002) proposed a within-channel account of CDD based on spread of peripheral excitation with variation in suppression enhancing the time periods of higher S/N ratios. Modeling absent of cross-channel processing has also been able to account for results obtained from subsets of CMR conditions. Berg (1996) applied Viemeister's (1979; see also Section 2 of this chapter) envelope-detection model to CMR conditions, finding that broad predetection filtering coupled with a decision statistic based on the amplitude spectrum of model output could account for some results. Focusing on within-channel cues processed through a modulation filterbank, Verhey et al. (1999) reported simulation of CMR for a subset of stimulus config-urations. CMR conditions requiring consideration of cross-channel processing have been modeled through two general approaches: detection of cross-spectral envelope decorrelation (Richards 1987) and cueing of times of high S/N ratio by masker coherence (Buus 1985). Across studies, most researchers find an explanation by cueing, often termed "listening in the valleys," able to account for a wider range of CMR results. However, van de Par and Kohlrausch (1998) observed that inclusion of the dc term in the calculation of cross-spectral envelope coherence (a distinction between cross covariance and cross correlation, with the latter including dc) mitigates some objection to this approach. Largely from the result of selectivity in the modulation domain, MDI is commonly assumed to reflect processing by modulation-specific channels. Limitation of this consid-eration is indicated by both the extent of interference obtained in the presence of multiple masker-modulation rates (Bacon et al. 1995) and the rate-dependent effect of second-order modulation (Sheft and Yost 1997b). Clearly for MDI, and most likely also CMR, thresholds often reflect a balance among potential detection cues, with one or another at times predominating in a specific stimulus condition. When multiple cues are present, marked individual differences are not uncommon.

In summary, though often indicating influence of envelope coherence, results do not rigorously support auditory grouping due to envelope coherence as a major basis of cross-spectral masking with modulated stimuli; many results are better understood as reflecting consequences of the modulation (e.g., modulation of S/N ratio or fine-structure periodicity). Overall, psychophysical results do show sensitivity to cross-spectral envelope coherence coupled with a degree of selectivity in the spectral domain that associates modulation with its appropriate carrier. However, the full extent of direct involvement and relevance of this ability to sound-source processing at present may be questioned.

## 3.6 Physiology and Modeling Cross-Spectral Processing of Envelope Modulation

All modeling of cross-spectral processing of envelope modulation assumes an initial stage of filtering that corresponds to the frequency selectivity of the auditory periphery. From there, two general approaches have been used. The first, introduced in Section 2, is based on the processing of a modulation filterbank, while the second utilizes the temporal response characteristics of frequency channels to code stimulus envelope. With selectivity to envelope modulation a central psychophysical result, brief consideration of the neural response is appropriate regarding either modeling approach.

Beginning at the level of the cochlear nucleus, neurons of the CANS can exhibit a bandpass response to modulation, with best frequency determined by either firing rate or temporal response measures. A thorough review of neural processing of AM signals is provided by Joris et al. (2004). In that there is no traveling wave for envelope modulation, the basis of modulation selectivity in the auditory system most likely reflects an interaction of temporal response character-istics. Considering the diversity of response characteristics both within and across levels of the CANS, some caution is warranted in applying a specific response class to a perceptual result. The basis of modulation-filterbank models comes from the findings of a topographical organization according to AM rate within the CANS (Schreiner and Langner 1988; Langner et al. 1992, 2002; Heil et al. 1995; Schulze and Langner 1997). Physiological results further indicate that the mapping of envelope periodicity is orthogonal to the tonotopic organization of the nuclei. Langner et al. (1997) demonstrated, by means of magenetoencephalog-raphy, this orthogonal organization in the human auditory cortex. With evidence of integration of periodicity information across spectral frequency (Biebel and Langner 2002), orthogonal mapping of spectral frequency and envelope period-icity offers a physiological basis for consideration of the modulation spectrogram in sound-source processing. Dependence of this topographical representation of AM rate on stimulus level is a modeling concern (Krishna and Semple 2000), though one that may be resolvable (Deligeorges and Mountain 2004). As Joris and coworkers (2004) comment, absence of topographical mapping in itself would not invalidate the existence of a modulation filterbank (p. 560), and even without fine tuning, neural selectivity to AM rate can still parse spectral components to allow for later grouping (p. 565).

Dau and Verhey (1999) attempted to account for CMR and MDI based on the outputs of modulation filterbanks, each analyzing the response of a separate peripheral auditory filter. Model predictions were based on integration of envelope fluctuation across spectral frequency. An alternative to the place mapping of a modulation filterbank is a temporal approach that represents stimuli in terms of some variant of a running autocorrelation of the outputs of multiple frequency channels (Meddis and Hewitt 1991; Patterson et al. 1992; Slaney and Lyon 1993; Strope and Alwan 2001). In this kind of temporal representation, the rate of change of the short-term spectral envelope reveals the modulation spectrum (Avendano and Hermansky 1997). In other words, the temporal representation is a transform away from directly expressing the same information as indicated by the processing of a modulation filterbank. The scale factor of the modeling of Carlyon and Shamma (2003) also shows this similarity. Retaining cross-channel timing information in their model, Carlyon and Shamma successfully simulated aspects of cross-spectral synchrony detection.

A processing scheme for source segregation that does not rely on place mapping of periodicity information was proposed by von der Malsburg and Schneider (1986). In their model, spectral segregation cued by cross-spectral envelope coherence is based on recognition of the synchronized temporal response across audio frequency. The scheme incorporates aspects of envelope detection by a synchronous receiver. Later work by Wang and colleagues (Wang 1996; Brown and Wang 2000; Hu and Wang 2004) has extended this correlation-based modeling approach to account for many aspects of stream segregation and spectral grouping by common AM. Sensitivity of cortical units to cross-spectral envelope coherence has been reported by Nelken et al. (1999) and Barbour and Wang (2002). Several recent physiological studies distinguish between the onset and sustained neural response to modulated stimuli (Sinex et al. 2002; Lu and Wang 2004; Wang et al. 2005). Wang et al. (2005) observed that at stimulus onset, a large population of cortical units responds with the subsequent response to ongoing modulation from a smaller neural subset. Though speculative, this distinction in neural response may contribute to the psychophysical results that indicate large effects of gating asynchrony contrasting with smaller effects attributable to coherence of ongoing modulation.

# 4. Information Processing of Envelope Modulation in Relation to Sound-Source Perception

In a most general sense, almost any study of AM perception could be considered in terms of information processing in relation to source. The following section discusses detection and discrimination results which most directly relate to source processing, and concludes with some reconsideration of aspects of cross-spectral masking.

Information is quantified by the level of stimulus uncertainty. Sheft and Yost (2001) measured AM detection using "dropped-cycle" modulators. The basis of the modulators was a sinusoidal function with a certain proportion of the cycles of the sinusoidal fluctuation eliminated over the course of the modulator duration. Two schemes were used for dropping modulator cycles. In the first, the interruption of modulator fluctuation followed a periodic pattern (e.g., only every fourth cycle of the underlying sinusoidal function was present). In the second scheme for dropping cycles, the same number of sinusoidal cycles was used as in each condition with periodic placement; however, on each presentation, the temporal locations of the sinusoidal fluctuations were randomized within the 500-ms modulator duration. With a WBN carrier, thresholds were comparable for the two schemes of omitting modulation cycles. That there was no effect of randomizing the temporal position of modulation cycles suggests that for AM, temporal uncertainty is not the appropriate metric of information-processing capability. Uncertainty in the spectral domain offers a better metric. Wright and Dai (1998) used a probe-signal procedure (see Sheft, Chapter 5) to measure detectability of unexpected rates of AM, finding that expectation had an effect only at low rates. The poor detection of low-rate AM when a high rate is expected can be accounted for by assuming that information is not uniformly integrated across the output of a modulation filterbank (Sheft and Yost 2001).

Sheft and Yost (2000) continued the study of AM detection with rate uncertainty, evaluating performance in the probe-signal paradigm, a 1-of-$m$ detection task, and with a joint detection–recognition procedure. Unlike the probe-signal paradigm, in which there is implicit uncertainty, uncertainty is explicit in the 1-of-$m$ detection task, with listeners aware that several potential signals will occur with equal probability during a block of trials. The joint detection–recognition procedure is simply 1-of-$m$ detection with a second response for signal identification. In all conditions, a WBN carrier was sinusoidally modulated at a rate of 4, 16, 64, or 256 Hz. Joint detection–recognition ability was measured for each pairwise combination of the four AM rates. On each trial there were two responses, one for detection and the other for recognition. Averaged results are shown in Figure 9.7. The open symbols at the left of each panel show mean detection ability for the two modulation rates used in that condition. Overall detection and recognition performance is shown by the first set of bars. The following two pairs of bars indicate performance levels on each task contingent on the response for the other being correct or incorrect, respectively. Unlike the results obtained in either probe-signal or 1-of-$m$ conditions, rate uncertainty led to only a slight decrement in detection ability, with little or no effect of the extent of the separation between the two rates used in a given condition. Recognition performance was always poorer than detection ability, with recognition near chance on trials in which the detection response was incorrect.

Starr et al. (1975) presented a theorem for predicting recognition accuracy based on detection performance in a joint detection–recognition task. When one of the AM rates was 4 Hz (Figure 9.7, three left panels), the theorem provides a reasonable prediction of recognition performance. For the three other pairwise combinations of AM rate, recognition performance was poorer, with the theorem failing to account for the extent of the decrement. The theorem is based on the assumptions of equal detectability and orthogonality of signals. The prediction by the theorem of recognition ability when the 4-Hz modulation rate was paired with one of the higher rates suggests independence of the 4-Hz versus higher rate processing. The inability of the theorem to predict recognition performance among the higher AM rates of 16, 64, and 256 Hz may then indicate correlation in the processing of these signals. The pattern of orthogonality across AM rate

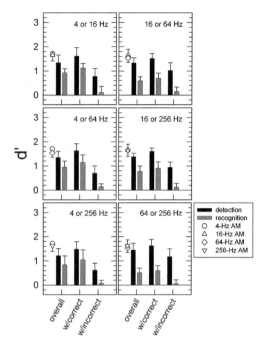

FIGURE 9.7. Mean performance of eight subjects in the joint detection–recognition task. The two AM rates used in each condition are indicated at the top of each panel. The open symbols indicate detection ability for the two rates used in that condition. The first set of bars shows overall detection and recognition performance. The following two pairs of bars indicate performance levels in each task contingent on the response for the other being correct or incorrect, respectively. Error bars represent one standard deviation from the mean. (From Sheft and Yost 2000.)

derived from the detection–recognition theorem suggests two rate-dependent cues for modulation detection, with one present at low rates and the other at higher rates. The physiological basis for this supposition is noted in Section 2. In terms of information processing of sound sources, the dominance of low-rate modulation in pattern discrimination and speech perception (see Section 5) may reflect the limited range of temporal coding of envelope fluctuation.

Envelope-pattern perception has been studied by several investigators using noise stimuli. Especially at narrow bandwidths, the stochastic fluctuation of noise stimuli is a dominant characteristic (see Section 2). Initial studies evaluated the ability of listeners to discriminate between reproducible and independent noise samples (Hanna 1984; Sheft and Yost 1994). Reproducible noise refers to presentation of the same noise sample across observation intervals of a trial. Though sharing a common bandwidth, CF, and duration, the modulation patterns of independent noises vary from sample to sample. Results indicate that with a stimulus duration of 400 ms or greater, discrimination ability improves with decreasing noise bandwidth, that is, as low-rate components begin to dominate the modulation spectrum. To eliminate potential cues due to either fine-structure modulation or variation in the long-term envelope amplitude spectrum across samples, Sheft and Yost (2002) evaluated discrimination of lowpass and bandpass noise modulators that differed only in terms of their phase spectra. Carriers were independent samples of WBN. Discrimination ability dropped rapidly with modulator bandwidth, exhibiting a lowpass characteristic notably more restricted than the lowpass TMTF for AM detection (Figure 9.8, triangles). That discrimination ability declined with bandwidth indicates both a loss of phase information with increasing rate and masking of the low-rate information by higher-rate modulation. If phase information was retained at higher rates, the function would show a positive slope with bandwidth. Without masking, the function would be horizontal. The function marked with circles in Figure 9.8 indicates ability to discriminate between a lowpass modulator and a ten-tone replica of the modulator constructed from the output of a simulated modulation filterbank. To generate the replica, the output of each filterbank channel was used to modulate a pure-tone carrier whose frequency was equal to the channel CF. Envelope-phase information was retained only for filterbank channels with CFs of 10 Hz or less. Except at the narrowest bandwidths, performance was poor, indicating that the ten-tone reconstruction was able to serve as a good replica of the stochastic modulators. Thus, along with inability of listeners to utilize the full range of phase information for envelope-pattern discrimination, spectral detail in the modulation domain does not appear to play a dominant role.

In the reproducible-noise discrimination task, use of concurrent noise bands that differ in CF allows for estimation of the extent of cross-spectral integration of envelope information. Sheft and Yost (1994) investigated multiband conditions in

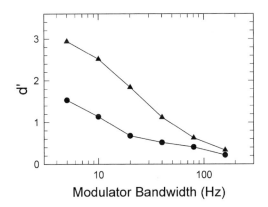

FIGURE 9.8. From Sheft and Yost 2002. In terms of raw or unadjusted cued-single-interval $d'$ values, mean envelope-discrimination ability of ten listeners as a function of the bandwidth of the lowpass-noise modulator. Triangles indicate conditions in which the modulator phase spectrum was randomized between samples. For the function indicated by circles, the discrimination was between a lowpass-noise modulator and a ten-tone replica of the modulator generated from the output of a simulated modulation filterbank.

which concurrent noise samples either shared a common bandwidth and pattern of envelope fluctuation, had a common bandwidth with asynchronous envelopes, or had different bandwidths. Performance in the multiband conditions was generally better than if the component bands were presented individually. Cross-spectral integration of modulation information was greater when the concurrent noise bands were synchronous rather than asynchronous. A similar result was reported by Bacon et al. (2002), who found that spectral integration for pure-tone detection was greater if the concurrent narrowband maskers were comodulated rather than incoherent.

Lutfi (1994) suggested that CMR may reflect a statistical constraint on masking due to the redundancy arising from the addition of cross-spectrally coherent modulation information. The following is an account of CMR based on statistical summation of information. When concerned with the combination of multiple sources of information, many psychophysical researchers have applied the established relationship that with independent sources, performance in terms of $d'$ is predicted to equal the square root of the sum of the squared $d'$ values associated with each source. More generally, Sorkin and Dai (1994) showed that the relationship between single- and multiband performances with a uniform interchannel correlation is

$$d'_n = \{(n\ \text{var}(d'))/(1-r) + (n(\text{mean}\ d')^2/(1-r+nr)\}^{1/2}, \qquad (9.2)$$

where $d'_n$ represents multiband performance, $d'$ is an array of the single-band performance levels for which variance and mean are calculated, $n$ is the number of channels and $r$ is the interchannel correlation.

Figure 9.9 shows, for the two-band case, three examples of predicted multiband performance as a function of interchannel correlation. The $d'$ associated with one band was always 1.6; for the other, $d'$ was 0.4, 0.8, or 1.6. With unequal single-band $d'$ values, the functions exhibit the unusual characteristic of

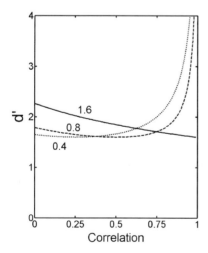

FIGURE 9.9. Predicted multiband performance in terms of $d'$ as a function of the correlation between the two individual information sources. The value of $d'$ associated with one band was always 1.6, and for the other was 0.4, 0.8, or 1.6. (Based on Sorkin and Dai 1994.)

predicted multiband performance increasing drastically with interchannel correlation. Assuming minimal spread of excitation from signal to masker channel, a CMR stimulus configuration represents this situation. A comodulated off-signal-frequency masker provides no information regarding the signal yet is highly correlated with the on-frequency band. Using Richards's (1987) values of cross-channel decorrelation associated with CMR and estimates of CMR psychometric-function slope from Moore et al. (1990), the relationship of equation (2) predicts roughly an 8-dB CMR. Application of Eq. (9.2) also predicts, as found empirically, a progressively diminishing increment in masking release with number of off-signal-frequency maskers. With multiple maskers, however, interchannel correlation is not constant, so that the application in this case is only approximate.

No assumption is made or needed concerning the actual mechanism in the above derivation. The intention is not to present an alternative to other hypotheses (e.g., "listening in the valleys"; see Section 3.5), but rather to illustrate compatibility of CMR with aspects of information processing. Buus et al. (1996) showed that either a correlation or signal-detection analysis can be applied to CMR results obtained with trial-to-trial random variation in signal level. Their analyses allowed for estimation of the time-varying weights listeners apply to the signal channel during the detection task. Findings were compatible with an explanation for CMR based on either "listening in the valleys" or cross-spectral envelope comparison if envelope compression is assumed.

As nonenergetic masking at the peripheral level with similarity of probe and masker a component, MDI exhibits characteristics used to describe informational masking (see Kidd, Mason, Richards, Gallun, and Durlach, Chapter 6). Sheft and Yost (2006) evaluated MDI in the context of informational masking, extending the definition of energetic masking to the modulation domain. Interactions among components of the modulation spectrum can be considered in some way analogous to interactions that occur within the audio-frequency spectrum, with both types of interaction representing energetic masking. To allow for consideration of auditory grouping and segregation effects, envelope slope and concurrency of modulation were manipulated. The task was to detect either 4- or 10-Hz SAM of the probe carrier. Masker modulators were either sinusoidal or complex waveforms defined by envelope slope and duty cycle (see Section 3.3). Across conditions with concurrent probe and masker modulation, significant departures from energetic masking in the modulation domain were obtained, with in some cases thresholds invariant over a 30-dB range. Conditions utilizing interrupted patterns of sinusoidal modulation compared concurrent versus sequential presentation of probe and masker envelope fluctuation. With either synchronous or asynchronous gating of the probe and masker carriers, concurrency of modulation had no effect. Additional conditions demonstrated that this result was not attributable to temporal masking of AM detection. The exception to energetic masking, along with the absence of an effect of modulation concurrency, was interpreted as indicative of a basis of MDI in informational masking. Commonly, stimulus uncertainty is associated with informational masking. The authors suggested that for MDI, the requisite uncertainty is present in the difficulty listeners have in associating near-threshold modulation with the appropriate carrier when several are present (Hall and Grose 1991). Effects of perceptual segregation on MDI are discussed in Section 3.5. In the context of an informational-masking basis of MDI, involvement of segregation enhances perceptual structure to reduce the uncertainty underlying the masking effect. To return to the argument of Garner (1974), enhancing structure increases the potential for information transmission.

## 5. Envelope Processing and Speech Perception

Direct evidence for a role of modulation processing in source-information extraction comes from speech studies (see Chapter 10 of this volume for consideration of speech as a sound source). While full discussion is beyond the scope of the present chapter, brief consideration is warranted. Several studies have performed information-transfer analysis on speech processed to retain only or primarily temporal-envelope cues (Van Tasell et al. 1987; Souza and Turner 1996; van der Horst et al. 1999; Apoux and Bacon 2004; Xu et al. 2005; Christiansen et al. 2006). That is, these studies formally measured the amount of information conveyed by the amplitude modulation of the speech signal. Analyses indicated that modulation information allowed for consonant recognition, with the extent of cross-spectral integration of modulation information dependent on phonetic feature and listening condition (e.g., in quiet or in noise).

Speech perception represents a high-level form of information processing. Houtgast and Steeneken (1973) proposed that for speech, information transmission is determined by the TMTF of the listening environment. Their basic premise is that speech intelligibility is closely related to the preservation of modulation information. Following this work, Plomp (1983) and Haggard (1985) argued that the modulation spectrum conveys crucial aspects of the speech signal. Their analyses indicate the importance of lower-rate modulation information for speech intelligibility. Speech-intelligibility studies in which the envelope modulation was filtered have confirmed the importance of low-rate modulation information (Van Tasell et al. 1987; Drullman et al. 1994; Hou and Pavlovic 1994; Shannon et al. 1995; Kanedera et al. 1997; Arai et al. 1999; van der Horst et al. 1999; Xu et al. 2005; Christiansen et al. 2006). Greenberg (1996; see also Greenberg and Kingsbury 1997) has further argued that the low-rate modulation spectrogram provides an invariant representation of speech. Invariance in this case refers to a consistent level of speech intelligibility despite large differences across talkers in the acoustic details of a specific speech message. Greenberg's argument is that by discarding spectrotemporal detail, the modulation spectrogram can reveal the stable structure of speech important for intelligibility.

Several groups have shown that by using the envelopes from four to five bands of filtered speech to modulate NBN carriers, good speech intelligibility can be achieved, though little spectral information is transmitted to the listener (e.g., Shannon et al. 1995; Dorman et al. 1997; Shannon et al. 1998; Loizou et al. 1999).

In practice, the number of channels required for asymptotic performance can vary with speech material and training, and is also higher in the presence of a masking noise (Friesen et al. 2001; Xu et al. 2005). Though affected by training, spectral transposition of envelope information degrades performance (Dorman et al. 1997; Shannon et al. 1998; Fu and Galvin 2003; Baskent and Shannon 2004). In this case, the listener receives speech-envelope information from a given audio-frequency region at a displaced spectral location. The performance decrement is consistent with the basic psychoacoustic finding that the ability to discriminate the pattern of envelope fluctuation diminishes with transposition of carrier frequency (Sheft and Yost 1992b; Takeuchi and Braida 1995). These results suggest that the processing of modulation information can depend on the spectral location of the carrier. A second aspect from envelope-speech studies relevant to sound-source processing concerns signal intelligibility in the presence of envelope-speech maskers. To the extent that envelope-speech perception represents solely temporal processing of the source (a case not true in limits in other than a single-band condition), masking of envelope speech by envelope speech indicates limitation on the ability to segregate sources while relying on only envelope cues, at least for a specific signal class. Analytically, Atlas (2003) demonstrated the separation of concurrent speech messages through modulation processing. Psychophysical results, however, show that the task is demanding; envelope-speech intelligibility is severely degraded in the presence of a competing envelope-speech message (Qin and Oxenham 2003; Stickney et al. 2004; Brungart et al. 2006).

## 6. Conclusions

The intent of the present chapter was to evaluate the role of envelope processing in sound-source perception in terms of both source determination and information processing. Regarding source determination, dominant effects of gating asynchrony have been reported across a wide range of stimulus configurations and experimental paradigms, with the effects of ongoing envelope modulation either relatively small, secondary to other aspects of source processing (e.g., stream segregation), or a reflection of a consequence of the modulation (e.g., modulation of S/N ratio) rather than an indication of direct involvement in auditory grouping. Apart from evaluation of envelope speech, far fewer studies have examined envelope processing in the context of information transmission from source to receiver. Despite this paucity, the inherent relationship between modulation and information transmission offers a basis for involvement. The central conclusion of this review is that though ongoing envelope modulation could play a role in both aspects of sound-source processing, current results and interpretation suggest greater potential involvement in information processing than source determination. This speculation is in part based on consideration of the efficacy of a system using the same stimulus attribute to bind and segregate a representation of the source as well as convey the source's message, and do this without leading to inaccuracy in either or both processes. Since modulation processing by definition is central to auditory information processing, a limited role for ongoing modulation is then suggested in source determination.

*Acknowledgments.* This work was supported by NIDCD Grant No. DC005423.

## *References*

Apoux F, Bacon SP (2004) Relative importance of temporal information in various frequency regions for consonant identification in quiet and in noise. J Acoust Soc Am 116:1671–1680.

Arai T, Greenberg S (1998) Speech intelligibility in the presence of cross-channel spectral asynchrony. In: Proceedings of the IEEE International Conference on Acoustics Speech Signal Process (ICASSP) 2:933–936.

Arai T, Pavel M, Hermansky H, Avendano C (1999) Syllable intelligibility for temporally filtered LPC cepstral trajectories. J Acoust Soc Am 105:2783–2791.

Atlas L (2003) Modulation spectral transforms—Applications to speech separation and modification. Technical Report, IEICE, Speech Dynamics by Ear, Eye, Mouth and Machine: An Interdisciplinary Workshop, Kyoto, Japan.

Atlas L, Janssen C (2005) Coherent modulation spectral filtering for single-channel music source separation. In: Proceedings of the IEEE International Conference Acoustics Speech Signal Process (ICASSP) 4:461–464.

Atlas L, Li Q, Thompson J (2004) Homomorphic modulation spectra. In: Proceeding of the IEEE International Conference Acoustics Speech Signal Process (ICASSP) 2:761–764.

Attias H, Schreiner CE (1997) Temporal low-order statistics of natural sounds. In: Mozer MC, Jordan MI, Petsche T (eds) Advances in Neural Information Processing Systems 9. Cambridge, MA: MIT Press, pp. 27–33.

Avendano C, Hermansky H (1997) On the properties of temporal processing for speech in adverse environments. In: Proceedings of WASPA'97,Mohonk, October 19–22, NY.

Bacon SP, Grantham DW (1989) Modulation masking: Effects of modulation frequency, depth, and phase. J Acoust Soc Am 85:2575–2580.

Bacon SP, Konrad DL (1993) Modulation detection interference under conditions favoring within- or across-channel processing. J Acoust Soc Am 93:1012–1022.

Bacon SP, Viemeister NF (1985) Temporal modulation transfer functions in normal-hearing and hearing-impaired listeners. Audiology 24:117–134.

Bacon SP, Grimault N, Lee J (2002) Spectral integration in bands of modulated or unmodulated noise. J Acoust Soc Am 112:219–226.

Bacon SP, Moore BCJ, Shailer MJ, Jorasz U (1995) Effects of combining maskers in modulation detection interference. J Acoust Soc Am 97:1847–1853.

Bacon SP, Lee J, Peterson DN, Rainey D (1997) Masking by modulated and unmodulated noise: Effects of bandwidth, modulation rate, signal frequency, and masker level. J Acoust Soc Am 101:1600–1610.

Barbour DL, Wang X (2002) Temporal coherence sensitivity in auditory cortex. J Neurophysiol 88:2684–2699.

Baskent D, Shannon RV (2004) Frequency-place compression and expansion in cochlear implant listeners. J Acoust Soc Am 116:3130–3140.

Beauvois MW, Meddis R (1996) Computer simulation of auditory stream segregation in alternating-tone sequences. J Acoust Soc Am 99:2270–2280.

Berg BG (1996) On the relation between comodulation masking release and temporal modulation transfer functions. J Acoust Soc Am 100:1013–1023.

Biebel UW, Langner G (2002) Evidence for interactions across frequency channels in the inferior colliculus of awake chinchilla. Hear Res 169:151–168.

Borrill SJ, Moore BCJ (2002) Evidence that comodulation detection interferences depend on within-channel mechanisms. J Acoust Soc Am 111:309–319.

Bregman AS (1990) Auditory Scene Analysis. Cambridge, MA: MIT Press.

Bregman AS, Pinker S (1978) Auditory streaming and the building of timbre. Can J Psychol 32:19–31.

Bregman AS, Abramson J, Doehring P, Darwin CJ (1985) Spectral integration based on common amplitude modulation. Percept Psychophys 37:483–493.

Bregman AS, Levitan R, Liao C (1990) Fusion of auditory components: Effects of the frequency of amplitude modulation. Percept Psychophys 47:68–73.

Brown GJ, Wang DL (2000) An oscillatory correlation framework for computational auditory scene analysis. In: Solla SA, Leen TK, Müller KR (eds) Advances in Neural Information Processing Systems 12. Cambridge, MA: MIT Press, pp. 747–753.

Bruckert L, Hermann M, Lorenzi C (2006) No adaptation in the amplitude modulation domain in trained listeners. J Acoust Soc Am 119:3542–3545.

Brungart DS, Iyer N, Simpson BD (2006) Monaural speech segregation using synthetic speech signals. J Acoust Soc Am 119:2327–2333.

Brunstrom JM, Roberts B (2001) Effects of asynchrony and ear of presentation on the pitch of mistuned partials in harmonic and frequency-shifted complex tones. J Acoust Soc Am 110:391–401.

Buss E, Richards VM (1996) The effects on comodulation masking release of systematic variations in on- and off- frequency masker modulation patterns. J Acoust Soc Am 99:3109–3118.

Buss E, Hall JW, Grose JH (2003) Effect of amplitude modulation coherence for masked speech signals filtered into narrow bands. J Acoust Soc Am 113:462–467.

Buss E, Hall JW, Grose JH (2004) Spectral integration of synchronous and asynchronous cues to consonant identification. J Acoust Soc Am 115:2278–2285.

Buus S (1985) Release from masking caused by envelope fluctuations. J. Acoust Soc Am 78:1958–1965.

Buus S, Zhang L, Florentine M (1996) Stimulus-driven, time-varying weights for comodulation masking release. J Acoust Soc Am 99:2288–2297.

Carlyon RP (1994) Detecting mistuning in the presence of synchronous and asynchronous interfering sounds. J Acoust Soc Am 95:2622–2630.

Carlyon RP, Shamna S (2003) An account of monaural phase sensitivity. J Acoust Soc Am 114:333–348.

Carlyon RP, Buus S, Florentine M (1989) Comodulation masking release for three types of modulator as a function of modulation rate. Hear Res 42:37–46.

Carrell TD, Opie JM (1992) The effect of amplitude comodulation on auditory object formation in sentence perception. Percept Psychophys 52:437–445.

Christiansen TU, Dau T, Greenberg S (2006) Spectro-temporal processing of speech—an information-theoretic framework. International Symposium on Hearing, Cloppenburg, August 18–23, Germany.

Ciocca V, Darwin CJ (1993) Effects of onset asynchrony on pitch perception: Adaptation of grouping? J Acoust Soc Am 93:2870–2878.

Cohen MF, Schubert ED (1987) The effect of cross-spectrum correlation on the detectability of a noise band. J Acoust Soc Am 81:721–723.

Cusack R, Deeks J, Aikman G, Carlyon RP (2004) Effects of location, frequency region, and time course of selective attention on auditory scene analysis. J Exp Psychol [Hum Percept Perform] 30:643–656.

Dai H, Green DM (1991) Effect of amplitude modulation on profile detection. J Acoust Soc Am 90:836–845.

Dannenbring GL, Bregman AS (1976) Stream segregation and the illusion of overlap. J Exp Psychol [Hum Percept Perform] 2:544–555.

Dannenbring GL, Bregman AS (1978) Streaming vs. fusion of sinusoidal components of complex tones. Percept Psychophys 24:369–376.

Darwin CJ (1981) Perceptual grouping of speech components differing in fundamental frequency and onset-time. Q J Exp Psychol 33A:185–207.

Darwin CJ (1992) Listening to two things at once. In: Schouten MEH (ed) The Auditory Processing of Speech. Berlin: Mouton de Gruyter, pp. 133–147.

Darwin CJ, Carlyon RP (1995) Auditory grouping. In: Moore BCJ (ed) Hearing. San Diego: Academic Press, pp. 387–424.

Darwin CJ, Ciocca V (1992) Grouping in pitch perception: Effects of onset asynchrony and ear of presentation of a mistuned component. J Acoust Soc Am 91:3381–3390.

Darwin CJ, Sutherland NS (1984) Grouping frequency components of vowels: When is a harmonic not a harmonic? Q J Exp Psychol 36A:193–208.

Darwin CJ, Ciocca V, Sandell GJ (1994) Effects of frequency and amplitude modulation on the pitch of a complex tone with a mistuned harmonic. J Acoust Soc Am 95:2631–2636.

Dau T, Verhey JL (1999) Modeling across-frequency processing of amplitude modulation. In: Dau T, Hohmann V, Kollmeier B (eds) Psychophysics, Physiology and Models of Hearing. Singapore: World Scientific, pp. 229–234.

Dau T, Kollmeier B, Kohlrausch A (1997) Modeling auditory processing of amplitude modulation. I. Detection and masking with narrow-band carriers. J Acoust Soc Am 102:2892–2905.

Dau T, Verhey J, Kohlrausch A (1999) Intrinsic envelope fluctuations and modulation-detection thresholds for narrow-band noise carriers. J Acoust Soc Am 106:2752–2760.

Dau T, Ewert SD, Oxenham AJ (2005) Effects of concurrent and sequential streaming in comodulation masking release. In: Pressnitzer D, de Cheveigné A, McAdams S, Collet L (eds) Auditory Signal Processing: Physiology, Psychoacoustics, and Models. New York: Springer, pp. 335–343.

Deligeorges S, Mountain DC (2004) Frequency, periodicity, and the ICC: A simple model for examining tonotopic and periodotopic axes in the inferior colliculus. Association for Research in Otolaryngology, Daytona Beach, FL.

Dorman MF, Loizou PC, Rainey D (1997) Simulating the effect of cochlear-implant electrode insertion depth on speech understanding. J Acoust Soc Am 102:2993–2996.

Drullman R, Festen JM, Plomp R (1994) Effect of reducing slow temporal modulations on speech reception. J Acoust Soc Am 95:2670–2680.

Durlach NI, Mason CR, Shinn-Cunningham BG, Abrogast TL, Colburn HS, Kidd G (2003) Informational masking: Counteracting the effects of stimulus uncertainty by decreasing target-masker similarity. J Acoust Soc Am 114:368–379.

Eddins DA (1993) Amplitude modulation detection of narrow-band noise: Effects of absolute bandwidth and frequency region. J Acoust Soc Am 93:470–479.

Eddins DA, Wright BA (1994) Comodulation masking release for single and multiple rates of envelope fluctuation. J Acoust Soc Am 96:3432–3442.

Ewert SD, Dau T (2000) Characterizing frequency selectivity for envelope fluctuations. J Acoust Soc Am 108:1181–1196.

Ewert SD, Verhey JL, Dau T (2002) Spectro-temporal processing in the envelope-frequency domain. J Acoust Soc Am 112:2921–2931.

Fantini DA, Moore BCJ (1994) Profile analysis and comodulation detection differences using narrow bands of noise and their relation to comodulation masking release. J Acoust Soc Am 95:2180–2191.

Fantini DA, Moore BCJ, Schooneveldt GP (1993) Comodulation masking release as a function of type of signal, gated or continuous masking, monaural or dichotic presentation of flanking bands, and center frequency. J Acoust Soc Am 93:2106–2115.

Friesen LM, Shannon RV, Baskent D, Wang X (2001) Speech recognition in noise as a function of the number of spectral channels: Comparison of acoustic hearing and cochlear implants. J Acoust Soc Am 110:1150–1163.

Fu QJ, Galvin JJ (2001) Recognition of spectrally asynchronous speech by normal-hearing listeners and Nucleus-22 cochlear implant users. J Acoust Soc Am 109:1166–1172.

Fu QJ, Galvin JJ (2003) The effects of short-term training for spectrally mismatched noise-band speech. J Acoust Soc Am 113:1065–1072.

Füllgrabe C, Moore BCJ, Demany L, Ewert SD, Sheft S, Lorenzi C (2005) Modulation masking produced by second-order modulators. J Acoust Soc Am 117:2158-2168.

Füllgrabe C, Berthommier F, Lorenzi C (2006) Masking release for consonant features in temporally fluctuating background noise. Hear Res 211:74–84.

Garner WR (1974) The Processing of Information and Structure. Potomac, MD: Lawrence Erlbaum.

Garner WR, Hake HW (1951) The amount of information in absolute judgments. Psychol Rev 58:446–459.

Ghitza O (2001) On the upper cutoff frequency of the auditory critical-band envelope detectors in the context of speech perception. J Acoust Soc Am 110:1628–1640.

Gilbert G, Lorenzi C (2006) The ability of listeners to use recovered envelope cues from speech fine structure. J Acoust Soc Am 119:2438–2444.

Giraud A, Lorenzi C, Ashburner J, Wable J, Johnsrude I, Frackowiak R, Kleinschmidt A (2000) Representation of the temporal envelope of sounds in the human brain. J Neurophysiol 84:1588–1598.

Gockel H, Carlyon RP, Deeks JM (2002) Effect of modulator asynchrony of sinusoidal and noise modulators on frequency and amplitude modulation detection interference. J Acoust Soc Am 112:2975–2984.

Goebl W, Parncutt R (2003) Asynchrony versus intensity as cues for melody perception in chords and real music. In: Proceedings of the 5th Triennial Conference of the European Society for Cognitve Science and Music (ESCOM5), Hanover, September 8–13, Germany, pp. 376–380

Gordon JW (1987) The perceptual attack time of musical tones. J Acoust Soc Am 82:88–105.

Gordon PC (2000) Masking protection in the perception of auditory objects. Speech Commun 30:197–206.

Green DM (1993) Auditory intensity discrimination. In: Yost WA, Popper AN, Fay RR (eds) Human Psychophysics. New York: Springer-Verlag, pp. 13–55.

Green DM, Dai H (1992) Temporal relations in profile comparison. In: Cazals Y, Horner K, Demany L (eds) Auditory Physiology and Perception. Oxford: Pergamon Press, pp. 471–478.

Green DM, Nguyen QT (1988) Profile analysis: Detecting dynamic spectral changes. Hear Res 32:147–164.

Greenberg S (1996) Understanding speech understanding: Towards a unified theory of speech perception. In: Proceedings of the ESCA Workshop on Auditory Basis of Speech Perception, Keele, England, pp. 1–8.

Greenberg S, Arai T (2001) The relation between speech intelligibility and the complex modulation spectrum. 7th European Conf Speech Comm Tech (Eurospeech-2001), pp. 473–476.

Greenberg S, Kingsbury BED (1997) The modulation spectrogram: In pursuit of an invariant representation of speech. In: Proceedings of the IEEE Int Conf Acoustics Speech Signal Process (ICASSP) 3:1647–1650.

Greenberg S, Arai T, Silipo R (1998) Speech intelligibility derived from exceedingly sparse spectral information. In: Proceedings of International Conference on Spoken Language Process (ICSLP) 6:2803–2806.

Grimault N, Bacon SP, Micheyl C (2002) Auditory stream segregation on the basis of amplitude-modulation rate. J Acoust Soc Am 111:1340–1348.

Grose JH, Hall JW (1993) Comodulation masking release: Is comodulation sufficient? J Acoust Soc Am 93:2896–2902.

Grose JH, Hall JW (1996) Across-frequency processing of multiple modulation patterns. J Acoust Soc Am 99:534–541.

Grose JH, Hall JW, Buss E (2001) Signal detection in maskers with multiple modulations. In: Breebaart DJ, Houtsma AJM, Kohlrausch A, Prijs VF, Schoonhoven R (eds) Physiological and Psychophysical Bases of Auditory Function. Maastricht NL: Shaker Publishing, pp. 258–265.

Grose JH, Hall JW, Buss E, Hatch DR (2005) Detection of spectrally complex signals in comodulated maskers: Effect of temporal fringe. J Acoust Soc Am 118:3774–3782.

Gustafsson HÅ, Arlinger SD (1994) Masking of speech by amplitude-modulated noise. J Acoust Soc Am 95:518–529.

Haggard M (1985) Temporal patterning in speech: The implications of temporal resolution and signal-processing. In: Michelsen A (ed) Time Resolution in Auditory Systems. Berlin: Springer-Verlag, pp. 215–237.

Hall JW, Grose JH (1988) Comodulation masking release: Evidence for multiple cues. J Acoust Soc Am 84:1669–1675.

Hall JW, Grose JH (1990) Comodulation masking release and auditory grouping. J Acoust Soc Am 88:119–125.

Hall JW, Grose JH (1991) Some effects of auditory grouping factors on modulation detection interference (MDI). J Acoust Soc Am 90:3028–3035.

Hall JW, Cokely JA, Grose JH (1988) Combined monaural and binaural masking release. J Acoust Soc Am 83:1839–1845.

Hall JW, Grose JH, Mendoza L (1995) Across-channel processes in masking. In: Moore BCJ (ed), Hearing. San Diego: Academic Press, pp. 243–266.

Hall JW, Grose JH, Hatch DR (1996) Effects of masker gating for signal detection in unmodulated and modulated bandlimited noise. J Acoust Soc Am 100:2365–2372.

Hall JW, Buss E, Grose JH (2006) Comodulation detection differences for fixed-frequency and roved-frequency maskers. J Acoust Soc Am 119:1021–1028.

Hanna TE (1984) Discrimination of reproducible noise as a function of bandwidth and duration. Percept Psychophys 36:409–416.

Hartmann WM (1988) Pitch perception and the segregation and integration of auditory entities. In: Edelman GM, Gall WE, Cowan WM (eds) Auditory Function: Neurobiological Bases of Hearing. New York: John Wiley & Sons, pp. 623–645.

Hartmann WM (1998) Signals, Sound, and Sensation. New York: Springer, pp. 412–429.

Hartman WM, Johnson D (1991) Stream segregation and peripheral channeling. Music Percept 9:155–184.

Hatch DR, Arné BC, Hall JW (1995) Comodulation masking release (CMR): Effects of gating as a function of number of flanking bands and masker bandwidth. J Acoust Soc Am 97:3768–3774.

Hawkins S (1995) Arguments for a nonsegmental view of speech perception. In: Proceedings of the XIIIth International Congress on Phonetic Sciences 3:18–25.

Heil P, Schulze H, Langner G (1995) Ontogenetic development of periodicity coding in the inferior colliculus of the Mongolian gerbil. Aud Neurosci 1:363–383.

Hill NI, Bailey PJ (1997) Profile analysis with an asynchronous target: Evidence for auditory grouping. J Acoust Soc Am 102:477–481.

Hou Z, Pavlovic CV (1994) Effects of temporal smearing on temporal resolution, frequency selectivity, and speech intelligibility. J Acoust Soc Am 96:1325–1340.

Houtgast T (1989) Frequency selectivity in amplitude-modulation detection. J Acoust Soc Am 85:1676–1680.

Houtgast T, Steeneken HJM (1973) The modulation transfer function in room acoustics as a predictor of speech intelligibility. Acustica 28:66–73.

Howard-Jones PA, Rosen S (1993) Uncomodulated glimpsing in "checkerboard" noise. J Acoust Soc Am 93:2915–2922.

Hu G, Wang D (2004) Monaural speech segregation based on pitch tracking and amplitude modulation. IEEE Trans Neural Net 15:1135–1150.

Hukin RW, Darwin CJ (1995) Comparison of the effect of onset asynchrony on auditory grouping in pitch matching and vowel identification. Percept Psychophys 57:191–196.

Iverson P (1995) Auditory stream segregation by musical timbre: Effects of static and dynamic acoustic attributes. J Exp Psychol [Hum Percept Perform] 21:751–763.

Joris PX, Schreiner CE, Rees A (2004) Neural processing of amplitude-modulated sounds. Physiol Rev 84:541–577.

Kanedera N, Arai T, Hermansky H, Pavel M (1997) On the importance of various modulation frequencies for speech recognition. In: Proceedings of Eurospeech '97: Rhodes, September 22–25, Greece. pp. 1079–1082.

Kay RH (1982) Hearing of modulation in sounds. Physiol Rev 62:894–975.

Kidd G, Mason CR, Abrogast TL (2002) Similarity, uncertainty, and masking in the identification on nonspeech auditory patterns. J Acoust Soc Am 111:1367–1376.

Kidd G, Mason CR, Richards VM (2003) Multiple bursts, multiple looks, and stream coherence in the release from informational masking. J Acoust Soc Am 114:2835–2845.

Kohlrausch A, Fassel R, Dau T (2000) The influence of carrier level and frequency on modulation and beat-detection thresholds for sinusoidal carriers. J Acoust Soc Am 108:723–734.

Krishna BS, Semple MN (2000) Auditory temporal processing : Responses to sinusoidally amplitude-modulated tones in the inferior colliculus. J Physiol 84:255–273.

Kubovy M (1987) Concurrent pitch segregation. In: Yost WA, Watson CS (eds) Auditory Processing of Complex Sounds. Hillsdale, NJ: Lawrence Erlbaum, pp. 299–313.

Kubovy M, Daniel JE (1983) Pitch segregation by interaural phase, by momentary amplitude disparity, and by monaural phase. J Audio Eng Soc 31:630–635.

Kwon BJ, Turner CW (2001) Consonant identification under maskers with sinusoidal modulation: Masking release or modulation interference? J Acoust Soc Am 110:1130–1140.

Langner G, Schreiner C, Albert M (1992) Tonotopy and periodotopy in the auditory midbrain of cat and guinea fowl. In: Cazals Y, Horner K, Demany L (eds) Auditory Physiology and Perception. Oxford: Pergamon Press, pp. 241–248.

Langner G, Sams M, Heil P, Schulze H (1997) Frequency and periodicity are represented by orthogonal maps in the human auditory cortex: Evidence from magnetoencephalography. J Comp Physiol A 181:665–676.

Langner G, Albert M, Briede T (2002) Temporal and spatial coding of periodicity information in the inferior colliculus of awake chinchilla (Chinchilla laniger). Hear Res 168:110–130.

Lawson JL, Uhlenbeck GE (1950) Threshold Signals, Radiation Laboratory Series, Vol 24. New York: McGraw-Hill.

Lentz JJ, Leek MR, Molis MR (2004) The effect of onset asynchrony on profile analysis by normal-hearing and hearing-impaired listeners. J Acoust Soc Am 116:2289–2297.

Loizou PC, Dorman M, Tu Z (1999) On the number of channels needed to understand speech. J Acoust Soc Am 106:2097–2103.

Lorenzi C, Simpson MIG, Millman RE, Griffiths TD, Woods WP, Rees A, Green GGR (2001) Second-order modulation detection thresholds for pure-tone and narrow-band noise carriers. J Acoust Soc Am 110:2470–2478.

Loughlin PJ, Tacer B (1996) On the amplitude- and frequency-modulation decomposition of signals. J Acoust Soc Am 100:1594–1601.

Lu T, Wang X (2004) Information content of auditory cortical responses to time-varying acoustic stimuli. J Neurophysiol 91:301–313.

Lu T, Liang L, Wang X (2001) Temporal and rate representations of time-varying signals in the auditory cortex of awake primates. Nat Neurosci 4:1131–1138.

Lutfi RA (1990) Informational processing of complex sounds. II. Cross-dimensional analysis. J Acoust Soc Am 87:2141–2148.

Lutfi RA (1994) Discrimination of random, time-varying spectra with statistical constraints. J Acoust Soc Am 95:1490–1500.

McCabe SL, Denham MJ (1997) A model of auditory streaming. J Acoust Soc Am 101:1611–1621.

McFadden D (1975) Beat-like interaction between periodic waveforms. J Acoust Soc Am 57:983.

McFadden D (1986) Comodulation masking release: Effects of varying the level, duration, and time delay of the cue band. J Acoust Soc Am 80:1658–1667.

McFadden D, Wright BA (1990) Temporal decline of masking and comodulation detection differences. J Acoust Soc Am 88:711–724.

McFadden D, Wright BA (1992) Temporal decline of masking and comodulation masking release. J Acoust Soc Am 92:144–156.

McNally KA, Handel S (1977) Effect of element composition of streaming and the ordering of repeating sequences. J Exp Psychol Hum Percept Perf 3:451–460.

Meddis R, Hewitt MJ (1991) Virtual pitch and phase sensitivity of a computer model of the auditory periphery. I: Pitch identification. J Acoust Soc Am 89:2866–2882.

Mendoza L, Hall JW, Grose JH (1995) Modulation detection interference using random and sinusoidal amplitude modulation. J Acoust Soc Am 97:2487–2492.

Miller GA, Licklider JCR (1950) The intelligibility of interrupted speech. J Acoust Soc Am 22:167–173.

Miyoshi T, Goto T, Doi T Ishida T, Arai T, Murahara Y (2004) Modulation cepstrum discriminating between speech and environmental noise. Acoust Sci Tech 25: 66–69.

Moore BCJ, Alcántara JI (1996) Vowel identification based on amplitude modulation. J Acoust Soc Am 99:2332–2343.

Moore BCJ, Borrill SJ (2002) Tests of a within-channel account of comodulation detection differences. J Acoust Soc Am 112:2099–2109.

Moore BCJ, Emmerich DS (1990) Monaural envelope correlation perception, revisited: Effects of bandwidth, frequency separation, duration, and relative level of the noise bands. J Acoust Soc Am 87:2628–2633.

Moore BCJ, Gockel H (2002) Factors influencing sequential stream segregation. Acta Acust 88:320–332.

Moore BCJ, Jorasz U (1992) Detection of changes in modulation depth of a target sound in the presence of other modulated sounds. J Acoust Soc Am 91:1051–1061.

Moore BCJ, Schooneveldt GP (1990) Comodulation masking release as a function of bandwidth and time delay between on-frequency and flanking-band maskers. J Acoust Soc Am 88:725–731.

Moore BCJ, Sek A (2000) Effect of relative phase and frequency spacing on the detection of three-component amplitude modulation. J Acoust Soc Am 108:2337–2344.

Moore BCJ, Shailer MJ (1994) Effects of harmonicity, modulator phase, and number of masker components on modulation discrimination interference. J Acoust Soc Am 95:3555–3560.

Moore BCJ, Hall JW, Grose JH, Schooneveldt GP (1990) Some factors affecting the magnitude of comodulation masking release. J Acoust Soc Am 88:1694–1702.

Moore BCJ, Glasberg BR, Gaunt T, Child T (1991) Across-channel masking of changes in modulation depth for amplitude- and frequency-modulated signals. Q J Exp Psychol 43A:327–347.

Moore BCJ, Sek A, Shailer MJ (1995) Modulation discrimination interference for narrow-band noise modulators. J Acoust Soc Am 97:2493–2497.

Nelken I, Rotman Y, Yosef OB (1999) Responses of auditory-cortex neurons to structural features of natural sounds. Nature 397:154–157.

Nelson PB, Jin SH, Carney AE, Nelson DA (2003) Understanding speech in modulated interference: Cochlear implant users and normal-hearing listeners. J Acoust Soc Am 113:961–968.

Oxenham AJ, Dau T (2001) Modulation detection interference: Effects of concurrent and sequential streaming. J Acoust Soc Am 110:402–408.

Pastore RE, Harris LB, Kaplan JK (1982) Temporal order identification: Some parameter dependencies. J Acoust Soc Am 71:430–436.

Patterson JH, Green DM (1970) Discrimination of transient signals having identical energy spectra. J Acoust Soc Am 48:894–905.

Patterson RD, Robinson K, Holdsworth J, McKeown D, Zhang C, Allerhand M (1992) Complex sounds and auditory images. In: Cazals Y, Horner K, Demany L (eds) Auditory Physiology and Perception. Oxford: Pergamon Press, pp. 429–446.

Plomp R (1983) The role of modulation in hearing. In: Klinke R, Hartmann R (eds) Hearing—Physiological Bases and Psychophysics. Berlin: Springer-Verlag, pp. 270–276.

Pollack I (1952) The information of elementary auditory displays. J Acoust Soc Am 24:745–749.

Pressnitzer D, McAdams S (1999) An effect of the coherence between envelopes across frequency regions on the perception of roughness. In: Dau T, Hohmann V, Kollmeier B (eds) Psychophysics, Physiology and Models of Hearing. Singapore: World Scientific, pp. 105–108.

Qin MK, Oxenham AJ (2003) Effects of simulated cochlear-implant processing on speech reception in fluctuating maskers. J Acoust Soc Am 114:446–454.

Rappold PW, Mendoza L, Collins MJ (1993) Measuring the strength of auditory fusion for synchronously and nonsynchronously fluctuating narrow-band noise pairs. J Acoust Soc Am 93:1196–1199.

Rasch RA (1978) The perception of simultaneous notes such as in polyphonic music. Acust 40:21–33.

Rasch RA (1979) Synchronization in performed ensemble music. Acustica 43:121–131.

Repp BH (1996) Patterns of note onset asynchronies in expressive piano performance. J Acoust Soc Am 100:3917–3932.

Richards VM (1987) Monaural envelope correlation perception. J Acoust Soc Am 82:1621–1630.

Richards VM (1990) The role of single-channel cues in synchrony perception: The summed waveform. J Acoust Soc Am 88:786–795.

Richards VM, Buss E, Tian L. (1997) Effects of modulator phase for comodulation masking release and modulation detection interference. J Acoust Soc Am 102:468–476.

Roberts B, Holmes SD (2006) Asynchrony and the grouping of vowel components: Captor tones revisited. J Acoust Soc Am 119:2905–2918.

Roberts B, Glasberg BR, Moore BCJ (2002) Primitive stream segregation of tone sequences without differences in fundamental frequency or passband. J Acoust Soc Am 112:2074–2085.

Schooneveldt GP, Moore BCJ (1987) Comodulation masking release (CMR): Effects of signal frequency, flanking-band frequency, masker bandwidth, flanking-band level, and monotic versus dichotic presentation of the flanking band. J Acoust Soc Am 82:1944–1956.

Schooneveldt GP, Moore BCJ (1989) Comodulation masking release (CMR) as a function of masker bandwidth, modulator bandwidth, and signal duration. J Acoust Soc Am 85:273–281.

Schreiner CE, Langner G (1988) Periodicity coding in the inferior colliculus of the cat. II. Topographical organization. J Neurophysiol 60:1823–1840.

Schulze H, Langner G (1997) Periodicity coding in the primary auditory cortex of the Mongolian gerbil (Meriones unguiculatus): Two different coding strategies for pitch and rhythm? J Comp Physiol A 181:651–663.

Shailer MJ, Moore BCJ (1993) Effects of modulation rate and rate of envelope change on modulation discrimination interference. J Acoust Soc Am 94:3138–3143.

Shannon RV, Zeng FG, Kamath V, Wygonski J, Ekelid M (1995) Speech recognition with primarily temporal cues. Science 270:303–304.

Shannon RV, Zeng FG, Wygonski J (1998) Speech recognition with altered spectral distribution of envelope cues. J Acoust Soc Am 104:2467–2476.

Sheft S (2000) Adaptation to amplitude modulation. Association for Research in Otolaryngology, St. Petersburg, FL.

Sheft S, Yost WA (1990) Cued envelope-correlation detection. J Acoust Soc Am (Suppl 1) 88:S145.

Sheft S, Yost WA (1992a) Concurrent pitch segregation based on AM. J Acoust Soc Am 92:2361.

Sheft S, Yost WA (1992b) Spectral transposition of envelope modulation. J Acoust Soc Am 91:2333.

Sheft S, Yost WA (1994) Reproducible noise discrimination with concurrent narrowband noises. J Acoust Soc Am 95:2964.

Sheft S, Yost WA (1997a) Binaural modulation detection interference. J Acoust Soc Am 102:1791–1798.

Sheft S, Yost WA (1997b) Modulation detection interference with two-component masker modulators. J Acoust Soc Am 102:1106–1112.

Sheft S, Yost WA (2000) Joint detection–recognition of amplitude modulation. J Acoust Soc Am 107:2880.

Sheft S, Yost WA (2001) AM detection with interrupted modulation. In: Breebaart DJ, Houtsma AJM, Kohlrausch A, Prijs VF, Schoonhoven R (eds) Physiological and Psychophysical Bases of Auditory Function. Maastricht NL: Shaker Publishing, pp. 290–297.

Sheft S, Yost WA (2002) Envelope phase-spectrum discrimination. AFRL Prog Rep 2, contract no SPO700-98-D-4002.

Sheft S,Yost WA (2004) Envelope-phase discrimination. Assoc Res Otolaryngol, Daytona Beach, Florida.

Sheft S, Yost WA (2005) Modulation masking of synchrony detection. Assoc Res Otolaryngol, New Orleans, Louisiana.

Sheft S, Yost WA (2006) Modulation detection interference as informational masking. In: International Symposium on Hearing, Cloppenburg, August 18–23, Germany.

Singh PG, Bregman AS (1997) The influence of different timbre attributes on the perceptual segregation of complex-tone sequences. J Acoust Soc Am 102:1943–1952.

Singh NC, Theunissen FE (2003) Modulation spectra of natural sounds and ethological theories of auditory processing. J Acoust Soc Am 114:3394–3411.

Sinex DG, Henderson J, Li H, Chen GD (2002) Responses of chinchilla inferior colliculus neurons to amplitude-modulated tones with different envelopes. J Assoc Res Otolaryngol 3:390–402.

Slaney M, Lyon RF (1993) On the importance of time—a temporal representation of sound. In: Cooke M, Beet S, Crawford M (eds) Visual Representations of Speech Signals. New York: John Wiley & Sons, pp. 95–116.

Smith ZM, Delgutte B, Oxenham AJ (2002) Chimaeric sounds reveal dichotomies in auditory perception. Nature 416:87–90.

Sorkin RD, Dai H (1994) Signal detection analysis of the ideal group. Org Behav Hum Decis Process 60:1–13.

Souza PE, Turner CW (1996) Effect of single-channel compression on temporal speech information. J Speech Hear Res 39:901–911.

Starr SJ, Metz CE, Lusted LB, Goodenough DJ (1975) Visual detection and localization of radiographic images. Radiol 116:533–538.

Stickney GS, Zeng FG, Litovsky R, Assmann P (2004) Cochlear implant speech recognition with speech maskers. J Acoust Soc Am 116:1081–1091.

Strickland EA, Viemeister NF (1996) Cues for discrimination of envelopes. J Acoust Soc Am 99:3638–3646.

Strickland EA, Viemeister NF (1997) The effects of frequency region and bandwidth on the temporal modulation transfer function. J Acoust Soc Am 102:1799–1810.

Strickland EA, Viemeister NF, Fantini DA, Garrison MA (1989) Within- versus cross-channel mechanisms in detection of envelope phase disparity. J Acoust Soc Am 86:2160–2166.

Strope BP, Alwan AA (2001) Modeling the perception of pitch-rate amplitude modulation in noise. In: Greenberg S, Slaney M (eds) Computational Models of Auditory Function. Amsterdam: IOS Press, pp. 315–327.

Summerfield Q, Culling JF (1992) Auditory segregation of competing voices: absence of effects of FM or AM coherence. Phil Trans R Soc Lond B 336:357–366.

Summerfield Q, Haggard M, Foster J, Gray S (1984) Perceiving vowels from uniform spectra: Phonetic exploration of an auditory aftereffect. Percept Psychophys 35:203–213.

Summerfield Q, Sidwell A, Nelson T (1987) Auditory enhancement of changes in spectral amplitude. J Acoust Soc Am 81:700–708.

Takeuchi AH, Braida LD (1995) Effect of frequency transposition on the discrimination of amplitude envelope patterns. J Acoust Soc Am 97:453–460.

Tansley BW, Suffield JB (1983) Time course of adaptation and recovery of channels selectively sensitive to frequency and amplitude modulation. J Acoust Soc Am 74: 765–775.

Terhardt E (1974) On the perception of periodic sound fluctuations (roughness). Acust 30:201–213.

van der Horst R, Leeuw AR, Dreschler WA (1999) Importance of temporal-envelope cues in consonant recognition. J Acoust Soc Am 105:1801–1809.

van Noorden LPAS (1975) Temporal coherence in the perception of tone sequences. PhD thesis, Eindhoven: Eindhoven Univ Tech.

van de Par S, Kohlrausch A (1998) Comparison of monaural (CMR) and binaural (BMLD) masking release. J Acoust Soc Am 103:1573–1579.

Van Tasell DJ, Soli SD, Kirby VM, Widin GP (1987) Speech waveform envelope cues for consonant recognition. J Acoust Soc Am 82:1152–1161.

Verhey JL, Dau T, Kollmeier B (1999) Within-channel cues in comodulation masking release (CMR): Experiments and model predictions using a modulation-filterbank model. J Acoust Soc Am 106:2733–2745.

Viemeister NF (1979) Temporal modulation transfer functions based upon modulation thresholds. J Acoust Soc Am 66:1364–1380.

Viemeister NF, Bacon SP (1982) Forward masking by enhanced components in harmonic complexes. J Acoust Soc Am 71:1502–1507.

Viemeister NF, Stellmack MA, Byrne AJ (2005) The role of temporal structure in envelope processing. In: Pressnitzer D, de Cheveigné A, McAdams S, Collet L (eds) Auditory Signal Processing. New York, Springer, pp. 221–229.

Vogel A (1974) Roughness and its relation to the time-pattern of psychoaccoustical excitation. In: Zwiker E, Terhardt E (eds) Facts and Modeling in Hearing. New York: Springer-Verlag, pp. 241–250.

von der Malsburg C, Schneider W (1986) A neural cocktail-party processor. Biol Cybern 54:29–40.

Vos J, Rasch R (1981) The perceptual onset of musical tones. Percept Psychophys 29:323–335.

Wakefield GH (1987) Detection of envelope phase disparity. J Acoust Soc Am (Suppl 1) 82:S34.

Wakefield GH, Viemeister NF (1985) Temporal interactions between pure tones and amplitude-modulated noise. J Acoust Soc Am 77:1535–1542.

Wang D (1996) Primitive auditory segregation based on oscillatory correlation. Cogn Sci 20:409–456.

Wang X, Lu T, Liang L (2003) Cortical processing of temporal modulations. Speech Comm 41:107–121.

Wang X, Lu T, Snider RK, Liang L (2005) Sustained firing in auditory cortex evoked by preferred stimuli. Nature 435:341–346.

Watson CS (2005) Some comments on informational masking. Acta Acust 91:502–512.

Wilson JP (1970) An auditory after-image. In: Plomp R, Smoorenberg GF (eds) Frequency Analysis and Periodicity Detection in Hearing. Leiden: AW Sijthoff, pp. 303–318.

Wojtczak M, Viemeister NF (2005) Forward masking of amplitude modulation: Basic characteristics. J Acoust Soc Am 118:3198–3210.

Wright BA (1990) Comodulation detection differences with multiple signal bands. J Acoust Soc Am 87:292–303.

Wright BA, Dai H (1998) Detection of sinusoidal amplitude modulation at unexpected rates. J Acoust Soc Am 104:2991–2996.

Xu L, Pfingst BE (2003) Relative importance of temporal envelope and fine structure in lexical-tone perception. J Acoust Soc Am 114:3024–3027.

Xu L, Thompson CS, Pfingst BE (2005) Relative contributions of spectral and temporal cues for phoneme recognition. J Acoust Soc Am 117:3255–3267.

Yost WA (1991) Auditory image perception and analysis: The basis for hearing. Hear Res 56:8–18.

Yost WA, Sheft S (1989) Across-critical-band processing of amplitude-modulated tones. J Acoust Soc Am 85:848–857.

Yost WA, Sheft S (1990) A comparison among three measures of cross-spectral processing of amplitude modulation with tonal signals. J Acoust Soc Am 87:897–900.

Yost WA, Sheft S (1993) Auditory perception. In: Yost WA, Popper AN, Fay RR (eds) Human Psychophysics. New York: Springer-Verlag, pp. 193–236.

Yost WA, Sheft S (1994) Modulation detection interference: Across-frequency processing and auditory grouping. Hear Res 79:48–58.

Yost WA, Sheft S (1997) Temporal modulation transfer functions for tonal stimuli: Gated versus continuous conditions. Aud Neurosci 3:401–414.

Yost WA, Sheft S, Opie J (1989) Modulation interference in detection and discrimination of amplitude modulation. J Acoust Soc Am 86:2138–2147.

Yost WA, Dye RH, Sheft S (1995) The synthetic-analytic listening task for modulated signals. J Acoust Soc Am 98:652–655.

Zeng FG, Nie K, Liu S, Stickney G, Del Rio E, Kong YY, Chen H (2004) On the dichotomy in auditory perception between temporal envelope and fine structure cues. J Acoust Soc Am 116:1351–1354.

Zera J, Green DM (1993) Detecting temporal onset and offset asynchrony in multicomponent complexes. J Acoust Soc Am 93:1038–1052.

# 10
# Speech as a Sound Source

Andrew J. Lotto and Sarah C. Sullivan

## 1. What Is the Sound Source for Speech?

Speech is one of the most salient and important sound sources for the human listener. As with many other natural sound sources, a listener can localize the direction from which a signal originated and can even determine some of the physical characteristics of the sound-producing object and event. But the real value of the speech signal lies not just in where the sound came from or by whom the sound was created, but in the linguistic message that it carries. The intended message of the speaker is the real sound source of speech, and the ability of listeners to apprehend this message in spite of varying talker and communication characteristics is the focus of this chapter.

This is not to say that the "where" and "by whom" questions related to the speech sound source are inconsequential. Localizing a speaker can be important for the segregation of the speech stream from competing speakers or noise (see Chapter 8 in this volume). Given the continuously varying nature of the speech signal, the segregation of speech from a particular talker is nontrivial, and there is a long history of research into this problem (see Chapters 5 and 7 of this volume). In addition to perceiving the location of a speaker, listeners can learn quite a bit about the speaker from his or her productions. The information in the signal that specifies characteristics of the speaker, such as gender, size, or affect, is referred to as *indexical* information. The indexical information is similar to the shape, size, and material composition information for other sound-producing objects/events (see Chapters 2 and 3 in this volume). It is clear that listeners can identify particular talkers from their speech (e.g., Bachorowski and Owren 1999), and this knowledge can color the interpretation of the incoming message. In the end, however, when one refers to *speech perception*, the task that comes to mind is the determination of the linguistic message intended by the speaker.[1]

Even if one accepts that the true source perception problem for speech is identification of the message carried by the signal, it is still unclear what the unit

---

[1] It will become clear in the remainder of this chapter that talker-specific characteristics play a role in speech perception. Here we are describing indexical identification as an outcome of sound source perception.

of identification is. Words may seem to be a reasonable candidate, since they are the smallest units carrying semantic information. Some theorists have suggested that coherent theories of speech perception can be developed from the assumption of the word as the fundamental unit (e.g., Lindblom et al. 1984; Stevens 1986, 2002; Kluender and Lotto 1999). However, the vast majority of research in speech perception is focused on the identification of phonemes or phonetic categories. In fact, the study of speech perception and spoken word recognition do not overlap as much as one might expect, and the fields generally cleave at the level of the phoneme. Most theories and models of speech perception explicitly state or implicitly assume that the goal or outcome of speech perception is a mapping from acoustics onto a phoneme or phonetic category representation. It is this mapping on which we will focus this review. Whether the fundamental unit of speech perception turns out to be the phoneme or the word or the syllable (or the di-phone or tri-phone), it is likely that the concepts and results summarized here will apply generally.

In this chapter, we present speech perception as a specific case of sound source identification. As with other source identification tasks, speech sound identification is based on the integration of multiple acoustic cues into a decision. However, the actual mapping from acoustic dimensions to phonetic categories is complicated by variability arising from speaker-specific characteristics, phonetic context, and the vicissitudes of listening conditions. After reviewing studies that explore the mechanisms by which listeners accommodate this variability, we will attempt to synthesize the results by describing auditory perception as "relative." That is, the perception of a particular sound is influenced by preceding (and following) sounds over multiple temporal windows. These effects of context (both temporally local and global) are likely to be important for any real-world perception of complex sounds (i.e., sound source perception).

## 2. Phonetic Categorization

Much of the tradition of speech perception research can be summarized as the study of *phonetic categorization*. That is, it has been focused on the ability of humans (and in some cases nonhuman animals) to map a set of sounds onto a discrete response typically corresponding to a phonetic segment (or minimal pair of syllables or words). For example, listeners are presented a synthesized series of syllables varying along a single acoustic dimension and are asked to press buttons labeled "da" and "ga" to identify the sound. While it has not been established that this mapping is a necessary step in normal speech perception (Lotto and Holt 2000; Scott and Wise 2003), robust phonetic categorization in the face of many sources of acoustic variance remains one of the most remarkable achievements of human auditory perception.

The task in phonetic categorization studies is quite similar to the nonspeech sound source identification tasks discussed by Lutfi (Chapter 2 of this volume). For example, Lutfi and Oh (1997) presented participants with synthesized approximates of struck clamped bars that differed in material. The participants pressed a button to indicate which of two intervals contained the sound produced by a target material (e.g., iron versus glass). The sounds varied along the acoustic attributes that distinguished the two materials. In traditional *phonetic categorization* tasks, listeners are asked to identify a phonetic category (typically from a closed set) based on sounds varying on just those dimensions that distinguish the categories. In fact, both of these tasks would be correctly referred to as *categorization* tasks. That is, a set of exemplars that vary in one or more physical dimensions are mapped onto a single response or label. The listener must be able to discriminate between members of each category but also to generalize their response across members of the same category.

Another similarity between phonetic categorization and other sound source identification tasks is that the category distinction is defined by a number of acoustic attributes or cues. Lutfi (Chapter 2) enumerates a number of acoustic cues that are related to the material or length of clamped struck bars, including the amplitude, frequency, and decay of different partials. Likewise, phonetic categories typically differ in a number of acoustic dimensions. For example, Lisker (1986) catalogued 16 different acoustic cues to the English voicing distinction, e.g., /b/ versus /p/, in syllable-initial position.[2] These include measures of relative amplitude, frequency, and duration of various components.

Even vowel categories are distinguished by a large number of acoustic attributes. Most people are familiar with the defining nature of formant frequencies for vowels. Formants are peaks in the spectral envelope corresponding to resonances of the vocal tract. The center frequencies of the first two formants (F1 and F2, labeled in order of increasing frequency) do a fairly good job of segregating categories for steady-state vowels.[3] However, in natural speech, the steady-state vowel is a bit of a mythical creature. Vowel categories can also be distinguished by overall duration (Peterson and Lehiste 1960; Strange 1989; Hillenbrand et al. 2000), by the extent and direction of changes in formant values within the vowel or *vowel-inherent spectral change* (Nearey and Assmann 1986), and by the steepness of the spectral envelope slope or tilt (Kiefte and Kluender 2005).

Of course, the fact that sounds differ in a number of acoustic dimensions as a function of their category membership does not mean that all dimensions

---

[2]It should be noted that the acoustic cues (and their perceptual weighting) differ for sounds that we label with the same phoneme when they appear in different positions in a syllable. For example, the acoustic cues that best distinguish English /l/ and /r/ described later in the text are relevant only when these sounds appear in a syllable-initial position. When the sounds occur in the syllable-final position, the relative importance of the cues changes (Sato et al. 2003). Whereas we label these sounds with the same phoneme and orthographic symbols regardless of position, they may be most appropriately considered different phonetic categories that are provided the same labels when we learn to read.

[3]Patterson et al. (Chapter 3 of this volume) provide a description of the acoustic characteristics of vowels as developed from the source—filter theory. In this chapter, we have opted to omit an overview of speech acoustics in favor of providing specific acoustic descriptions for phonetic distinctions as they are discussed.

are equally informative for the category distinction. For example, the English distinction between /l/ and /r/ in the syllable-initial position is realized, in part, by the starting frequencies of F2 and F3 (which then transition to the formant values of the following vowel). If one plots the initial F2 and F3 frequencies for exemplars of /l/ and /r/ produced by a number of speakers with a variety of following vowels, the resulting distributions show very little overlap in the F3 dimension and quite a bit of overlap in F2 (Dalston 1975; Lotto et al. 2004). That is, initial F3 is a far more reliable cue for distinguishing /l/ and /r/ than is initial F2. And in fact, native English speakers rely much more on initial F3 than on F2 when categorizing these sounds (Yamada and Tohkura 1990; Iverson et al. 2003).

A particularly interesting aspect of speech perception is the salient effects of differential experience on phonetic categorization and discrimination. It is well documented that native Japanese speakers have difficulty perceiving and producing the English /r/–/l/ distinction. One reason for this difficulty is that Japanese listeners appear to apply ineffective weighting functions to the cues for this distinction. That is, Japanese listeners tend to rely on initial F2 (as opposed to F3) when categorizing /l/ and /r/ exemplars (Yamada and Tohkura 1990; Iverson et al. 2003). Japanese productions of this contrast also result in distributions that are differentiated more by initial F2 than by F3 (Lotto et al. 2004). This weighting strategy appears to be a result of learning Japanese, which contains a distinction between /w/ and a flap consonant that is similar to /l/ and /r/ but is distinguished by F2 (Lotto et al. 2004). Thus, experience with a particular phonetic system can result in the application of suboptimal weighting strategies for nonnative contrasts (see also Francis and Nusbaum 2002; Kim and Lotto 2002).

In summary, phonetic categorization is a process by which a listener determines a sound's category by integrating and weighting multiple cues, and these weighting functions are not always optimal. This description should strike the reader as equally applicable to categorizing sounds on the basis of whether it was the result of a struck iron bar or a dropped wooden dowel; that is, it is a general description of sound source identification. One of the concerns in sound source identification is determining whether listeners are using optimal decision and weighting rules for a given task. Lutfi (2001), for example, derives optimal weighting functions for hollowness detection analytically from equations describing the acoustic outputs of vibrating hollow and solid bars. Such an approach is unlikely to be feasible for determining optimal weighting strategies for phonetic categorization. While there are good models for predicting the acoustic output for different vocal tract configurations, it is doubtful that one will be able to develop analytical solutions that capture all of the variability inherent in different productions of the same phonetic segment. In fact, it is this variability in the mapping between acoustics and phonetic categories (or intended gestures) that is the bugaboo for the understanding and modeling of human speech perception.

The sources of the variability range from perturbations common to all sound sources, such as room acoustics, channel transmission characteristics, and competing sources, to changes that are characteristic of speech such as coarticulation and differences between talkers. Several of the other chapters in this volume that review the particular challenges of competing sources include discussion of speech signals (e.g., Chapters 5, 6, and 8 in this volume). Here, we will concentrate on how the auditory system accommodates acoustic variation due to surrounding phonetic environment and talker-specific characteristics in phonetic categorization tasks. After reviewing some of the relevant empirical results, we will suggest that it is useful to conceptualize this accommodation as being the result of adaptive encoding by the auditory system working on multiple time scales.

## 3. Phonetic Context Effects

The acoustic pattern that is associated with a particular phonetic segment is notoriously context-dependent. One reason for this context dependence is that articulation is constrained by the physics of mass and inertia. At reasonable rates of speech production, it is difficult to move the articulators quickly enough to fully reach the targets that would characterize an articulation produced in isolation. For example, the vowel /ʌ/ (as in *but*) is produced in isolation with the tongue body relatively retracted. However, when producing *dud*, the tongue moves anterior to produce the initial and final /d/ and may not completely retract for the vowel, leading to a "fronted" articulation of /ʌ/. However, a more retracted version of the vowel will occur in a /g_g/ context, where the /g/ articulation requires the tongue to make a more posterior occlusion. That is, the articulation of the vowel is assimilated to the articulations of the surrounding context consonants; it is *coarticulated*.

Coarticulation is not just the result of physical constraints on articulators. The articulation of a phoneme can be influenced by following phonemes (*anticipatory coarticulation*) and coarticulation occurs even when there is relatively little interdependence of the articulators involved in the target and context phonemes. It appears that coarticulation is in part a result of the motor plan for speech (Whalen 1990). In fact, some cases of coarticulation or context-dependent production may be specified at the level of linguistic rules (e.g., regressive place assimilation; see Gaskell and Marslen-Wilson 1996).

Whatever the underlying causes, the result of coarticulation is context-dependent acoustics for phonetic categories. This acoustic variability is not evident simply as noise on inessential dimensions, but is present in those very dimensions that serve as substantial cues to phoneme identification. This provides a difficulty for simple template- or feature-matching models of phonetic categorization because there are few acoustic invariants that one can point to as defining a particular category. In the vowel coarticulation example provided above, the result of coarticulation is that the formant frequency values during the "vowel portion" vary as a function of the surrounding consonants (see Figure 10.1). At quick speaking rates, the formant values for /ʌ/ in "dud" resemble the values

**Isolated**           **/dʌd/**

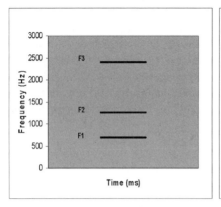 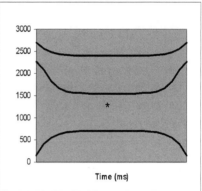

FIGURE 10.1. Formant tracks for a vowel spoken in isolation and in a /d_d/ context. The star indicates the frequency for the second formant when produced in isolation. The shift in frequency for this formant in context demonstrates coarticulation.

for the vowel /ɛ/ (the vowel in *bet*) spoken in isolation (Lindblom 1963; Nearey 1989). Thus, the approach of defining vowels simply by their formant frequencies is thwarted.

Another example, from Mann (1980), demonstrates coarticulation effects that cross over a syllable boundary between two consonants. As mentioned above, /d/ is articulated by creating an occlusion of the air stream relatively anterior in the mouth (at the alveolar ridge when produced in isolation). The exact placement of the tongue tip in creating this occlusion can be influenced by the context phonemes being produced. Producing /d/ after the matching anterior production of /l/ leads to a more anterior place of articulation. Producing /d/ after /r/ (produced with a more retracted tongue) results in a more posterior place of articulation. You can demonstrate this by producing /al da/ and /ar da/ with a natural or quick speaking rate. (You may try this at home, even without direct supervision.) The same coarticulation effects occur for /g/, which is produced with a relatively posterior occlusion. These articulation changes result in acoustic changes along the very dimensions that best distinguish /d/ from /g/. Figure 10.2 is a schematic of the formants for the four context-target conditions /al da/, /ar da/, /al ga/, /ar ga/ (based loosely on measures from Mann 1980). One can see in the context-consistent conditions (consistent in anterior or posterior articulation), /al da/ and /ar ga/, that the main distinction between /d/ and /g/ is the onset frequency of F3. However, in the context-inconsistent conditions, /al ga/ and /ar da/, the F3 of /ga/ is drawn higher and the F3 of /da/ is drawn lower. The resulting syllables are nearly indistinguishable. How do listeners deal with this ambiguity?

Mann (1980) demonstrated that listeners accommodate context-dependent acoustics through context-sensitive perception. She presented listeners with a

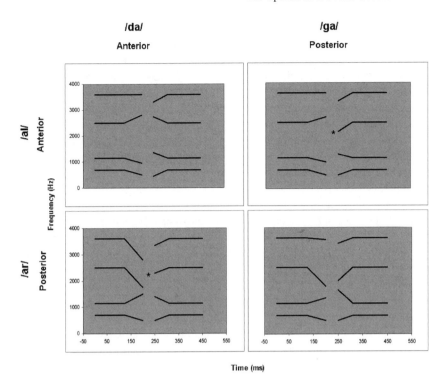

FIGURE 10.2. Schematic spectrograms representing the frequencies of the first four formants for productions of /ga/ and /da/ in the context of /al/ and /ar/. The stars indicate third-formant onset values that are nearly equivalent, resulting in syllables that are acoustically similar but represent different phonetic categories.

synthesized series of consonant–vowel (CV) stimuli that varied acoustically in initial F3 frequency and, consequently, varied perceptually from /da/ to /ga/. These stimuli were preceded by recordings of /al/ and /ar/ (with a 50-ms silent gap between syllables). Listeners' identifications of the target CVs indicated that the perceived identity of a phoneme was dependent on the preceding context. Following /al/ stimuli were identified as /ga/ more often than when the same stimuli followed /ar/. That is, identical acoustics lead to different perceptions depending on the preceding context. Note that the direction of this context-dependent shift is opposite the direction of coarticulation. In production, a preceding /al/ makes the CV more anterior or /da/-like. In perception, a preceding /al/ is identified as more /ga/-like. It appears that perception is compensating for the acoustic effects of coarticulation. In this way, the intended phoneme can be perceived despite variability in the acoustic form. This perceptual constancy is a hallmark of adaptive object or event perception (Brunswik 1956). It should stand as a particularly informative case of successful sound source identification.

A similar compensation for coarticulation is demonstrable for the case of a vowel coarticulated with, preceding, and following consonants, such as presented in Figure 10.1. Lindblom and Studdert-Kennedy (1967) first demonstrated this context-sensitive perception for Swedish vowels with liquid ("w" and "y") contexts. To protect the average English reader from hurting themselves while attempting to produce Swedish vowels in /w_w/ frames, we describe here similar results obtained by Nearey (1989) and Holt et al. (2000). Listeners were presented vowels varying in F2 midpoint frequency, from a good /ʌ/ to a good /ɛ/, in either isolation or /d_d/ context. More /ʌ/ responses were made to the vowels in /d_d/ context than in isolation. This again reverses the effects of coarticulation, which would result in vowel acoustics more appropriate for /ɛ/ in this context. Several other examples of apparent compensation for coarticulation have been examined (see Repp 1982 for a review).

These demonstrations leave one wondering what aspect of the context is used by the auditory system to derive context-specific identifications of the target. Does the perceptual system recognize the phonemic content of the context and then shift identification based on this context identity? While this is an explanation preferred by some word-recognition models (e.g., TRACE, Elman and McClelland 1988), it is inconsistent with some of the data. Infants as young as four months old demonstrate shifts in responses to /da/-/ga/ stimuli as a function of /al/ or /ar/ context despite not having a developed phonological system (Fowler et al. 1990), and native Japanese speakers show similar effects to those of English speakers despite not being able to discriminate /al/ and /ar/ (Mann 1986). Perhaps even more damaging to an account that relies on phonemic content of the context, birds trained to peck keys in response to presentations of /da/ or /ga/ show the same context-sensitive shift in responses that humans show for /al/ and /ar/ contexts (Lotto et al. 1997). Thus, it does not appear that the listener needs to access the phonemic identity of the context in order to compensate for coarticulation.

Lotto and Kluender (1998) proposed that it may be the spectral makeup of the context that determines the context effects in perception as opposed to its status as a phonemic entity. They presented listeners with a /ga/-/da/ series preceded by a frequency glide that tracked the transition of F3 for /r/ or /l/ (with a 50-ms silent interval). This context, which was not identifiable as a speech sound, was sufficient to result in a target categorization shift in the same direction as if the CVs were preceded by /al/ and /ar/. In a similar demonstration, Holt et al. (2000) replaced the /d_d/ context for a /ʌ/ to /ɛ/ series with tonal glides that matched the trajectory of F2. Once again, these nonphonetic contexts resulted in similar shifts as obtained for speech contexts. Nonspeech context effects can be obtained from bandpass noise, sine-wave tones, or single formants filtered from speech (Holt 1999; Lotto 2004).

Regardless of whether the context is speech or nonspeech, the effects on target identification can be described as contrastive. For example, /al/ with its *high*-frequency F3 offset (see Figure 10.2) results in more /ga/ responses, which is characterized by a *low*-frequency F3 onset. Alternatively, an /ar/ context with a low-frequency F3 offset results in more high-frequency F3 onset, or /da/, responses. Tone or noise-band contexts centered on the F3 offsets for

/al/ and /ar/ also result in contrastive shifts in target identification. Similarly, the high-frequency F2 onset and offset of /d/ contexts (or FM glide analogues) result in a lower perceived F2 for vowels in /d_d/ context compared to isolation (/ʌ/ being the vowel typically containing a lower F2). Thus, the effects of context can be predicted by the relative distribution of acoustic energy across frequencies regardless of the source of the context sound.

This constellation of findings implicates a rather general auditory process, which is insensitive to whether the sounds involved are speech. In addition to the demonstrations of nonspeech contexts affecting speech-target perception, one can obtain contrastive effects of speech contexts on the perception of target nonspeech sounds (Stephens and Holt 2003) and nonspeech context effects on nonspeech targets (Aravamudhan 2005). If, in fact, a general auditory process is partly responsible for compensation for coarticulation, then it is not surprising that the effects are present in infants, or nonnative language listeners (e.g., Japanese listeners and English stimuli), or even birds. One may also conclude that this process would play a role in sound source identification for sources that are not speech, that the identification of any complex sound may be affected by its acoustic context.

The original descriptions of these speech–nonspeech context effects referred to the results as demonstrations of *frequency contrast* (Lotto and Kluender 1998). However, this is a misnomer, because the "frequencies" present in the speech contexts don't change, but the relative energy present at each frequency does change. The /al/ and /ar/ contexts contain harmonics at the same frequencies when produced with the same fundamental frequency. The difference between them is the distribution of energy amplitude across those harmonics, with the peaks in energy defining the formants. Likewise, the targets /da/ and /ga/ differ in the relative amplitude of the harmonics in the F3 region. It is the amplitude differences between the spectral patterns that are being enhanced. *Spectral contrast* is a more appropriate description of these effects. Thus, one should be able to predict the effect of a context by the frequency regions of its spectral prominences. Conversely, one should be able to predict a complementary effect for contexts that have spectral troughs. Coady et al. (2003) preceded a CV series varying in F2 onset (/ba/-/da/) with a harmonic spectrum (rolling off at –6 dB/octave approximating the spectral tilt of speech) that contained either a low-frequency or high-frequency trough (zero energy at several consecutive harmonics) in the F2 region. The results demonstrated a contrastive effect of context. A context with a low-frequency *trough* leads to more target identifications consistent with a low-frequency *prominence* (i.e., /ba/).

The Coady et al. (2003) experiment is reminiscent of experiments conducted by Summerfield and colleagues on vowel "negative" aftereffects (Summerfield et al. 1984; Summerfield and Assmann 1987). They presented a uniform harmonic spectrum composed of equal-amplitude harmonics preceded by a spectral complement for a particular vowel (with troughs replacing formant prominences). Listeners reported hearing the vowel during presentation of the uniform spectrum. This result is in line with predictions of spectral contrast.

Regions that are relatively prominent in the context are attenuated in the target, and troughs in the contexts are enhanced in the target, in this case, leading to a pattern that resembles a vowel. Summerfield et al. (1984) note that the results are also consistent with the psychoacoustic phenomenon of auditory enhancement (Green et al. 1959; Viemeister 1980; Viemeister and Bacon 1982). Auditory enhancement can be demonstrated by presenting an equal-amplitude harmonic complex with one of the harmonics omitted followed by the same complex with the harmonic included. The replaced harmonic will stand out perceptually, and its auditory representation appears to be enhanced, since it can lead to increased forward masking of a tone relative to the complex being presented without the context (Viemeister and Bacon 1982). It is quite possible that the mechanisms underlying auditory enhancement produce some of the spectral contrast witnessed for speech sounds.

However, it does not appear that auditory enhancement can provide the complete story. Enhancement seems to be a largely monaural effect. Summerfield and Assmann (1989) failed to find effects of a precursor stimulus in their vowel experiments when the precursor was presented to the contralateral ear to the target. On the other hand, spectral contrast effects for speech are maintained even when the context (/al/ or /ar/) is presented to the opposite ear from the target (/da/-/ga/, Holt and Lotto 2002). Nonspeech effects are also present for dichotic presentation of context and target (Lotto et al. 2003). For both speech and nonspeech contexts, the effect is smaller for dichotic presentation versus diotic. These results suggest that peripheral mechanisms such as VIIIth-nerve adaptation or adaptation of suppression may play a partial role, but that interactions are occurring more centrally as well. More evidence for nonperipheral mechanisms comes from examining the time course of speech effects. Holt and Lotto (2002) varied the duration of the silent gap between /al/-/ar/ contexts and CV targets from 25 to 400 ms (50 ms being the value used in all previously described experiments). There was a monotonic decrease in effect size with increasing interval duration, but the effect was still significant at 275 ms. Lotto et al. (2003) demonstrated an effect of nonspeech context with a gap of 175 ms. These results again are consistent with both peripheral and more central mechanisms, because the effect is strongest for short intervals of 25 ms, within the temporal window of peripheral interactions, but lasts several hundred milliseconds, which is an unlikely window for purely peripheral mechanisms. Viemeister and Bacon (1982) reported no appreciable auditory enhancement for their masking study beyond about 100 ms of silent gap.

Perhaps the best evidence that speech effects cannot be accounted for solely by peripheral interactions is that context can affect preceding targets. Wade and Holt (2005a) had subjects identify words as "got" or "dot" with an embedded tone following the vowel. The tone was either high or low frequency. When the tone followed the consonant by 40 ms, it resulted in contrastive shifts in consonant identity (more "got" responses for embedded high-frequency tone). Whether the mechanisms responsible for "forward" and "backward" contrast effects are the same remains an open question. But it is clear that the identification of a complex sound can be heavily influenced by its surrounding context.

Another question that is unanswered is how sound source segregation influences context effects. The fact that sounds obviously originating from different sources (e.g., speech and tones) can affect each other in perception suggests that context effects may precede or be independent of source segregation. However, strict tests of the priority of segregation and context effects have not been conducted. Whereas nonspeech can affect speech when presented to opposite ears (Lotto et al. 2003), no one has tested whether a context that is localized to a specific region of exterior space will affect a target perceived as coming from a different location. Nor have there been attempts to manipulate the segregation of context and target by providing alternative perceptual organizations such as in an auditory streaming paradigm (Bregman 1990). It wouldn't be surprising if segregation influenced context effects. Empirical results from the visual modality demonstrate that context effects are malleable in relation to perceptual organization. For example, Gilchrist (1977) has reported that brightness contrast occurs only for luminances that are perceived as coplanar (see also Gogel 1978). Source segregation may explain a finding from Lotto and Kluender (1998). They preceded a /da/-/ga/ series modeled on a male voice with /al/-/ar/ contexts produced by the same male or a female. The female contexts did result in a significant shift in target identification, but the effect was significantly smaller than that obtained for the male contexts. Whether this difference was due to the listener perceiving the change in sources or because the spectral patterns for the female were not optimal for shifting the targets was not investigated.

Further investigation will also be required to resolve how spectral contrast interacts with linguistic information such as lexical status and phonological rules to determine a speech sound's identity. It is interesting to note that perceptual accommodation of linguistically determined assimilation does not appear to require that one has experience with the particular language being presented (Gow and Im 2004). General perceptual mechanisms and principles may be involved even in these cases, which previously were accounted for by appealing to linguistic-specific knowledge (e.g., Gaskell and Marslen-Wilson 1996).

## 4. Talker Normalization

As discussed in Section 2, an examination of the distributions of phonetic categories in acoustic space allows one to determine an optimal weighting and decision strategy for distinguishing contrasts in a particular language. Several theorists have proposed that language learners derive phonetic categories from these distributions averaged over many encountered talkers (Kuhl 1993; Jusczyk 1997; Lotto 2000). However, whereas average distributions will provide a best guess as to phonetic identity across all talkers, they will be suboptimal for any particular talker. While the acoustic variability associated with different talkers is useful when one's task is indexical identification (e.g., distinguishing the gender of the speaker), it can be a challenge for the robust identification of the intended phoneme. In order to effectively identify phonemes in all communication settings, listeners must be able to "tune" their auditory representations to the particular talker. This accommodation of talker-specific characteristics is referred to as *talker normalization* and has been a focus of speech-perception research since the inception of the field (Potter and Steinberg 1950).

Peterson and Barney (1952) presented an early description of talker differences in vowel acoustics that has structured much of the work on talker normalization over the past 50 years. They measured formant frequency values for adult males and females and children for the vowels of English spoken in /h_d/ context. Despite the lack of context-induced variability, the distributions for the vowel categories show a great deal of dispersion and overlap (see Hillenbrand et al. 1995 for an updated data set). However, listeners can still identify the phoneme intended by the speaker. One way to account for this ability is to propose that listeners are using less-variable ratios of formants rather than treating each formant as an independent informative dimension (Fujisaki and Kawashima 1968; Traunmüller 1981; Syrdal and Gopal 1986; Miller 1989). This approach is exemplified by the suggestion of Potter and Steinberg (1950) that a vowel may be equivalent to a pattern of stimulation on the basilar membrane regardless of its location along the membrane. Sussman (1986) proposes that columns of combination-sensitive neurons could encode the same formant ratios across changes in absolute frequencies. Talker normalization solutions, such as these, that rely on information contained solely within the vowel (or, more generally, the phonetic segment) have been referred to as *intrinsic* by Ainsworth (1975).

Whereas formant ratios can decrease some of the talker variability, there is clear evidence that listeners also apply normalization strategies that utilize information *extrinsic* to the target vowel. In 1957, Ladefoged and Broadbent conducted a classic experiment in which they presented listeners with a synthesized context sentence (*Please say what this word is __.*) followed by a synthesized vowel embedded in the frame /b_t/. Listeners identified the final target word as *bit*, *bet*, *bat*, or *but*. Using a synthesizer, the researchers varied the spectral characteristics of the context sentence. For example, in one condition they lowered the range of F1 frequencies and in another raised it. Manipulations of the context sentence had large effects on the perceived vowel category. A vowel that was categorized as /ɪ/ (*bit*) by 88% of listeners following the unaltered context was categorized as /ɛ/ (*bet*) by 90% of listeners following a context with lowered F1 range. That is, the categorization of the vowel sound was strongly dependent on the characteristics of the context sentence.

Ladefoged and Broadbent (1957) also report that the manipulated context sentences maintained intelligibility but sounded as if they were produced by different speakers. From this view, the results can be interpreted as indicative of talker normalization. Listeners' responses appeared to be the result of tuning phonetic categories given information about typical values of formant frequencies for that speaker. In general, productions of the vowels /ɪ/ and /ɛ/ differ in F1 frequency, with /I/ having a lower-frequency F1 (Peterson and Barney 1952). However, the actual value that corresponds to a "lower-frequency F1" is relative to speech produced by a talker. When the range of F1 in the context sentence is lowered, a moderate F1 value appears "high" and encourages an /ɛ/ categorization. Similar demonstrations of context sentences on target identifications have been made by a number of researchers (Broadbent and Ladefoged 1960; Ainsworth 1975; Assmann et al. 1982; Remez et al. 1987; Darwin et al. 1989; Ladefoged 1989; Nearey 1989; Johnson 1990; Watkins and Makin 1994).

Ladefoged and Broadbent (1957) proposed that their findings are consistent with a proposal by Joos (1948) that listeners make vowel identifications by referencing a talker-specific formant space created from the context material. That is, the identity of a vowel is dependent on its formant values relative to other vowels produced by the same talker. A vowel token positioned with respect to a low range of F1 values will be perceived differently from the same token positioned relative to high F1 values. This explanation is consistent with a number of theories of normalization in which it is proposed that the listener recalibrates his or her perception of a segment by making reference to talker-specific information (e.g., Ainsworth 1975; Nordstrom and Lindblom 1975; Nearey 1989). Common to these approaches is the requirement that the listener retain some distributional information about the speech of the particular talker, whether that information is just the average F3 values for back vowels (Nordstrom and Lindblom 1975), the ranges of the formant frequencies (Joos 1948), or an entire mapping of the vowel space. Results from word-recognition and memory studies make it clear that some talker-specific information is retained in the speech representation (Goldinger 1998).

Much of the investigation of talker normalization has been concentrated on perceptual compensation for anatomical differences between talkers, such as gender differences. One approach to these differences is to rescale speech based on an estimate of vocal-tract length (e.g., Nordstrom and Lindblom 1975). Alternatively, Patterson et al. (Chapter 3 of this volume) present a transform that normalizes vowels by extracting variance related to vocal-tract length. The problem with approaches that explicitly relate to vocal-tract length is that many speaker-specific differences are not due strictly to length. Even the difference between males and females is partly due to differences in the proportions of different regions of the vocal tract, in addition to overall length. And even when one accounts for these structural differences, there is not complete overlap between male and female vowel spaces, suggesting a difference in articulation style (Fant 1966; Nordstrom 1977). In fact, there are many individual differences in production (Johnson et al. 1993b) that are unrelated to anatomy, including dialect and accent. It is likely that a problem as complicated as talker normalization for speech will be accomplished by a number of different processes working sequentially or in parallel.

The fact that general auditory processes appear to play some role in compensation for coarticulation may lead one to question whether there are general processes that aid in talker normalization. The size normalization process proposed by Patterson et al. (Chapter 3) may be an example of an auditory process not specialized for human speech that is involved in talker normalization. Recently, Holt (2005) described a new auditory phenomenon that may also play an important role in normalization. The stimulus paradigm appears to be a mix of the normalization experiment of Ladefoged and Broadbent (1957) and the nonspeech/speech-context-effect experiments of Lotto and Kluender (1998). The target that listeners had to identify was a member of a /da/-/ga/ series (modified from natural speech tokens). The target was preceded by a 70-ms tone situated at a frequency that was shown to be a neutral context (set at a frequency that was in the middle of the F3 range for the CV). This *standard tone* was, in turn, preceded by a series of 21 70-ms tones varying in frequency. The 21 tones were randomly sampled from a rectangular distribution of tone frequencies that either had a low or high mean. The low mean corresponded to the F3 offset-frequency of /ar/, and the high mean corresponded to the F3 for /al/ from the experiments by Lotto and Kluender (1998). The context tones are referred to as the *acoustic history*. Representations of the stimuli are presented in Figure 10.3. As in the context-effects experiments, listeners were asked to identify the final syllable as /da/ or /ga/. However, in contrast to the previous experiments, the context was not adjacent to the syllable (the neutral standard tone always directly preceded the target), and the difference in the context conditions cannot be described in terms of a specific spectral pattern (the order of tones in the acoustic histories changed on each trial). Nevertheless, the results resemble those obtained in the context-effects experiments. Listeners identified the target as /ga/ more often following the high-mean history and as /da/ more often following the low-mean history.

This is a contrastive response pattern, except that the contrast is not with a particular spectral pattern but with the spectral energy averaged over a relatively long (more than 2 s) temporal window. In support of the conclusion that this is another contrast effect, Holt (2006) demonstrated that complementary results can be obtained when the acoustic history is a series of noise bursts with troughs at sampled frequencies instead of tones.

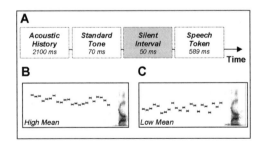

FIGURE 10.3. Description of the stimulus paradigm of Holt (2005). (**A**) Time interval for each stimulus event; (**B**) and (**C**) spectrograms of two stimuli with tones sampled from either a high-mean (**B**) or a low-mean (**C**) distribution.

The acoustic histories of Holt (2005) resemble in some respect the carrier sentences of Ladefoged and Broadbent (1957). Both are extended contexts that differ in the range of frequencies that contain amplitude peaks (tones or first formants). Given this correspondence, one may propose that a similar process plays a role in both demonstrations. The results of Ladefoged and Broadbent (1957) can also be redescribed in contrastive terms. If one lowers the average frequency of F1 in the carrier sentence, then the F1 for the vowel in the target word is perceived as higher (i.e., more /ɛ/). That is, it may be that talker normalization is, in part, another example of spectral contrast influencing speech perception.

There are several other studies that demonstrate that perception of a target syllable is influenced by the spectral makeup of the carrier phrase, and in each case the effect can be described as contrast with the average spectral pattern of the precursor. Watkins (1988, 1991) applied a filter to a carrier phrase and demonstrated that a target vowel was perceived as if it were filtered with an inverse of the phrase filter (see also Watkins and Makin 1994, 1996). Similarly, Kiefte and Kluender (2001) presented carrier phrases that varied in the slope of their spectral tilt (the slope of the amplitude falloff for higher-frequency harmonics). Steeper spectral tilts led to target vowel identifications that were more consistent with a shallow spectral tilt. One can conceive of these demonstrations as examples of talker normalization or normalizing for the effects of filtering by a transmission channel. Whatever the cause of these deviations, the effect appears to be that target identification is made relative to the preceding (and following, Watkins and Makin 1996) spectral patterns.

The demonstrations of context-based perception discussed thus far are related to spectral differences in the context, but what of temporal differences? One salient difference between talkers is speaking rate. Given that temporal cues (such as voice onset time) are important for phonetic categorization, it would appear necessary that listeners compensate for inherent temporal variations among talkers. As an example, the distinction between /ba/ and /wa/ in English is, in part, defined by the duration of the formant transitions from onset to the vowel; short-duration transitions are associated with /b/. (Think of the production in each case as movement away from approximated lips. This movement is faster for /ba/.) However, these transition durations also vary with speaking rate (Miller and Baer 1983). Listeners appear to accommodate speaking-rate variation by perceiving the transition duration relative to the following vowel duration, which could be considered a correlate of speaking rate. A synthesized CV that is perceived as /wa/ when the vowel is short will be perceived as /ba/ when the vowel is lengthened (Miller and Liberman 1979). This is again a contrastive response pattern in phonetic categorization. The effective perceived transition duration is shortened when the vowel is lengthened. The same pattern can be witnessed in nonspeech categorization. Pisoni et al. (1983) reported analogous shifts for sine-wave analogues of /ba/ and /wa/ that were categorized as beginning with an "abrupt" or "gradual" transition (see also Diehl and Walsh 1989). The implication that a general contrast process may underlie this context effect is consistent with the findings of vowel-length effects for infants (Jusczyk et al. 1983) and nonhuman animals (macaques: Stevens et al. 1988; budgerigars: Dent et al. 1997).

As with spectral effects, one can demonstrate that changing the average durations for segments (speaking rate) in a carrier phrase will affect target identification (Diehl et al. 1980; Summerfield 1981; Kidd 1989; Wayland et al. 1994). Wade and Holt (2005b) utilized the acoustic histories paradigm described above to examine whether carrier phrase effects could be induced with nonspeech precursors. They preceded members of a /ba/-/wa/ series (varying in formant transition duration) with a series of tones sampled from a single rectangular distribution with a range from F1 to F2. The context conditions differed in terms of the duration of these tones, with short (30 ms) and long (110 ms) conditions. The precursors had a reliable contrastive effect on the categorization of the target CV (more /ba/ responses for long condition). Thus, it appears that rate normalization shares much in common with the other versions of talker normalization reviewed above.

Whereas the correspondence of speech and nonspeech effects presented here implicates general auditory processes in talker normalization, it should be noted again that normalization is a complex problem that likely requires a multitude of mechanisms. Many of the differences between talkers cannot be summarized as overall changes in rate or average spectra. For example, the perturbations of production resulting from a foreign accent or dialect difference are often specific to individual phonetic categories. Yet listeners appear to be able to adjust their categorization on the basis of a talker's dialect (Evans and Iverson 2004). Listeners also appear to make use of information from vision when normalizing for a talker (Johnson et al. 1999; Glidden and Assmann 2004). In order to account for the entire constellation of findings, it is likely that the proposal of many perceptual processes will need to be entertained. However, a subset of these processes appear to be of a general auditory nature and are likely to play a role in any real-world sound source identification task.

## 5. A Synthesis: Relative Perception

In this review, we have proposed that phonetic categorization is an example of a sound source identification task. As such, the results of investigations into perceptual weighting strategies, source segregation, auditory attention, and memory, etc. discussed in the other chapters of this volume may be applied to the complex problem of speech perception. Another implication of this proposal is that phenomena in speech perception may provide insights into the auditory processes that are active for categorization of any complex sound. The demonstrations of phonetic context effects (or compensation for coarticulation) and talker normalization reviewed here indicate that the identification of a target sound can be influenced by the acoustic makeup of surrounding context sounds. To the extent that sound sources are not perceived in isolation, contextual sounds may be an important determiner of behavior in many nonspeech identification tasks.

The effects of context on identification can be described as contrastive. For example, energy in a particular frequency region is perceived as less intense in contrast to a preceding (or following) peak of energy in that region. What general mechanisms in the auditory system lead to this type of perceptual contrast? There are a number of candidate neural mechanisms that emphasize the difference between sounds. Delgutte and his colleagues (1996; Delgutte 1997) have established a case for a broad role for neural adaptation in perception of speech, noting that the adaptation may enhance spectral contrast between sequential segments. This contrast is predicted to arise because neurons adapted by stimulus components close to their preferred (characteristic) frequency are relatively less responsive to subsequent energy at that frequency, whereas components not present (or weakly present) in a prior stimulus are encoded by more responsive unadulterated neurons. Adaptation of suppression is another possible contrast-inducing mechanism that has been implicated in auditory enhancement (Palmer et al. 1995). Clearly, neural adaptation is a mechanism that would be active in both speech and nonspeech source identification tasks.

Recent studies have provided strong evidence that the auditory system, like the visual system (e.g., Movshon and Lennie 1979; Saul and Cynader 1989), exhibits another form of adaptation—known as stimulus-specific adaptation (SSA)—that has intriguing parallels to the spectral contrast effects reviewed above. Ulanovsky et al. (2003, 2004) have demonstrated SSA in primary auditory cortex using a version of the "oddball" paradigm common to mismatch negativity studies (Näätänen et al. 1978). In particular, they presented a repeating tone as a standard that was sporadically replaced by a deviant tone with a different frequency. The response to the deviant tone was enhanced relative to when the tones were presented equally often in a sequence. That is, the cortical neurons provide an enhanced response to acoustic novelty. This is a contrastive response pattern. The effects of context in speech can also be viewed as an enhancement to change from the prevailing acoustic environment. The acoustic histories of Holt (2005) establish a context with energy centered in high- or low-frequency regions, and the introduction of components outside of those regions leads to a perceptual emphasis of those components.

Abrupt changes in sound waves or light are indicative of novel forces working on an object or of the presence of multiple sources. Emphasis of change, whether it is spectral contrast or brightness contrast, can help the perceiver in directing attention to new information or to segregate different sources. Thus, contrast appears to be not just a single process or the result of a single mechanism, but is instead an operating characteristic of adaptive perceptual systems.

In order to detect change, perceptual systems need to retain information about context stimuli. This retention appears to operate over multiple time scales. In phonetic context effects, the time scale is on the order of tens to hundreds of milliseconds. In the carrier phrase and acoustic history experiments, the time scale appears to be seconds. One could consider this retention to be an example of *auditory memory* (see Demany and Semal, Chapter 4). However, memory is a term that is usually associated with cognition as opposed to perception. We prefer to think of the tracking of statistical regularities in the input and the encoding of targets relative to those regularities as fundamental to perception.

Given the purported importance of tracking statistics to source perception, it is incumbent on us to determine what "statistics" are computed and over what temporal windows they are computed. Data from carrier-phrase and acoustic-history experiments suggest that the average spectra of contexts are likely computed. In the carrier-phrase experiments of Kiefte and Kluender (2001), listeners appear to extract the average spectral tilt of the precursor and perceive the target relative to that average spectrum. In Holt's (2005) acoustic history experiments, the mean of the tone distributions seem to be extracted for comparison with the target. In a follow-up study, Holt (2006) demonstrated that repeated presentation of a tone with the mean frequency had the same effect on identification as presentation of the entire distribution and that in general, the variance of the distribution plays little or no role in the effect.

The extraction of the average spectrum by the auditory system provides a possible means of normalizing for talker differences. Work on speech production models by Story and Titze (2002; Story 2005) has provided evidence that individual talker differences are apparent in the vocal-tract shape used in the production of a neutral or average vowel. The productions of other vowels and consonants can be considered as perturbations of this neutral vowel shape. These perturbations are remarkably consistent across talkers, so that much of the talker variability is captured by the differences in the neutral vowel shape. If the auditory system is extracting an average spectrum and then enhancing deviations from that average (contrast), then one can think of the perceiver as extracting the acoustics of the neutral (average) vocal-tract shape and enhancing the perturbations from this average, which result from the phonetic articulations of the speaker. This tuning of perception to the average spectrum (and by extension neutral vocal tract) of a speaker would drastically lower the variability associated with talker differences. Again, this beneficial result for robust speech communication is only a specific case of the general processes involved in auditory source perception.

It is likely that the auditory system can track regularities beyond mean spectra. The results of temporal contrast studies, such as those involving /ba/ versus /wa/, indicate that average duration of segments or something like it is tracked. The work on SSA in audition and vision (e.g., Fairhall et al. 2001; Ulanovsky et al. 2004) suggests that there is a variety of statistical regularities in sensory signals that can be tracked. Similar conclusions come from the literature on mismatch negativity studies, which have demonstrated that the auditory system reacts to deviants from a standard stimulus repetition based on amplitude, intensity, spatial location, and even phonetic category (see Näätänen and Winkler 1999 for a review). Certainly, there must be some restrictions on the types of regularities that are extracted, but to date, there has not been systematic study of these constraints. The context studies reviewed here provide a possible paradigm for testing the limits of the auditory system's abilities in this regard.

The concept of perceiving a target sound with respect to previous statistical or distributional information can be extended to the entire process of categorization as discussed in Section 2 of this review. We presented the idea of optimal cue-weighting strategies as determinable from the category distributions described in acoustic space. If listeners do develop weighting strategies based on the distributions of experienced exemplars, then they must retain some description of these distributions that is created over time. It is unclear what exactly is retained. It could be something as detailed as a full representation of each exemplar (e.g., Goldinger 1997; Johnson 1997) or a "tally" of the values of experienced exemplars on a constrained set of acoustic attributes. Whatever the answer turns out to be, it is becoming clear that listeners retain a fairly good representation of the distributions of experienced sounds. Sullivan et al. (2005) presented bands of noise varying in center frequency from two overlapping distributions that were arbitrarily labeled as categories "A" and "B" to listeners who learned to categorize the sounds with feedback. Within six minutes of training (one repetition of the 50 stimuli in each distribution), the participants were able to categorize the sounds with near-optimal performance. In order to do this, they had to calculate the crossover point of the two distributions and use it as a decision criterion. Listeners appeared to do this with notable precision. Obviously, phonetic categories and categories for other sound sources are developed over a longer time interval than a single experimental session, but the parallels between phonetic context effects and the formation of phonetic categories are intriguing. In each case, the perception of a target is made relative to a larger context, whether it is a carrier phrase or all experienced tokens of different phonemes. There is even evidence for contrastive effects at the category level. The exemplars of vowel categories that are judged as "best" members of the category or result in the strongest responses are not those exemplars that are most typical but those that are most different from competing categories (Johnson et al. 1993a; Kluender et al. 1998). Also, a vowel that is ambiguous between two categories preceded by a good exemplar from one of the categories will be perceived as a member of the contrasting vowel category (Repp et al. 1979; Healy and Repp 1982; Lotto et al. 1998). Thus, there appear to be similarities in response patterns and importance of context that extends from peripheral neural adaptation to categorization, across time scales differing in many orders of magnitude.

Whether these similarities are superficial or whether they reveal something fundamental about auditory perception remains to be seen. But as hearing scientists move toward an understanding of sound source perception in the environment, it is clear that it will not be sufficient to examine the ability of listeners to detect an acoustic feature or register a value along an acoustic dimension in isolation. Perception in the real world is about perception in context.

*Acknowledgments:* Preparation of this chapter was supported in part by grants from NIH-NIDCD and NSF.

## References

Ainsworth WA (1975) Intrinsic and extrinsic factors in vowel judgments. In: Fant G, Tatham M (eds) Auditory Analysis and Perception of Speech. London: Academic Press, pp. 103–113.

Aravamudhan R (2005) Perceptual overshoot with speech and nonspeech sounds. Ph.D. thesis, Kent State University, Kent, OH.

Assmann PF, Nearey TM, Hogan JT (1982) Vowel identification: Orthographic, perceptual and acoustic aspects. J Acoust Soc Am 71:975–989.

Bachorowski JA, Owren MJ (1999) Acoustic correlates of talker sex and individual talker-identity are present in a short vowel segment produced in running speech. J Acoust Soc Am 106:1054–1063.

Bregman AS (1990) Auditory Scene Analysis: The Perceptual Organization of Sound. Cambridge, MA: Bradford Books, MIT Press.

Broadbent DE, Ladefoged P (1960) Vowel judgments and adaptation level. Proc Biol Sci 151:384–399.

Brunswik E (1956) Perception and the Representative Design of Psychological Experiments. Berkeley, CA: University of California Press.

Coady JA, Kluender KR, Rhode WS (2003) Effects of contrast between onsets of speech and other complex spectra. J Acoust Soc 114:2225–2235.

Dalston RM (1975) Acoustic characteristics of English /w, r, l/ spoken correctly by young children and adults. J Acoust Soc 57:462–469.

Darwin CJ, McKeown JD, Kirby D (1989) Perceptual compensation for transmission channel and speaker effects on vowel quality. Speech Commun 8:221–234.

Delgutte B (1997) Auditory neural processing of speech. In: Hardcastle WJ, Laver J (eds) The Handbook of Phonetic Sciences. Oxford: Blackwell, pp. 507–538.

Delgutte B, Hammond BM, Kalluri S, Litvak LM, Cariani P (1996) Neural encoding of temporal envelope and temporal interactions in speech. In: Ainsworth W, Greenberg S (eds) Proceedings of the ESCA Research Workshop on the Auditory Basis of Speech Perception. Keele, UK; 15–19 July. pp. 1–11.

Dent ML, Brittan-Powell EF, Dooling RJ, Pierce A (1997) Perception of synthetic /ba/ /wa/ speech continuum by budgerigars (*Melopsittacus undulatus*). J Acoust Soc Am 102:1891–1897.

Diehl RL, Walsh MA (1989) An auditory basis for the stimulus-length effect in the perception of stops and glides. J Acoust Soc Am 85:2154–2164.

Diehl RL, Souther AF, Convis CL (1980) Conditions on rate normalization in speech perception. Percept Psychophys 27:435–443.

Elman JL, McClelland JL (1988) Cognitive penetration of the mechanisms of perception: Compensation for coarticulation of lexically restored phonemes. J Mem Lang 27: 143–165.

Evans BG, Iverson P (2004) Vowel normalization for accent: An investigation of best exemplar locations in northern and southern British English sentences. J Acoust Soc Am 115:352–361.

Fairhall AL, Lewen GD, Bialek W, de Ruyter van Steveninck RR (2001) Efficiency and ambiguity in an adaptive neural code. Nature 412:787–792.

Fant G (1966) A note on vocal tract size factors and non-uniform F-pattern scalings. Speech Trans Lab Q Prog Stat Rep 7:22–30.

Fowler CA, Best CT, McRoberts GW (1990) Young infants' perception of liquid coarticulatory influences on following stop consonants. Percept Psychophys 48:559–570.

Francis AL, Nusbaum HC (2002) Selective attention and the acquisition of new phonetic categories. J Exp Psychol [Hum Percept] 28:349–366.

Fujisaki H, Kawashima T (1968) The roles of pitch and higher formants in the perception of vowels. IEEE Trans Audio Elect AU-16:73–77.

Gaskell G, Marslen-Wilson WD (1996) Phonological variation and inference in lexical access. J Exp Psychol [Hum Percept] 22:144–158.

Gilchrist A (1977) Perceived lightness depends on perceived spatial arrangement. Science 195:185–187.

Glidden CM, Assmann PF (2004) Effects of visual gender and frequency shifts on vowel category judgments. Acoust Res Let Online 5:132–138.

Gogel WC (1978) The adjacency principle in visual perception. Sci Am 238:126–139.

Goldinger SD (1997) Words and voices: Perception and production in an episodic lexicon. In: Johnson K, Mullennix JW (eds) Talker Variability in Speech Processing. San Diego, CA: Academic Press, pp. 33–66.

Goldinger SD (1998) Echoes of echoes? An episodic theory of lexical access. Psychol Rev 105:251–279.

Gow DW, Im AM (2004) A cross-linguistic examination of assimilation context effects. J Mem Lang 51:279–296.

Green DM, McKay MJ, Licklider JCR (1959) Detection of a pulsed sinusoid in noise as a function of frequency. J Acoust Soc Am 31:1446–1452.

Healy AF, Repp BH (1982) Context independence and phonetic mediation in categorical perception. J Exp Psychol [Hum Percept] 8:68–80.

Hillenbrand JM, Getty L, Clark MJ, Wheeler K (1995) Acoustic characteristics of American English vowels. J Acoust Soc Am 97:3099–3111.

Hillenbrand JM, Clark MJ, Houde RA (2000) Some effects of duration on vowel recognition. J Acoust Soc Am 108:3013–3022.

Holt LL (1999) Auditory constraints on speech perception: An examination of spectral contrast. Diss Abstr Int (Sci) 61:556.

Holt LL (2005) Temporally non-adjacent non-linguistic sounds affect speech categorization. Psychol Sci 16:305–312.

Holt LL (2006) The mean matters: Effects of statistically-defined non-speech spectral distributions on speech categorization. J Acoust Soc Am 120:2801–2817.

Holt LL, Lotto AJ (2002) Behavioral examinations of the level of auditory processing of speech context effects. Hear Res 167:156–169.

Holt LL, Lotto AJ, Kluender KR (2000) Neighboring spectral content influences vowel identification. J Acoust Soc Am 108:710–722.

Iverson P, Kuhl PK, Akahane-Yamada R, Diesch E, Tohkura Y, Kettermann A, Siebert C (2003) A perceptual interference account of acquisition difficulties for non-native phonemes. Cognition 87:B47–B57.

Johnson K (1990) The role of perceived speaker identity in F0 normalization of vowels. J Acoust Soc Am 88:642–654.

Johnson K (1997) Speech perception without speaker normalization: An exemplar model. In: Johnson K, Mullennix JW (eds) Talker Variability in Speech Processing. San Diego: Academic Press, pp. 145–166.

Johnson K, Flemming E, Wright R (1993a) The hyperspace effect: Phonetic targets are hyperarticulated. Language 69:505–528.

Johnson K, Ladefoged P, Lindau M (1993b) Individual differences in vowel production. J Acoust Soc Am 94:701–714.

Johnson K, Strand EA, D'Imperio M (1999) Auditory-visual integration of talker gender in vowel perception. J Phonet 27:359–384.

Joos M (1948) Acoustic Phonetics. Language 24:1–136.

Jusczyk PW (1997) The Discovery of Spoken Language. Cambridge, MA: MIT Press.

Jusczyk PW, Pisoni DB, Reed M, Fernald A, Myers M (1983) Infants' discrimination of the duration of a rapid spectrum change in nonspeech signals. Science 222:175–177.

Kidd GR (1989) Articulatory-rate context effects in phoneme identification. J Exp Psychol [Hum Percept] 15:736–748.

Kiefte M, Kluender KR (2001) Spectral tilt versus formant frequency in static and dynamic vowels. J Acoust Soc Am 109:2294–2295.

Kiefte M, Kluender KR (2005) The relative importance of spectral tilt in monopthongs and diphthongs. J Acoust Soc Am 117:1395–1404.

Kim M-RC, Lotto AJ (2002) An investigation of acoustic characteristics of Korean stops produced by non-heritage learners. The Korean Language in America 7:177–188.

Kluender KR, Lotto AJ (1999) Virtues and perils of an empiricist approach to speech perception. J Acoust Soc Am 105:503–511.

Kluender KR, Lotto AJ, Holt LL, Bloedel SL (1998) Role of experience for language specific functional mappings of vowel sounds. J Acoust Soc Am 104:3568–3582.

Kuhl PK (1993) Early linguistic experience and phonetic perception: Implications for theories of developmental speech perception. J Phonet 21:125–139.

Ladefoged P (1989) A note on "Information conveyed by vowels." J Acoust Soc Am 85:2223–2224.

Ladefoged P, Broadbent DE (1957) Information conveyed by vowels. J Acoust Soc Am 29:98–104.

Lindblom B (1963) Spectrographic study of vowel reduction. J Acoust Soc Am 35:1773–1781.

Lindblom B, Studdert-Kennedy M (1967) On the role of formant transitions in vowel recognition. J Acoust Soc Am 42:830–843.

Lindblom B, MacNeilage P, Studdert-Kennedy M (1984) Self-organizing processes and the explanation of language universals. In: Butterworth B, Comrie B, Dahl Ö (eds) Explanations for Language Universals. Berlin: Walter de Gruyter and Co, pp. 181–203.

Lisker L (1986) "Voicing" in English: A catalogue of acoustic features signaling /b/versus /p/ in trochees. Lang Speech 29:3–11.

Lotto AJ (2000) Language acquisition as complex category formation. Phonetica 57:189–196.

Lotto AJ (2004) Perceptual compensation for coarticulation as a general auditory process. In: Agwuele A, Warren W, Park S-H (eds) Proceedings of the 2003 Texas Linguistic Society Conference. Sommerville, MA: Cascadilla Proceedings Project, pp. 42–53.

Lotto AJ, Holt LL (2000) The illusion of the phoneme. In: Billings SJ, Boyle JP, Griffith AM (eds) CLS 35 The Panels. Chicago: Chicago Linguistic Society, pp. 191–204.

Lotto AJ, Kluender KR (1998) General contrast effects of speech perception: Effect of preceding liquid on stop consonant identification. Percept Psychophys 60:602–619.

Lotto AJ, Kluender KR, Holt LL (1997) Perceptual compensation for coarticulation by Japanese quail (Coturnix coturnix japonica). J Acoust Soc Am 102:1134–1140.

Lotto AJ, Kluender KR, Holt LL (1998) The perceptual magnet effect depolarized. J Acoust Soc Am 103:3648–3655.

Lotto AJ, Sullivan SC, Holt LL (2003) Central locus for nonspeech context effects on phonetic identification. J Acoust Soc Am 113:53–56.

Lotto AJ, Sato M, Diehl RL (2004) Mapping the task for the second language learner: The case of Japanese acquisition of /r/ and /l/. In: Slifka J, Manuel S, Matthies M (eds) From Sound to Sense: 50+ Years of Discoveries in Speech Communication. Electronic Conference Proceedings, Boston; 12 June. pp. C181–C186.

Lutfi RA (2001) Auditory detection of hollowness. J Acoust Soc Am 110:1010–1019.

Lutfi RA, Oh EL (1997) Auditory discrimination of material changes in a struck-clamped bar. J Acoust Soc Am 102:3647–3656.

Mann VA (1980) Influence of preceding liquid on stop-consonant perception. Percept Psychophys 28:407–412.

Mann VA (1986) Distinguishing universal and language-dependent levels of speech perception: Evidence from Japanese listeners' perception of English /l/ and /r/. Cognition 24:169–196.

Miller JD (1989) Auditory-perceptual interpretation of the vowel. J Acoust Soc Am 85:2114–2134.

Miller JL, Baer T (1983) Some effects of speaking rate on the production of /b/ and /w/. J Acoust Soc Am 73:1751–1755.

Miller JL, Liberman AM (1979) Some effects of later-occurring information on the perception of stop consonant and semivowel. Percept Psychophys 25:457–465.

Movshon JA, Lennie P (1979) Pattern-selective adaptation in visual cortical neurons. Nature 278:850–852.

Näätänen R, Winkler I (1999) The concept of auditory stimulus representation in cognitive science. Psychol Bull 125:826–859.

Näätänen R, Gaillard AW, Mantysalo S (1978) Early selective attention effect on evoked potential reinterpreted. Acta Psychol 42:313–329.

Nearey TM (1989) Static, dynamic, and relational properties in vowel perception. J Acoust Soc Am 85:2088–2113.

Nearey TM, Assmann PF (1986) Modeling the role of inherent spectral change in vowel identification. J Acoust Soc Am 80:1297–1308.

Nordstrom PE (1977) Female and infant vocal tracts simulated from male area functions. J Phonet 5:81–92.

Nordstrom PE, Lindblom B (1975) A normalization procedure for vowel formant data. In: Proceedings of the 8th International Congress of Phonetic Sciences, Leeds, England, p. 212.

Palmer AR, Summerfield Q, Fantini DA (1995) Responses of auditory-nerve fibers to stimuli producing psychophysical enhancement. J Acoust Soc Am 97:1786–1799.

Peterson GE, Barney HL (1952) Control methods used in a study of the vowels. J Acoust Soc Am 24:175–184.

Peterson GE, Lehiste I (1960) Duration of syllable nuclei in English. J Acoust Soc Am 32:693–703.

Pisoni DB, Carrell TD, Gans SJ (1983) Perception of the duration of rapid spectrum changes in speech and nonspeech signals. Percept Psychophys 34:314–322.

Potter RK, Steinberg JC (1950) Toward the specification of speech. J Acoust Soc Am 22:807–820.

Remez RE, Rubin PE, Nygaard LC, Howell WA (1987) Perceptual normalization of vowels produced by sinusoidal voices. J Exp Psychol [Hum Percept] 13:40–61.

Repp BH (1982) Phonetic trading relations and context effects: New evidence for a speech mode of perception. Psychol Bull 92:81–110.

Repp BH, Healy AF, Crowder RG (1979) Categories and context in the perception of isolated steady-state vowels. J Exp Psychol [Hum Percept] 5:129–145.

Sato M, Lotto AJ, Diehl RL (2003) Patterns of acoustic variance in native and non-native phonemes: The case of Japanese production of /r/ and /l/. J Acoust Soc Am 114:2392.

Saul AB, Cynader MS (1989) Adaptation in single units in visual cortex: The tuning of aftereffects in the spatial domain. Vis Neurosci 2:593–607.

Scott SK, Wise RJS (2003) Functional imaging and language: A critical guide to methodology and analysis. Speech Commun 41:7–21.

Stephens JDW, Holt LL (2003) Preceding phonetic context affects perception of nonspeech sounds. J Acoust Soc Am 114:3036–3039.

Stevens KN (1986) Models of phonetic recognition II: A feature-based model of speech recognition. In: Mermelstein P (ed) Proceedings of the Montreal Satellite Symposium on Speech Recognition, pp. 67–68.

Stevens KN (2002) Toward a model for lexical access based on acoustic landmarks and distinctive features. J Acoust Soc Am 111:1872–1891.

Stevens EB, Kuhl PK, Padden DM (1988) Macaques show context effects in speech perception. J Acoust Soc Am 84(Suppl. 1):577.

Story BH (2005) A parametric model of the vocal tract area function for vowel and consonant simulation. J Acoust Soc Am 117:3231–3254.

Story BH, Titze IR (2002) A preliminary study of vowel quality transformation based on modifications to the neutral vocal tract area function. J Phonet 30:485–509.

Strange W (1989) Evolving theories of vowel perception. J Acoust Soc Am 85:2081–2087.

Sullivan SC, Lotto AJ, Diehl RL (2005) Optimal auditory categorization on a single dimension. In: Forbus K, Gentner D, Regier T (eds) Proceedings of the Twenty-Sixth Annual Conference of the Cognitive Science Society. Mahwah, NJ: Lawrence Erlbaum, p. 1639.

Summerfield Q (1981) Articulatory rate and perceptual constancy in phonetic perception. J Exp Psychol [Hum Percept] 7:1074–1095.

Summerfield Q, Assmann PF (1987) Auditory enhancement in speech perception. In: Schouten MEH (ed) NATO Advanced Research Workshop on the Psychophysics of Speech Perception. Dordrecht, Netherlands: Martinus Nijhoff Publishers, pp. 140–150.

Summerfield Q, Assmann PF (1989) Auditory enhancement and the perception of concurrent vowels. Percept Psychophys 45:529–536.

Summerfield Q, Haggard M, Foster J, Gray S (1984) Perceiving vowels from uniform spectra: Phonetic exploration of an auditory aftereffect. Percept Psychophys 35:203–213.

Sussman HM (1986) A neuronal model of vowel normalization and representation. Brain Lang 28:12–23.

Syrdal AK, Gopal HS (1986) A perceptual model of vowel recognition based on the auditory representation of American English vowels. J Acoust Soc Am 79:1086–1100.

Traunmüller H (1981) Perceptual dimension of openness in vowels. J Acoust Soc Am 69:1465–1475.

Ulanovsky N, Las L, Nelken I (2003) Processing of low-probability sounds by cortical neurons. Nat Neurosci 6:391–398.

Ulanovsky N, Las L, Farkas D, Nelken I (2004) Multiple time scales of adaptation in auditory cortex neurons. J Neurosci 24:10440–10453.

Viemeister NF (1980) Adaptation of masking. In: van den Brink G, Bilsen FA (eds) Psychophysical, Physiological, and Behavioural Studies in Hearing. Delft, Netherlands: Delft University Press, pp. 190–199.

Viemeister NF, Bacon SP (1982) Forward masking by enhanced components in harmonic complexes. J Acoust Soc Am 71:1502–1507.

Wade T, Holt LL (2005a) Effects of later-occurring nonlinguistic sounds on speech categorization. J Acoust Soc Am 118:1701–1710.

Wade T, Holt LL (2005b) Perceptual effects of preceding non-speech rate on temporal properties of speech categories. Percept Psychophys 67:939–950.

Watkins AJ (1988) Spectral transitions and perceptual compensation for effects on transmission channels. In: Ainsworth W, Holmes J (eds) Proceedings of the 7th Symposium of the Federation of Acoustical Societies of Europe: Speech' 88, Edinburgh, England. pp. 711–718.

Watkins AJ (1991) Central, auditory mechanisms of perceptual compensation for spectral-envelope distortion. J Acoust Soc Am 90:2942–2955.

Watkins AJ, Makin SJ (1994) Perceptual compensation for speaker differences and for spectral-envelope distortion. J Acoust Soc Am 96:1263–1282.

Watkins AJ, Makin SJ (1996) Some effects of filtered contexts on the perception of vowels and fricatives. J Acoust Soc Am 99:588–594.

Wayland SC, Miller JL, Volaitis LE (1994) The influence of sentential speaking rate on the internal structure of phonetic categories. J Acoust Soc Am 95:2694–2701.

Whalen DH (1990) Coarticulation is largely planned. J Phonet 18:3–35.

Yamada RA, Tohkura Y (1990) Perception and production of syllable-initial English /r/ and /l/ by native speakers of Japanese. In: Proceedings of the International Conference on Spoken Language Processing, Kobe, Japan; 19–22 November. pp. 757–760.

# 11
# Sound Source Perception and Stream Segregation in Nonhuman Vertebrate Animals

RICHARD R. FAY

## 1. Introduction

Since the appearance of Bregman's (1990) influential book *Auditory Scene Analysis*, research on both human and nonhuman animals has been influenced to seriously consider sound sources and their perceptions. The new human psychoacoustic work has come together in this volume under the phrase "sound source perception" because the kernel of Bregman's lessons has been interpreted as an emphasis on sound sources and the factors that promote and determine their perception in human listeners. Modern psychoacoustic research on these concepts has expanded in a more general direction, and has included experiments and topics that were suggested by, but go beyond, the original notions of stream segregation and scene analysis. Researchers interested in animal hearing have been motivated by the terms Bregman originally used, "auditory scene analysis" and "stream segregation," to argue that animal listeners, too, are faced with the fundamental biological problem of source segregation and determination. The animal work is at an earlier stage of development, however, and has focused on demonstrating the core phenomena of sound source perception using the terminology of "scene analysis" and "stream segregation." These terms have come to stand for the general idea that individual sound sources in mixtures must be "heard out," or segregated, or determined, before they can play a useful biological role in communication, imaging the local world, and the other functions of hearing. The titles of the works reviewed here reflect this. This application of concepts that are self-evident in human perception to nonhuman animals has been a very difficult undertaking because of the essentially human-centered experimental designs that characterized the early work on scene analysis; for example, most experiments require language of the participants, and many of the experimental designs focus on aspects of human hearing that focus on speech and music perception (see also Handel 1989). This volume is the first to focus on some of the second-generation experiments and ideas that come together under the title "sound source perception." Nevertheless, the general arguments made in Bregman's (1990) volume, especially in Chapter 1, have been so compelling that one interested in animal sensory behavior cannot help but ask whether this general description of human perception must apply to other species as well.

Auditory scene analysis is set up at one point as follows: "Dividing evidence between distinct perceptual entities (visual objects or auditory streams) is useful because there really are distinct physical objects and events in the world that we humans inhabit. Therefore, the evidence that is obtained by our senses really ought to be untangled and assigned to one or another of them" (Bregman 1990, p. 13). It is pointed out in the first chapter that the visual problem is easier to think about than the auditory, but that these capabilities must be about the same for audition. It is not emphasized that these capabilities and problems are probably the same for all the senses (except, perhaps, for the vestibular sense, since the source of stimulation is seldom an important question), for all species with a sense of hearing, and for all time. Bregman talks in evolutionary terms at times, e.g., "The internal organs of animals evolve to fit the requirements of certain constant factors in their environments. Why should their auditory systems not do likewise?" (Bregman 1990, p. 39). But such evolutionary statements might be read to suggest that these adaptations have taken hundreds of millions of years to be "perfected," and so appear only among humans and other recent species with large brains and special cognitive abilities, or among species with special or obvious requirements for sound source perception, such as songbirds needing to recognize a conspecific's call among a mixture of bird calls (e.g., Hulse et al. 1997, Bee and Klump 2004, 2005).

It is suggested here that an auditory system could not have been an advantage to any organism (i.e., could not have evolved among all vertebrate species) had it not had the primitive (shared) capacity for sound source perception from the start. In other words, if the sensory evidence obtained by our (ancestors') senses were not untangled and assigned to one or another auditory object or event, the sensory evidence itself would be all but useless, and could not contribute to fitness. Bregman's prime example is the perceptual segregation of a cradle's squeak from the simultaneous voice of the mother; this segregation must take place for the baby's language to develop normally. It is argued here that this sort of segregation must take place for useful "hearing" ever to have taken place at all, in any organism.

## 1.1 What Is Auditory Scene Analysis and Stream Segregation?

Sound source perception, including auditory scene analysis (ASA), has multiple definitions, some that workers in the field implicitly use in the design of experiments, and some primarily proceeding from Bregman's conceptions outlined and detailed in his book on the topic (Bregman 1990). Throughout this chapter, sound source perception and ASA will refer to an apparently self-evident capacity of auditory perception that results in the determination of sound sources (or of sources in general) when the sounds they produce are heard in the presence of other simultaneous sounds or sources. It is not the mere detection of sources in the presence of noise or distracters, but is the disentangling of components of one source from those of others. Nor is it the recognition of signals, such as conspecific vocalizations, but rather the determination that the signal in question arises from an independent source. The source of a conspecific vocalization is what is biologically relevant (an individual bird of the same species), not necessarily the message encoded in the sound quality itself. In this way, sound source perception is conceived as a necessary part of all auditory systems that permit the organism to behave appropriately with respect to sound sources and auditory events (e.g., Lewis and Fay 2004). Sound source perception can fail at times, and in the rare cases of failure in the everyday world, the organism may be at a severe risk of behaving inappropriately (and not surviving as a result). Thus, sound source perception is thought of as a fundamental, defining function of a sense of hearing, and would be expected to be part of the pressures on all auditory systems, from the time they first evolved to the present. The necessity for a kind of sound source perception as part of most sensory/perceptual systems probably determined many capacities and structures of the vertebrate ear and brain (and possibly invertebrate brains as well; see, for example, Schul and Sheridan 2006).

Auditory stream segregation, a component of sound source perception, is most easily understood in terms of human experience; it refers to the perception of a unitary source among multiple sources. An auditory stream also implies extension in time, or an ongoing source, and its ongoing nature seems to help define the source or sources. Auditory streams are self evident in human experience and can be demonstrated as well with simple tone sequences and more biologically significant sounds such as human speech and music. The segregation of alternating A-B tone sequences into two streams based on the pitch of the tones (Miller and Heise 1950) is an often-used stand-in for what we generally mean by stream segregation and scene analysis in everyday life. However, it must be recognized that the reduced or simplified nature of these stimuli and their sequential as opposed to simultaneous nature, properties advantageous for experimental and theoretical analysis, might mislead us in comparative and physiological studies of the broader phenomenon of sound source perception. Does this model stand in appropriately for what we mean by sound source perception among all animals, or is it unique on account of its simple and abstract nature? Human perception is very good when it comes to generalization from simple to complex situations, and vice versa. Most other animals generalize poorly, and most brain cells don't generalize at all.

## 2. Comparative Behavioral Studies

### 2.1 Studies on Starlings

There are only a few animal studies of sound source perception, but these are of relatively great interest because they focus on demonstrating and analyzing the capacity itself, and tend not to focus on those perceptual capacities that are required or assumed to exist in ASA, but are not identical with ASA itself. Hulse et al. (1997) were apparently the first to explicitly study auditory scene analysis in a nonhuman animal, the European starling (*Sturnus vulgaris*). The authors of this and subsequent studies explained their choice of the European starling as their subject because starlings seemed to be particularly competent at responding to their specific scenes of auditory sources, made up primarily of vocalizations of conspecifics and others. The method consisted in operant training methods (keypeck for food) in response to 10-second recordings of birdsong. Starlings were initially trained to discriminate between stimuli made up of sample mixtures one species' song combined with another (essentially, two concurrent sources to be perceived independently). One type of stimulus always included the examples of the starling's own song, and another type consisted of samples of songs of two species other than starlings. The birds were trained to discriminate examples of the first stimulus type from the second. Each stimulus type was represented by 15 unique examples. Starlings learned to discriminate with at least 85% accuracy. The authors concluded that a capacity for source segregation was the most likely explanation for the discrimination. By segregating the sources in each sound type, it would have been an easy matter to discriminate between a stimulus that contained songs of the starling (for example) mixed with songs of another species, and those that did not. All ($n = 6$) the starlings were equally adept at correctly identifying stimuli containing the starling song compared to stimuli that did not. An alternative explanation for the discrimination was that birds memorized acoustic features that were unique to each of the 30 stimulus samples. To test for this possibility, the starlings were transferred to novel stimulus examples of the two types in a discrimination test. The birds immediately transferred without further training, and the memorization hypothesis was rejected.

With the assumption of source segregation by starlings, the authors reasoned that the discrimination would transfer to a case in which the stimulus consisted of only one species song in isolation. If the discrimination were based on the fusion of two sources into a complex virtual source for the starling (a failure of source determination), the discrimination would not be expected to transfer to the case of a single song. The starlings classified all the isolated song types at levels above chance, but with a statistically significant deficit compared with the discrimination of the two-song examples. This means that the one-song stimuli were not treated as identical to the two-song mixtures (they were not identical, after all), but were treated as nearly equivalent.

Finally, Hulse et al. (1997) added the sound of a "dawn chorus" (songs of many individuals of many species) to all stimulus files, and the discrimination was tested again. Performance declined, but remained above chance. The decline in performance is expected because the addition of the dawn chorus made the stimuli quite different from the original ones that the birds had been trained on (the "other" sources were not a single species, but were a mixture of many species). In general, the authors concluded that stream segregation was the most parsimonious explanation for these results as a whole, and that starlings should be seen as being as capable of auditory scene analysis as human observers. The authors point out that source segregation apparently occurred easily in starlings, in spite of the fact that the various sources were delivered mixed through one loudspeaker. Thus, spatial separation of sources is not necessary for source segregation to occur.

Wisniewski and Hulse (1997) investigated the abilities of starlings to discriminate between individual birds on the basis of their song. After acquiring the discrimination between 10 examples of starling A's song and 10 examples of starling B's song, the authors showed that the discrimination was performed when both types of stimuli were heard in the presence of song from starling C. The discrimination remained even after four conspecific distracters were added. Finally, it was shown that the discrimination persisted after the distracting song of starling C was played backward. The results of all experiments were consistent with the hypothesis that starlings treated starling A's and B's songs as categories that they could easily discriminate and segregate on the basis of a set of cues that remain unknown. Further, the results of all experiments were well described and accounted for by the principles of auditory scene analysis as outlined by Bregman (1990).

MacDougall-Shackelton et al. (1998) very cleverly extended the above findings to the case of the segregation of A-B tone sequences. A common attitude among comparative researchers is that a functional capacity, such as source segregation, is probably a specific adaptation to species-specific conditions or constraints. One interpretation of the demonstration of source segregation in birds is that it is likely that this capacity is a very general phenomenon probably shared widely. A common alternative hypothesis is that it is a capacity specific to vocalization sounds, since it is obviously in the starling's interest to segregate its own vocalizations from others. There would be no such fitness advantage in segregating such biologically irrelevant sounds as tone patterns.

Starlings were trained as above to sequences of tone bursts that varied according to their temporal pattern. One pattern was the typical "galloping" pattern composed of three tone bursts in rapid succession separated by a time interval equal to one of the tone bursts (repetition period of 400 ms). The other patterns were two isochronous patterns of bursts, one with the 400 ms repetition period, and one with a 200 ms repetition period. The birds were trained to discriminate the galloping pattern from either one of the isochronous patterns. Within any one training session, the pulses were of a fixed frequency of 1000 Hz, but between sessions, the frequency was varied between 1000 and 4538 Hz. After the birds achieved at least 85% correct performance on this discrimination, novel probe stimuli were introduced on a small proportion of trials. The probe stimuli had the temporal structure of the galloping pattern but with a frequency difference between the second burst of the group of three, and the first and last burst. If this frequency difference is great enough for segregation to occur (as it does for human listeners), the probe stimuli is heard as two sources (or likely heard as one of two possible sources) with different temporal patterns and pitches. Therefore, if segregation occurred on probe trials, the probes were hypothesized to evoke a perception more like one of the isochronous stimuli, and less like the galloping stimuli. The frequency of responses indicating that the galloping probe stimuli (with frequency changes) were perceived as isochronous monotonically grew as the probe stimuli had frequency differences that grew from 50 Hz to 3538 Hz. Stream segregation of the kind experienced by human observers listening to the same sounds is the best explanation for the starlings' behaviors in this experiment. This experiment remains the only one in the literature to demonstrate that nonhuman animals can segregate streams evoked by arbitrary and abstract pure tone sequences having only frequency differences (but see Izumi 2002). The authors argue, on the basis of these and other data, that auditory scene analysis, in some form, is probably common among animals.

The MacDougall-Shackelton et al. (1998), Hulse et al. (1997), and Wisniewski and Hulse (1997) papers make a strong case that a nonhuman animal perceives sources and can segregate them similarly to human beings. The segregation of abstract and arbitrary tone sequences helps to argue that starling perception, in general, is similar to that of human beings and is not necessarily narrowly adapted to species-specific communication tasks. The segregation, also, of starling and other birdcalls argues that this capacity is not just a peculiar and biologically irrelevant response to meaningless tones, but has adaptive value. Taken as a whole, the starling experiments suggest to us that source segregation and auditory scene analysis is a general biological phenomenon, and that it doesn't require a mammalian brain.

In a related experiment, Benny and Braaten (2000) investigated auditory source segregation behavior in two estrilid finches using the same paradigm. The finches generally behaved like starlings in these tasks, and were shown capable of source segregation. Of special interest in these experiments were the detailed data on discrimination acquisition of the two finch species when the songs they were trained on included the species' own song, or not. It was hypothesized that the learning of the discrimination would be naturally easier for a species discriminating its own song possibly because of the experience it would have had with these sounds. The evidence for this hypothesis was mixed. Zebra finches (*Taeniopygia guttata*) tended to acquire the baseline discrimination more rapidly when one of the songs to be segregated was zebra finch song. However, this difference was noted only after one individual zebra finch was eliminated from the study because it failed to acquire the discrimination to criterion. Finally, zebra finches (but not Bengalese finches) showed the hypothesized effect. Bengalese finches (*Lonchura striata domestica*) took the same time to acquire the discrimination whether the discrimination involved the species' own song or not. Another datum relevant to this hypothesis is that zebra finches (but not Bengalese finches) trained to detect heterospecific song were distracted by the presence of conspecific song. Thus, the evidence supports the hypothesis that conspecific song is of "special" importance to zebra finches, but not to Bengalese finches. Note that these speculations on the special nature of conspecific song in perception do not relate in a consistent way to the occurrence or robustness of auditory stream segregation. For example, zebra finches were said to be "distracted" from the discrimination that defined stream segregation (showed poorer source segregation performance) when they were required to withhold a response to conspecific song. One might think that the presence of conspecific song would make the segregation task easier, not harder. Thus, the hypothesized "special" nature of conspecific song for perception has an effect on behavior, but seems not to determine whether or how well songs are segregated.

## 2.2 Studies on Goldfish

A series of studies of source segregation in goldfish was initially motivated by the hypothesis that source segregation is very widespread, possibly occurring in all animals that hear, and that it is not found exclusively in species that communicate acoustically and is not a special adaptation for communication. The first experiment to bear on the topic is a study by Fay (1992) on "analytic listening" by goldfish. Before the phrases "source segregation" or "stream segregation" became widely understood, the hearing out of an acoustic source among several was called *analytic listening*. Actually, the experiment was motivated by an early comment by van Bergeijk (1964, p. 296) that "given that a fish can discriminate between two sounds A and B when they are presented separately, can he still discriminate either one when both are presented simultaneously? Or, do the two sounds blend to form a new entity (such as a chord)." Here, van Bergeijk seems to have specified the fundamental requirement of an auditory system (as we understand it), and he was interested to know whether fishes had an equivalent auditory system and heard in this way. This important kind of question about what we know as source segregation could be considered trivial if hearing were defined by what we humans do and if this definition were extended to all vertebrate animals that have ears. But the question was not trivial in the context of fish hearing because, while a capacity for sound detection was well known, fishes were thought to be unable to localize sound sources (van Bergeijk 1964) and perhaps be otherwise unintelligent or challenged when it came to perceiving the nature of sound sources.

In Fay (1992), goldfish were classically conditioned to respond to a simple mixture of two tones (166 and 724 Hz at about 30 dB above threshold), and then tested for generalization to stimuli that consisted of a single pure tone of various frequencies. Stimulus generalization (Guttman 1963) is a simple phenomenon in animal discrimination learning that permits an estimate of the extent to which stimuli are perceived as equivalent or inequivalent. A full response to a novel probe sound, after conditioning to a given complex sound, indicates generalization from the training sound to the probe sound and can be interpreted as the degree of equivalence or perceptual similarity of one sound to another. The segregation of one or the other tone from the complex mixture would result in a generalization gradient (a plot of response strength as a function of tone frequency) that had a peak at 166 Hz, or 724 Hz, or both. The failure to segregate would have resulted in a small response strength (indicating inequivalence) and probably a flat generalization gradient. The result was a two-peaked generalization function (at 166 and 724 Hz) and a lesser response at other frequencies. The response strength was about 60% of the maximum obtainable, possibly indicating that while the two tones were heard out as individual sources when present in a mixture, a single tone of any frequency was still not equivalent to the mixture in its total perceptual effects.

Fay (1998) then carried out an experiment on goldfish specifically aimed at auditory source segregation. The stimuli were synthetic filtered tone pulses repeated at various rates. They were meant to mimic the sorts of sounds fishes make when communicating, but it is doubtful that a goldfish would ever have encountered these sorts of sounds in everyday life, and does not vocalize itself. The segregation hypothesis was that a simple mixture of pulse trains (a high-frequency pulse repeated at 85 pulses per second (pps), and a low-frequency pulse repeated at 19 pps) would be segregated into individual sources based on a different spectral envelopes and repetition rates. The pulses were pulsed sinusoids (238 and 625 Hz), creating a "low-frequency" and "high-frequency" pulse. Then the pulses could be repeated at various rates between 19 and 85 pps.

The main experiment consisted of two groups of eight animals conditioned to a simultaneous mixture of two pulse trains, the high-frequency pulse repeated at 85 pps, and the low-frequency pulse repeated at 19 pps. One of the groups was then tested for generalization to the low-frequency pulse repeated at rates between 19 and 85 Hz, and the other group was tested using the high-frequency pulse repeated at the same range of rates. The results for the two groups showed two, oppositely sloped, generalization gradients. The group tested with the low-frequency pulse produced a gradient that sloped downward as a function of repetition rate, with the most robust responses at the lowest repetition rates. The group tested with the high-frequency pulse had a gradient that sloped upward as a function of repetition rate (the largest response at the 85 Hz repetition rate). These results demonstrated that goldfish correctly associated a particular spectral envelope and repetition rate (i.e., information about the two mixed pulse trains was obtained independently). Auditory source segregation provides the best description of these results. Fay (1998) then went on to show that a 500-ms stimulus onset asynchrony between the two pulse trains resulted in more robust segregation, as it does in human perception.

These findings on goldfish were extended with additional experiments, demonstrating that segregation occurred more robustly when the spectral difference between pulse trains increased (Fay 2000). In these new experiments, 625-Hz filtered pulses alternated with pulses having different center frequencies (500 Hz to 240 Hz) for a total alternating pulse rate of 40 pps (Miller and Heise 1950). Animals were tested for generalization to the 625 Hz pulse alone repeated at a variety of rates between 20 and 80 pps. If these alternating pulses were segregated into two streams (a high frequency of 625 Hz, and low-frequency stream), then the generalization behavior should resemble that following conditioning to a pulse presented at one-half the repetition rate of the conditioning pulses (20 pps). The results were consistent with the segregation of the alternating pulses with large frequency separation (625–240 Hz), but not with the small frequency separation (625–500 Hz).

These experiments with goldfish demonstrate perceptual behaviors that are indistinguishable from what would be expected from human listeners under similar circumstances. The best description of these behaviors is that they are examples of concurrent auditory source segregation. Thus, we are fairly confident in believing that goldfish, and fishes in general, are capable of sound source segregation, as we understand it for human listeners. Since others have come to the same conclusion with respect to monkeys, starlings, and birds in general, it seems likely that all vertebrate animals have the capacity for source segregation and thus some form of auditory scene analysis. Therefore, scene analysis and source segregation are phenomena of biological interest and significance that can be studied in a comparative and evolutionary context.

## 2.3 Studies on Other Species

Izumi (2002) has investigated auditory source segregation using tone sequences in the Japanese macaque (*Macaca fuscata*). In an initial experiment, macaques were trained in a go/no-go paradigm to discriminate between rising tone sequences and nonrising sequences in isolation, and with added distracters either overlapping (in frequency range) the target sequences or not. It was hypothesized that the animals could not segregate or discriminate the target sequences under overlapping distracter conditions, but could discriminate the rising sequences when the targets and distracters were nonoverlapping. The results confirmed the hypothesis, suggesting that the monkeys segregated the sequences based on frequency proximity. In experiment 2, Izumi (2002) evaluated the hypothesis that cues other than "local pitch" were used to accomplish the discrimination. In this experiment, the starting frequencies of the target sequences were altered to determine whether the animals based their discrimination on the exact frequencies used, or whether they were responding to the global (rising) nature of the pattern. Results showed that the discrimination could be made only when the target and distracter sequences were nonoverlapping. It was concluded that the cues for discrimination were global ones (rising pattern) and not the local pitch of a component tone, indicating further that the animals segregated the target sequences similarly to human listeners. Izumi (2002) related this ability of the macaque to problems the species would encounter in responding to the species' own vocalizations in the usual environment, but it would seem that due to the abstract nature of the stimuli used, these results would apply (and be valuable) to all listening situations.

Hulse (2002) has reviewed the literature on auditory scene analysis with respect to animal communication, and has identified some observations and experiments on diverse species that are relevant to the question. Feng and Ratnam (2000) have similarly reviewed the literature regarding hearing in "real-world situations." Once arguments have been made and experiments done demonstrating stream segregation in one or two taxa, it is difficult not to generalize the result to diverse species, because the arguments and rationale are essentially the same for all taxa; it would be remarkable or even unbelievable that some species could be demonstrated incapable of source segregation. In this way, several papers have appeared that basically assume that scene analysis is a ubiquitous capacity among animals and then go on to offer examples and arguments for this assumption. For example, anurans (frogs and toads) find themselves in the same situation as birds in having to respond to an individual's call within a chorus of primarily conspecific advertisement calls. It seems impossible to consider this task without assuming some sort of source segregation among all anuran species that behave this way (Klump and Gerhardt 1992). In this context, the variation and evolution of advertisement calls with respect to frequency and temporal patterns can be regarded as attempts to match the acoustic dimensions that are most important for source segregation, so that studying the acoustic characteristics of these calls could reveal properties of the segregation capacities. The same argument could be made for fishes that produce advertisement calls (Bodnar et al. 2001) and the several other species that evaluate and find individual males that usually call in a chorus.

Echolocating animals are another example of the apparent necessity for source segregation; echoes are thought to inform the listener about the nature of the reflections from individual sources (e.g., Moss and Surlykke 2001; Barber et al. 2003; Kanwal et al. 2003). For these sorts of tasks, one cannot imagine confusing acoustic components from a source with those of the general background. In general, it appears to be self-evident that animals that use communication sounds must be able to segregate these biologically significant sounds from whatever acoustic background exists from moment to moment, and not to confuse acoustic components of the background with those of the communication sounds.

This logic is often used to suggest that capacities for stream segregation ought to be found among species that communicate acoustically and that these species are particularly well adapted for segregating their own calls or songs. But any sound source could be substituted for a communication sound source and the logic and apparent necessity for source segregation would be as compelling. Animals must behave appropriately with respect to all sources or objects in their environment, regardless of the "biological significance" of the source or of competing sources.

## 3. Comparative Neurophysiological Studies

Recently, there have been several attempts to discover some of the neural mechanisms or neural correlates of scene analysis or source segregation. Many of these experiments, by necessity, are performed on nonhuman animals, and thus assume that these processes exist in species other than the human. Some of these experiments make use of noninvasive electrophysiology (surface evoked potentials) with human subjects (e.g., Sussman et al. 1999; Alain et al. 2002; see also Chapters 4, 5, and 9 of this volume for further consideration of Sussman's

(1999) work). Thus, some human studies will be reviewed here to the extent that they could as well be applied to nonhuman animals (the fact that the experiment is on a human being is not critical or a necessary part of the design).

Clearly, however, to attempt a neurophysiological experiment on scene analysis is potentially a very difficult undertaking, primarily because what is essentially rather private and subtle in perception is not easily or obviously translated to the sorts of dependent variables monitored in neurophysiological experiments, particularly single-cell experiments in which one cell (or cell cluster) is often viewed as if it were a "homunculus," or autonomous decision-maker. In general, the strategies for experimental design have been to (1) focus on rather abstract examples of sequential stream segregation (e.g., A–B tone sequences), (2) identify stimulus continua along which stream segregation appears and disappears depending on the stimulus value (e.g., A–B frequency separation, A–B repetition rate, or presentation time), and (3) use spike count or probability as the dependent variable such that the pattern at one recording location transitions from responses to both tones (indicating synthesis) to a response to only one tone (indicating segregation) as one stimulus variable is manipulated. The correlation between the variable's effect on the spike count that signals segregation/integration and the segregation/integration of sources in perception is taken as evidence that the neural cause or correlate of source segregation has been identified. This experimental logic has been used for investigating the correlates of stream segregation in macaque cortex (Fishman et al. 2001, 2004; Micheyl et al. 2005) and in starling forebrain (Bee and Klump 2004, 2005). This logic has been applied only to the case of sequential segregation of simple A–B tone sequences.

Fishman et al. (2001, 2004) have investigated the cortical neural correlates of sequential stream segregation (alternating A–B tone sequences) in the macaque monkey (*Macaca fascicularis*). Using fixed electrode arrays, Fishman et al. (2001) recorded multiunit activity, averaged evoked potentials, and current source density from the same electrodes in primary auditory cortex of awake monkeys. The A-tone frequency was chosen to be maximally excitatory for the unit cluster, and the B-tone was off the best frequency, producing excitation that was about one-half that of the A tone. The fundamental observation was that when alternating A–B tone sequences were presented, the responses to B tones were suppressed relative to the A-tone responses when repetition rates were high (above 10 pulses per second). This suppression corresponds to the perception of segregated streams of A and B tones when the tone repetition rate grows above 10 Hz. Apparently, the A-tone response plays a role in suppressing the B-tone response, and conversely; the presence of the B tone tends to result in a larger response to the A tones than in controls (the same rep rate for A tones with no B tones). In addition, the A-tone suppression of B-tone responses tended to increase as the B-tone frequency deviated from the A-tone frequency and as tone duration increased (Fishman et al. 2004). These physiological observations correspond qualitatively with the appearance of segregated sources in perceptual experiments with humans (e.g., Miller and Heise 1950); segregation tends to occur more at higher repetition rates, and when the frequency difference between A and B tones increases. The multiunit activity responding nearly equally to alternating A–B sequences at low repetition rates corresponds to integration, while the suppression of B responses at higher rates or larger frequency differences corresponds to the segregation of the A source. Presumably, there will be clusters of units representing the segregated B tones that are differently tuned, such that the A-tone responses will be suppressed by the B-tone responses. Thus, the authors considered these observations as a neural analogue of sequential stream segregation, and concluded that synchronized transient responses in A1 cortex play a role in this sort of stream segregation. These findings are consistent with those of previous single-unit forward masking experiments in cat auditory cortex (Brosch and Schreiner 1997), suggesting that forward masking plays a role in sequential stream segregation generally.

In a similar study of units of the auditory forebrain in awake starlings, Bee and Klump (2004) found corresponding results. They have identified the same two stimulus continua that are closely related to sequential tone segregation in the starling: frequency and the temporal relations among brief tone pulses. Thus, frequency selectivity and physiological forward masking are the processes identified as underlying sequential stream segregation in the starling. Specifically, these authors evaluated the hypothesis that the degree of overlap in excitation along a tonotopic gradient decreases under stimulus conditions known to promote source perception. In other words, the differences in the spiking response to two tones (one at CF and the other off-CF) should increase (i.e., segregation should increase) as the frequency difference between the two tones (B–A Hz) increased, and should increase as the tone repetition time decreased (increased speed of repetition). In addition, Bee and Klump (2004) tested the hypothesis that differential responses to CF and off-CF tones (i.e., segregation) are influenced by the relatively stronger physiological forward masking of off-CF tones by preceding CF tones.

The results of careful and systematic experiments on numerous multiunit recording locations tended to support the hypotheses; the differences between responses to CF and off-CF tones increased with the frequency difference between the tones and with tone repetition rate. Thus, the same stimulus manipulations that promote perceptual source segregation also tend to differentiate the neural responses to CF and off-CF tones. It was concluded "that frequency selectivity and physiological forward masking can play a role in the perceptual segregation of interleaved tone sequences, and are thus likely mechanisms involved in sequential auditory stream segregation." (Bee and Klump 2004, p. 1100).

Most recently, Micheyl et al. (2005) have reinvestigated source segregation by cells in A1 of awake macaques in a very careful and quantitative study in which variables and hypotheses were clear and well defined. As in Fishman et al. 2001, 2004 and Bee and Klump 2004, 2005, stimuli were sequential AB tone sequences, and the dependent variable was the spike count from single cells. This experiment was unique, however, in looking for the neural correlates of the temporal buildup of stream segregation in human listeners (e.g., Carlyon et al. 2001) and in using statistical decision concepts to define the detection of one or more sources. In addition, these authors carefully and logically defined what neural patterns corresponded to the perception of one or two sources and the hypothetical neural mechanisms that were revealed. The most remarkable result was that the temporal evolution of the perception of two streams was so well accounted for by the neural data using simple definitions for the number of streams represented. The time course, function shape, effects of stimulus presentation rate, and the effects of frequency separation on the probability of hearing two streams were quantitatively and qualitatively mirrored by the neurometric functions based on spike counts of cortical neurons. This is a remarkable correspondence between psychophysics and neurophysiology.

Schul and Sheridan (2006) have argued for the demonstration of source segregation in the nervous system of a katydid (*Neocococephalus retusus*). They motivated their study by the apparent necessity for source segregation in perceiving predators (i.e., echolocating bats) in the presence of conspecific advertisement call vocalizations. These authors recorded auditory-evoked activity from the broadly tuned TN-1 interneuron, part of the circuit that responds to bat echolocation calls, but which also has sensitivity to the male advertisement calls. Advertisement calls and bat echolocation sounds are both impulsive sounds with different spectral regions (15 kHz and 40 kHz) and repetition rates (5–10 pulses per second, and 100–150 pulses per second) respectively. When the TN-1 neuron is stimulated by the long-duration advertisement calls (at a high repetition rate), it adapts completely; it responds with spikes only during the first three seconds of prolonged stimulation and is subsequently silent. Schul and Sheridan (2006) found that bat calls occurring during the period of complete adaptation to an advertisement call evoked a robust response over a wide range of intensities. This apparent release from adaptation to a sound of different spectral region was interpreted as evidence that the TN-1 interneuron segregated the two sources, one with the slow pulse rate from the other with the rapid pulse rate. This behavior occurred even when the spectra of the signals of different repetition rate were exchanged. While other interpretations of TN-1 behavior are possible, the authors interpreted this as a demonstration of a preattentive, primitive form of auditory stream segregation in an insect, and thus concluded that auditory scene analysis would have been a universal selective pressure during the evolution of all hearing systems.

## 3.1 Evoked Potential Studies

Other physiological investigations of auditory stream segregation have used evoked brain potentials. For example, in one such study of sequential stream segregation, Sussman et al. (1999) used the phenomenon of mismatch negativity (MMN) to investigate the role of attention in auditory stream segregation. MMN is an enhanced negative shift of the event-related evoked potential occurring around 140–220 ms after stimulus onset. It tends to occur to a novel or deviant event in the context of a homogeneous series of sounds. It is known to occur independently of attention or whether attention is focused, and is thus known as a preattentive or automatic phenomenon. Sussman et al. (1999) found that MMN occurred in parallel with source segregation of alternating high- and low-frequency tones, suggesting a preattentive locus for the segregation effect.

Alain et al. (2001, 2002) recorded auditory-event-related potentials from listeners presented with either harmonic sounds or a complex sound with one mistuned component. Listeners tend to report two sounds (or sources) in the mistuned case; the mistuned component tends to pop out in perception as if it were an independent source. These authors found that when stream segregation occurs for the case of the mistuned harmonic, the evoked potential contains a slow negative component (similar to the MMN) peaking at about 180 ms. They termed this phenomenon an "object-related negativity," or ORN. In addition, concurrent auditory object perception was also associated with a long-latency positive wave (P400) in the evoked potential. The authors found that the ORN was associated with the perception of the mistuned harmonic as a separate tone. They concluded that concurrent source segregation depends on a transient neural response triggered by the automatic (preattentive) detection of inharmonicity. The P400 component was associated with the participant's response (i.e., when active discrimination was required). In general, these and other results with long-latency evoked potentials were taken as evidence that the auditory cortex plays a role in scene analysis.

## 3.2 Critique

These physiological studies of auditory stream segregation are among the first attempts to determine the neural correlates of scene analysis, and will certainly not be the last. But it is not clear that all the assumptions underlying these experiments are useful ones and will advance our understanding. In the single-unit (or unit-cluster) experiments (Fishman et al. 2000, 2001; Bee and Klump 2003, 2004; Micheyl et al. 2005), the fundamental assumption is that a single forebrain cell, or cluster of cells, can represent a perceptual behavior in that the activity of these cells is somehow identical to the perceptual behavior of source segregation. The assumption is that source segregation occurs whenever brain cells of a given class respond in a particular way. This may be an unrealistic assumption in general. These studies demonstrate a correlation between a complex perceptual behavior and a particular neural response. The response seems to signal two sources or only one source depending on sound repetition rate, frequency separation, or sound duration, in parallel with the perception of human listeners. But to assume that these neural responses are responsible for the perception (i.e., are a cause or mechanism) is not warranted, as several of these authors themselves point out. Thus, these studies can only suggest the neural mechanisms of source segregation. These suggestions are rather dissatisfying.

In the search for the neural mechanisms of stream segregation, these experiments have pointed to forward masking or suppression (or the more neutral habituation; Micheyl et al. 2005) and frequency selectivity as the underlying processes. These are the processes required for the physiological results to correspond to the perceptual results. This would be plausible if source segregation applied only to abstract sequential tone sequences of the kind studied. Here, the most obvious stimulus dimensions are tone frequency and the times between pulses, and it would seem parsimonious that perceptual phenomena associated with these variables could be explained by invoking two fundamental, independently defined processes of hearing. But what is actually meant by scene analysis and stream segregation seems to be far more complex than this. These tone-sequence experiments have come to stand for all the phenomena of source segregation, and for at least some of the phenomena of auditory scene analysis. But scene analysis suggests more the notion of concurrent (as opposed to sequential) segregation of natural sound sources (e.g., a mixture of birdcalls from which one species' call is heard out). A natural argument could be made that the perception of A–B tone sequences as one or two sources is *like* the source or stream segregation and scene analysis required in the everyday world to respond appropriately to actual sound sources, but it cannot really be considered a good model of these processes; it is only a rather reduced and abstract analogy. So, the tentative explanations for this sort of stream segregation in terms of forward suppression and frequency analysis cannot be useful mechanistic explanations for what we really mean by the phenomena of source segregation and auditory scene analysis, and might actually be misleading in the sense that we could be led to evaluate unproductive hypotheses. Thus, the suggestions for mechanisms of source segregation arising from the single-unit physiological studies described above are dissatisfying, at best.

## 4. Conclusions

Auditory source perception and source segregation are well-known and well-studied aspects of human perception. This chapter reviews the experimental evidence on whether these phenomena occur in more diverse taxa, and could thus be studied in a comparative and evolutionary context. So far, behavior consistent with source segregation (broadly defined) has been observed in goldfish, starling, and macaque, and behavioral and communication contexts have been described for other species for which source segregation seems likely and necessary (e.g., in echolocating bats, frog and fish choruses, and in a katydid). Based on these observations, it is suggested that all animals that hear are likely capable of some sort of source segregation and scene analysis in perception. Thus, source segregation and auditory scene analysis are useful concepts, not only in human psychoacoustics, but in the comparative neurobiology of perception generally. It is argued that these fundamental processes of nervous system function are so useful and ubiquitous that their requirement has probably played an important role in specifying the organization, structure, and evolution of the vertebrate brain.

# References

Alain C, Arnott SR, Picton TW (2001) Bottom-up and top-down influences on auditory scene analysis: Evidence from event-related brain potentials. J Exp Psychol [Hum Percept Perform] 27:1072–1089.

Alain C, Schuler BM, McDonald KL (2002) Neural activity associated with distinguishing concurrent auditory objects. J Acoust Soc Am 111:990–995.

Barber JR, Razak KA, Fuzessery ZM (2003) Can two streams of auditory information be processed simultaneously? Evidence from the gleaning bat *Antrozous pallidus*. J Comp Physiol 189:843–855.

Bee MA, Klump GM (2004) Primitive auditory stream segregation: A neurophysiological study in the songbird forebrain. J Neurophysiol 92:1088–1104.

Bee MA, Klump GM (2005) Auditory stream segregation in the songbird forebrain: Effects of time intervals on responses to interleaved tone sequences. Brain Behav Evol 66:197–214.

Benney KS, Braaten RF (2000) Auditory scene analysis in estrildid finches (*Taenopygia guttata* and *Lonchura striata domestica*): A species advantage for detection of conspecific song. J Comp Psychol 114:174–182.

Bodnar DA, Holub AD, Land BR, Skovira J, Bass AH. (2001) Temporal population code of concurrent vocal signals in the auditory midbrain. J Comp Physiol 187:865–873.

Bregman AS (1990) Auditory Scene Analysis: The Perceptual Organization of Sound. Cambridge: MIT Press.

Brosch M, Schreiner CE (1997) Time course of forward masking tuning curves in cat primary auditory cortex. J Neurophysiol 77:923–943.

Carlyon RP, Cusak R, Foxton JM, Robertson IH (2001) Effects of attention and unilateral neglect on auditory stream segregation. J Exp Psychol [Hum Percept Perform] 27: 115–127.

Fay RR (1992) Analytic listening by the goldfish. Hear Res 59:101–107.

Fay RR (1998) Auditory stream segregation in goldfish (*Carassius auratus*). Hear Res 120:69–76.

Fay RR (2000) Frequency contrasts underlying auditory stream segregation in goldfish. J Assoc Res Otolaryngol 1:120–128.

Feng AS, Ratnam R (2000) Neural basis of hearing in real-world situations. Annu Rev Psychol 51:699–725.

Fishman YI, Reser DH, Arezzo JC, Steinschneider M (2001) Neural correlates of auditory stream segregation in primary auditory cortex of the awake monkey. Hear Res 151: 167–187.

Fishman YI, Arezzo JC, Steinschneider M (2004) Auditory stream segregation in monkey auditory cortex: Effects of frequency separation, presentation rate, and tone duration. J Acoust Soc Am 116:1656–1670.

Guttman N (1963) Laws of behavior and facts of perception. In: Koch S (ed) Psychology: A Study of a Science, Vol. 5, New York: McGraw-Hill, pp. 114–178.

Handel S (1989) Listening: An Introduction to the Reception of Auditory Events. Cambridge, MA: MIT Press.

Hulse SH (2002) Auditory scene analysis in animal communication. Adv Study Behav 31:163–200.

Hulse SH, MacDougall-Shackelton SA, Wisniewski AB (1997) Auditory scene analysis by songbirds: Stream segregation of birdsong by European starlings (*Sturnus vulgaris*). J Comp Psychol 111:3–13.

Izumi A (2002) Auditory stream segregation in Japanese monkeys. Cognition 82:B113–B122.

Kanwal JS, Medevdev AV, Micheyl C (2003) Neurodynamics for auditory stream segregation: Tracking sounds in the mustached bat's natural environment. Network Comput Neural Syst 14:413–435.

Klump GM, Gerhardt HC (1992) Mechanisms and functions of call-timing in male–male interactions in frogs. In: McGregor PK (ed) Playback and Studies of Animal Communication. New York: Plenum Press, pp. 153–174.

Lewis, ER, Fay RR (2004) Environmental variables and the fundamental nature of hearing. In: Evolution of the Vertebrate Auditory System. New York: Springer-Verlag, pp. 27–54.

MacDougall-Shackelton SA, Hulse SH, Gentner TQ, White W (1998) Auditory scene analysis by European starlings (*Sturnus vulgaris*): Perceptual segregation of tone sequences. J Acoust Soc Am 103:3581–3587.

Micheyl C, Tian B, Carlyon RP, Rauschecker JP (2005) Perceptual organization of tone sequences in the auditory cortex of awake macaques. Neuron 48:139–148.

Miller GA, Heise GA (1950) The trill threshold. J Acoust Soc Am 22:637–638.

Moss CF, Surlykke A (2001) Auditory scene analysis by echolocation in bats. J Acoust Soc Am 110:2207–2226.

Schul J, Sheridan RA (2006) Auditory stream segregation in an insect. Neuroscience 138:1–4.

Sussman E, Ritter W, Vaughn HG (1999) An investigation of the auditory streaming effect using event-related brain potentials. Psychophysiology 36:22–34.

van Bergeijk WA (1964) Directional and nondirectional hearing in fish. In: Tavolga WN (ed) Marine Bio-Acoustics, Oxford: Pergamon Press, pp. 281–299.

Wisniewski AB, Hulse SH (1997) Auditory scene analysis by European starlings (*Sturnus vulgaris*): Discrimination of song segments, their segregation from multiple and reversed conspecific songs, and evidence for conspecific song categorization. J Comp Psych 111:337–350.

# Index

For more information about the series, please visit www.springer.com

Printed in the United States of America